Biology of Perch

Biology of Perch

Editors

Patrice Couture
Professeur
Institut national de la recherche scientifique
INRS - Centre Eau Terre Environnement
Québec (Québec)
CANADA

Greg Pyle
Alberta Water and Environmental Sciences Building
Dept. of Biological Sciences
University of Lethbridge
Lethbridge
CANADA

CRC Press
Taylor & Francis Group
Boca Raton London New York

CRC Press is an imprint of the
Taylor & Francis Group, an **informa** business
A SCIENCE PUBLISHERS BOOK

Cover illustrations

For the yellow perch photo: reproduced by permission from Aquarium du Québec-Sépaq

For the Balkhash perch photo: reproduced by permission from Nadir Mamilov

CRC Press
Taylor & Francis Group
6000 Broken Sound Parkway NW, Suite 300
Boca Raton, FL 33487-2742

Printed on acid-free paper
Version Date: 20150707

International Standard Book Number-13: 978-1-4987-3032-7 (Hardback)

Library of Congress Cataloging-in-Publication Data

Biology of perch / editors, Patrice Couture, Greg Pyle.
pages cm
Includes bibliographical references and index.
ISBN 978-1-4987-3032-7 (hardcover : alk. paper) 1. Perch. I. Couture, Patrice. II. Pyle, Greg.

QL638.P4B575 2015
597'.73--dc23 2015018288

Visit the Taylor & Francis Web site at
http://www.taylorandfrancis.com

and the CRC Press Web site at
http://www.crcpress.com

Foreword

Perch *Perca* spp. are known to many due to their wide distribution in temperate climates, their attractive colouration and distinct body stripes and their robust body shape. Many angle them for sport and find them very palatable, especially smoked! I was fortunate to start my career in fish biology research by studying the population dynamics of the European perch *Perca fluviatilis* in a small freshwater lagoon in south-west England and then moved to participate in a long-term investigation on the same animal in Windermere. Following this I also had the opportunity to study the North American perch *Perca flavescens* in Canada. I can still remember being particularly struck by the brilliant golden yellow colour of these fish in northern Manitoba, probably resulting from a transfer of carotenes in their diet of gammarids. I have not had any experience of or even seen the third member of this genus, the Balkhash perch *Perca schrenkii*, which has a much more restricted distribution, although quite by chance I was recently given a picture of it.

My interest in these animals led me to write two books on percids, the family to which perch belong, published in 1987 and 2000. As time goes by technology changes and new techniques are applied in research. In fish research for example, this is particularly apparent in genetics and telemetry. In long-term studies further data points are added allowing new interpretations of the data as a whole. In addition continuing perturbations such as climate change and increased anthropogenic impacts (overfishing and competition for fresh water as a resource) have their effect on the fishes. Thus after a 15 year gap it is probably expedient to produce an update on what we know about these important animals. The pursuit of science requires continual self-correcting and taking on board new or modified concepts. Greg Pyle and Patrice Couture have undertaken this by editing a book containing ten distinct and authoritative chapters. These include a revision on *Perca* phylogeny using the latest molecular tools and more evidence is supplied to support the migration of *Perca* ancestors across the North Atlantic Land Bridge rather than the Bering Land Bridge. I was particularly interested in the chapter on the little known Balkhash perch. New interpretations on long-term studies such as those in the St. Lawrence River and Windermere indicate the importance of gathering data over a long time period; often the importance is ignored or the gathering curtailed by governments. Also covered in this book are advances in our knowledge of perch reproduction, parasites, behaviour and the use of perch in ecotoxicology research.

I continue to be concerned by the decline of fish stocks worldwide including those in fresh water. To manage these stocks, albeit not too successfully in many cases, we need to know more about the biology of these fishes and how they interact with their environment. The present update on our knowledge about perch will help to inform managers and hopefully aid conservation of this precious resource. I wonder what the next 15 years will yield.

John F. Craig
Craig Consultancy
Whiteside, Dunscore, Dumfries, DG2 0UU
Scotland
U.K.

Preface

The three freshwater fish species of the genus *Perca* are major components of freshwater systems all around the Northern hemisphere. Within their respective natural ranges, they are essential as both prey and predators and their presence shapes food webs. Their abundance supports important sport and commercial fisheries and a perch aquaculture industry is emerging. Outside of their range, they are formidable invasive species.

Given their abundance and importance, perch have also been the object of countless scientific investigations in ecology, stock management, ecotoxicology and in several other areas. For instance, years ago we adopted the North American yellow perch (*Perca flavescens*) as our model species during investigations into the effects of environmental contaminants on wild fish populations. Yellow perch was an obvious choice for us, given that it was a very common species in the metal-contaminated lakes that we were studying at the time. In fact, in some contaminated lakes, yellow perch was the only fish species present! Moreover, they were more or less ubiquitous across Canada and the northern United States and were relatively abundant where they occurred, which made them easy to sample and amenable to studies that generalized conclusions over broad geographic areas, even, perhaps, to their two cousins. Unfortunately, we soon realized that the basic literature relevant to our model organism—with a few notable exceptions—was somewhat diffuse and disparate. Yet perch—particularly the North American and European species—have recently seen an increase in research attention, in such diverse areas as population biology, ecology, biogeography, behaviour, and aquatic ecotoxicology.

This book endeavours to update and consolidate the perch literature in a manner that will be useful to fisheries managers, academics and students interested in perch biology. We wanted the book to focus exclusively on the three extant species comprising the genus *Perca*: *P. flavescens*, *P. fluviatilis*, and *P. schrenkii*. We also wanted to capture the range of research that is currently being conducted on *Perca* spp. and how that research is advancing our basic understanding of perch biology. We therefore approached internationally renowned perch specialists from every aspect of fish biology to contribute to this volume. We hope that the following pages meet that goal.

June 29, 2015

Patrice Couture, Québec City, Québec
Greg Pyle, Lethbridge, Alberta

Contents

1

Introduction to Biology of Perch

Greg Pyle[1,*] and Patrice Couture[2]

Recent estimates suggest that there are approximately 33,200 species of fish in the world, 7000 of which have been described since 1995 (Eschmeyer 2014). Of those, approximately 12,000 species strictly occupy fresh water habitats (Nelson 2006). The order Perciformes represents the largest group of vertebrates, consisting of some 18 suborders of fish, the largest of which is Percoidei, which contains some 71 families representing about 530 genera and 3000 species (Nelson 2006). One of those families, Percidae, consists of 10 genera of fishes limited to fresh water, particularly in the Northern Hemisphere. The focus of the present volume is on the three extant members comprising the genus *Perca*: *P. flavescens* Mitchill, 1814, *P. fluviatilis* Linnaeus, 1758, and *P. schrenkii* Kessler, 1874.

Perca flavescens is the yellow perch (sometimes referred to as the North American perch or "lake perch"). It is widely distributed across Canada and the northern United States, from just east of the Continental Divide to the eastern seaboard of North America, as far north as Great Slave Lake and south to Florida's panhandle. Its value as a sport fish varies throughout its range. Some of its life history traits make it amenable to angling, including its schooling habit in shallow areas during spring, its colourful appearance, and its delicious flesh. Where it occurs, it is relatively abundant relative to other freshwater species. Consequently, those that target *P. flavescens* value it as a sport fish.

[1] Dept. of Biological Sciences, Water Institute for Sustainable Environments (WISE), University of Lethbridge, Lethbridge, AB T1K 3M4 Canada.
[2] Institut national de la recherche scientifique Centre Eau Terre Environnement Québec QC G1K 9A9 Canada.
* Corresponding author: gregory.pyle@uleth.ca

Perca fluviatilis is the European perch (sometimes referred to as the Eurasian perch). It is widely distributed across Europe and Asia, ranging from northern Scandinavia to central Italy, and from the west coast of Ireland to the Kolmya River in eastern Siberia. Introduced populations occur in South Africa, New Zealand, and Australia (Nelson 2006). Together with *P. flavescens*, *Perca* spp. are truly circumpolar in their combined distribution in the Northern Hemisphere. *Perca fluviatilis* is somewhat larger than *P. flavescens*, and is prized as a sport fish throughout its range. However, given its opportunistic dietary habits, its tolerance to a range of environmental conditions from fresh to brackish (even salty) waters, and its prolific reproduction, *P. fluviatilis* has the potential to outcompete heterospecifics and become a pest to environmental managers.

In contrast to the widespread distribution of *P. fluviatilis*, *P. schrenkii* (the Balkhash perch) occupies a very limited geographic distribution around the Balkhash and Alakol lakes area of Kazakhstan. Despite its much closer geographic proximity to *P. fluviatilis*, molecular data suggest that it is more closely related to *P. flavescens* (Sloss et al. 2004). Although *P. schrenkii* formed the basis of an important commercial fishery in the Balkhash region, recent species introductions and over-exploitation have led to significant population declines.

Both *P. flavescens* and *P. fluviatilis* support important commercial and sport fisheries. In North America, most commercial fishing for yellow perch takes place in the Laurentian Great Lakes and the St. Lawrence River basin. In 2013, over 4.4 million kg (9.5 million lbs) of yellow perch was harvested from Lake Erie alone using both sport and commercial fishing gear. This catch represented an 11.2% reduction in harvest from 2012 despite a significant increase in fishing effort (GLFC 2014).

The biology of the species of *Perca* was well documented by Professor John Craig in his seminal work, *The Biology of Perch and Related Fish* (1987). Since the publication of his book, several advances in basic biological techniques have been realized, resulting in advances in general fish biology, including in the basic understanding of the biology of perch. This book does not attempt to repeat the efforts of Craig (1987), but rather extends them in several important ways.

In Chapter 2, Dr. Carol Stepien and her colleagues update the *Perca* phylogeny on the basis of recent molecular evidence and confirm its status as a monophyletic clade. New insights into perch biology are made possible by relatively recent advances in molecular tools. Researchers have reconstructed historic biogeographical distribution patterns that account for the current distribution of the three perch species, as well as their evolutionary relationships. As a group, perch have a comparatively low genetic diversity relative to other fishes, and demonstrate both broad-scale and fine-scale genetic differences among populations. Broad-scale differences are genetic, morphological, and behavioural among populations separated over large geographic areas (e.g., western yellow perch populations have a distinctly different morphological appearance than those from eastern populations). Fine-scale differences occur among spawning groups within a particular lake. These fine-scale genetic differences result from a fidelity to distinct spawning groups, which is a common trait among all three species. Understanding these genetic differences can help fisheries managers maintain perch stocks to meet competing demands between commercial and recreational fishing, as well as to inform conservation efforts to protect perch populations against large-scale environmental threats, such as environmental contamination or climate change.

In Chapter 3, Professor Nadir Mamilov from Al-Farabi Kazakh National University in Kazakhstan provides us with a glimpse of the rarest perch of all—the Balkhash perch. Although it is similar in many respects to the yellow perch of North America and the European perch, the Balkhash perch also shows several differences. Unlike its two cousins' broad geographic distributions, the Balkhash perch is restricted to a very small range in the Balkhash region in Southeast Asia. At one time, the Balkhash perch populations supported a large commercial fishery. However, over-fishing and the introduction of a number of exotic species in the region has since reduced Balkhash perch populations to a fraction of their earlier size. Increased predation pressure by newly introduced species has forced Balkhash perch into fast-flowing mountain streams, where they have recently been found at elevations over 1500 m. As Professor Mamilov discusses, understanding the profound effect that exotic species has on indigenous populations is critical to develop successful conservation strategies.

Canadian professor Joseph Rasmussen and Dr. Lars Brinkmann (Chapter 4) focus their attention on the distribution of yellow perch throughout the western part of their range. In a new analysis, they determine that mean annual air temperature and mean summer air temperature are key for understanding yellow perch distribution patterns, particularly in establishing the northern and southern boundaries of their range. However, temperature alone cannot explain their elevational limits in the Rocky Mountains, because there is an abundance of suitable perch habitat west of the Continental Divide that remains completely devoid of perch. Unlike Balkhash perch (Chapter 3), yellow perch are averse to fast moving water. Consequently, physical barriers, such as slope and current, limit their distribution west of the Continental Divide. This analysis reconciles with Stepien et al. (Chapter 2), who propose that perch colonized North America from Europe over a north Atlantic land bridge, as opposed to a Beringia land bridge from East Asia to western North America. Given that perch do not occur naturally west of the North American Continental Divide, it seems that the north Atlantic route is most plausible.

Moving from west to east, Yves Mailhot and colleagues (Chapter 5) turn our attention to the eastern yellow perch populations of the St. Lawrence River. The St. Lawrence River ranks among the world's largest rivers, flowing for over 3000 km from its headwaters in northern Minnesota into the Atlantic Ocean. Chapter 5 describes over 20 years of sampling data from a 350 km stretch of the St. Lawrence River through the Canadian province of Québec, which includes a range of different freshwater habitats, including long, narrow flowing stretches of river, archipelagos, and three large, shallow fluvial lakes. These extensive datasets clearly demonstrate distinctive traits within and among yellow perch subpopulations along the extent of the study area likely owing to the high degree of spawning group fidelity described in Chapter 2. The data presented in Chapter 5 also serve as an excellent case study in yellow perch conservation. The authors advise future yellow perch fisheries managers to consider the distinct perch population characteristics to maximize the likelihood of success. Experience with some St. Lawrence perch fisheries has shown that management decisions based solely on political or commercial considerations are destined to fail.

Just as Mailhot and colleagues (Chapter 5) showcase the knowledge to be gained from long-term fisheries datasets for understanding yellow perch stock dynamics in the St. Lawrence River system, Dr. John Craig and colleagues (Chapter 6) provide a similar case study on European perch from Lake Windermere, UK using data collected over 70 years! No perch population has been as comprehensively studied as the population of Lake Windermere. Long-term research of the Windermere perch population was originally motivated to inform policies designed to protect against worldwide overfishing by understanding the factors that influence population dynamics. Since then, the Windermere perch population has undergone considerable changes, including losing up to 98% of the population owing to a disease outbreak. Many of the selective pressures acting on this population, such as disease and predation by introduced species, can have surprising and unexpected effects on long-term population dynamics. Understanding these dynamics becomes extremely important, especially when facing large-scale environmental disturbances such as climate change.

Perch reproductive biology is discussed in varying degrees of detail in several chapters throughout this book. The reproductive biology of Balkhash perch is described in some detail in Chapter 3. Case studies are provided both for European perch (Chapter 6) and yellow perch (Chapters 4 and 5), while perch reproductive behaviours are described in Chapter 9. In Chapter 7, however, Dr. P. Fontaine and colleagues from the Université de Lorraine, France, highlight the similarities between the European and yellow perch in their reproductive biology, and summarize over 20 years of studies describing the influences of environmental variables, such as temperature and photoperiod, on gonad maturation and spawner quality. The authors of Chapter 7 present convincing evidence that perch aquaculture is a promising avenue in the near future, which will allow supporting restoration efforts as well as restocking of populations threatened by overfishing. Linking information provided in Chapter 7 with our knowledge of perch genetics (Chapter 2), spawning group fidelity and good practices in perch stock management (Chapters 5 and 6), the chapter contains invaluable information for managers to develop sound strategies for the protection of these important species

In Chapter 8, Drs. J. Behrmann-Godel and A. Brinker provide a detailed account of European and yellow perch parasites and host-parasite interactions. Perch parasites tend to be very diverse owing to generalized dietary preferences of perch, their widespread geographic distribution, and broad habitat preferences. This parasite diversity appears to hold true for individuals in a population that tend to migrate in and out of different habitats as well as at larger scales and among populations. The authors suggest that understanding host-parasite interactions not only advances our knowledge of parasite ecology, but it can also be used to glean insights into co-evolutionary mechanisms over a much longer timescale than any individual fish or parasite. An interesting outcome of some recent molecular-level research into perch-parasite interactions may be that we can learn about the evolutionary history of perch by simply studying their parasites. However, more work needs to be done, especially by conducting comparative studies between European and yellow perch.

Similar to its reproductive biology, the three perch species show remarkable similarities in their behaviours, according to Dr. C. Semeniuk and colleagues in Chapter 9. As we have already seen (Chapter 2), sympatric perch can form distinct local subpopulations owing partly to their tendency to form mating groups to which they have a high perennial fidelity. These subpopulations can show differences in genetics, morphology, and behaviour. Three possible mechanisms that can lead to these distinct subpopulations include homing, kin group fidelity, and cryptic population isolation. In this chapter, Semeniuk et al. capture the remarkable range and plasticity of perch behaviours that allow it to live a solitary or shoaling existence, in benthic or littoral habitats, as part of either the prey or predator guilds, or even as a cannibal. Given the relatively low genetic variability among perch populations, the range of behaviours demonstrated by these species is truly impressive and is a reflection of local adaptations. Moreover, perch behaviours are controlled by local environmental conditions, including both biotic and abiotic factors. Consequently, as environmental conditions change so too does individual perch behaviour. Understanding how those behaviours change in response to a changing environment, and the ecological consequences of those changing behaviours are important to fisheries resource managers and environmental conservation efforts.

In the final chapter, Chapter 10, Couture et al. demonstrate the value of perch in aquatic ecotoxicological research. The literature is dominated by studies involving yellow perch and European perch; comparatively little ecotoxicological work has been done on Balkhash perch. Perch are a useful species in ecotoxicological studies because their circumpolar geographic distribution overlaps with large-scale industrial activities, such as metal mining and smelting. The authors provide two case studies of their own work, summarizing approximately 15 years of research on wild perch populations from industrial regions in Ontario and Québec, Canada. The research described in that chapter makes use of both field and laboratory techniques in studies investigating the physiological, immunological, and toxicological effects of metals in wild perch populations. New molecular tools have been developed, such as a novel yellow perch cDNA microarray, that allow researchers the ability to explore questions related to long-term adaptation to environmental contamination.

The biology of perch has seen a number of fundamental advances since Professor Craig's original work. New research tools are bridging the gap between results obtained under the controlled conditions of the laboratory and natural perch populations in the wild. Several of our chapters have common themes that demonstrate the similarity among the three different perch species, but also highlight their differences—especially among conspecific subpopulations. We have strived to capture some of the exciting advances that have occurred recently in perch biology and hope that this volume contributes towards the next generation of new advances in the field.

References

Craig, J.F. 1987. The Biology of Perch and Related Fish. Croom Helm, UK p. 333.

Eschmeyer, W.N. 2014. Online Catalog of Fishes [Datebase]. Available from http://research.calacademy.org/redirect?url=http://researcharchive.calacademy.org/research/Ichthyology/catalog/fishcatmain.asp [accessed 08/13/14 2014].

Great Lakes Fisheries Commission (GLFC). 2014. Report of the Yellow Perch Task Group. Report to the Standing Technical Committee, GLFC. p. 41.

Nelson, J.S. 2006. Freshwater Fishes of the World. 4th Edition ed. John Wiley and Sons, Ltd., New York, NY.

Sloss, B.L., N. Billington and B.M. Burr. 2004. A molecular phylogeny of the Percidae (Teleostei, Perciformes) based on mitochondrial DNA sequence. Mol. Phylogenet. Evol. 32(2): 545–562. doi:10.1016/j.ympev.2004.01.011.

2

Evolutionary Relationships, Population Genetics, and Ecological and Genomic Adaptations of Perch (*Perca*)

Carol A. Stepien,[1,*] Jasminca Behrmann-Godel[2] and Louis Bernatchez[3]

ABSTRACT

The latest results about the evolutionary, biogeographic, and population genetic relationships of the three species comprising the percid fish genus *Perca* are presented, explained, and discussed. New analyses from new data dated the origin of the genus to an estimated 19.8 million years ago (mya) during the early Miocene Epoch, and the distribution of ancestral *Perca* likely extended across the North Atlantic Land Bridge until the mid-Miocene. The earliest evolutionary bifurcation led to the diversification of the European perch *P. fluviatilis* from the lineage shared by the common ancestor of the North American yellow perch

[1] Great Lakes Genetics/Genomics Laboratory, Lake Erie Center and Department of Environmental Sciences, The University of Toledo, Toledo, OH, 43615, USA.

[2] Limnological Institute, University of Konstanz, Mainaustrasse 252, 78457 Konstanz, Germany. E-mail: Jasminca.Behrmann@uni-konstanz.de

[3] Département de Biologie, Institut de Biologie Intégrative et des Systèmes (IBIS), Pavillon Charles-Eugène-Marchand, 1030, Avenue de la Médecine, Local 1145 Université Laval, Québec (Québec) G1V 0A6 Canada.
E-mail: Louis.Bernatchez@bio.ulaval.ca

* Corresponding author: carol.stepien@utoledo.edu

P. flavescens and the Eurasian Balkash perch *P. schrenkii*. The latter two species diverged during the later Miocene, after the Land Bridge was closed. The European and yellow perches are both widely distributed across their respective continents, with biogeographic areas housing high genetic distinctiveness. Population genetic structure in their northern regions were shaped by post-glacial colonization patterns from multiple refugia, whose admixture increased diversity. Today's spawning groups are modest in genetic diversity yet very divergent from one another, which may reflect an apparent tendency of perch to live with relatives throughout their lives. There is a disconnect between the genetic divisions among populations and the delineation of fishery management units in the yellow perch, which is of concern. Employing a combined fisheries management and genetics/genomic approach will provide further understanding to help maintain the genetic diversity and unique adaptations of perch populations in the face of increasing anthropogenic influences, including climate change.

Keywords: Balkhash perch, biogeography, conservation genetics, evolutionary patterns, European perch, fisheries management, genomics, *Perca flavescens*, *Perca fluviatilis*, *Perca schrenkii*, phylogenetics, phylogeography, population genetics, yellow perch

2.1 Evolutionary and Biogeographic History of *Perca*

The percid genus *Perca* contains three economically and ecologically important species, which are top piscivores in North America and Eurasia. The yellow perch *P. flavescens* Mitchill, 1814 is endemic to North America, whereas two *Perca* species are native to Eurasia—the European perch *P. fluviatilis* Linnaeus, 1758 and the Balkhash perch *P. schrenkii* Kessler, 1874. Both the yellow perch and the European perch are widespread across much of their respective continents (Figs. 1 and 2), where they support popular recreational and commercial fisheries. They each have been widely introduced for angling outside their native ranges. Perch also serve as an important model species in ecotoxicology studies (e.g., Chapter 10, this volume). The goals of this chapter are to summarize the current knowledge and recent study results about their evolutionary and population genetic diversification, with implications and suggestions for further research.

The Balkhash perch is native to Lakes Balkhash and Alakolin Kazakhstan, where it supported abundant fisheries during the 1930s and 1940s (Sokolovsky et al. 2000). The Lake Balkhash fishery collapsed, however, contemporary harvests remain in the Lake Alakol region (Sokolovsky et al. 2000); the species also occurs in associated catchments extending into China (Berg 1965). During the 1960s and 1970s, the Balkhash perch was introduced to various water bodies in Uzbekistan, Kazakhstan, and Middle Asia (Kamilov 1966; Nuriyev 1967; Pivnev 1985). The Balkhash perch is classified in "The IUCN of Threatened Species" (v2013.2) as "data deficient" (http://www.iucnredlist.org/). Chapter 3 (this volume) is dedicated to this rare species.

As the phylogenetic relationships of the Balkhash perch to the other two *Perca* species was unresolved, we harvested sequence data from NIH GenBank, obtained new samples to sequence (by the Stepien laboratory), analyzed two mitochondrial genes

A

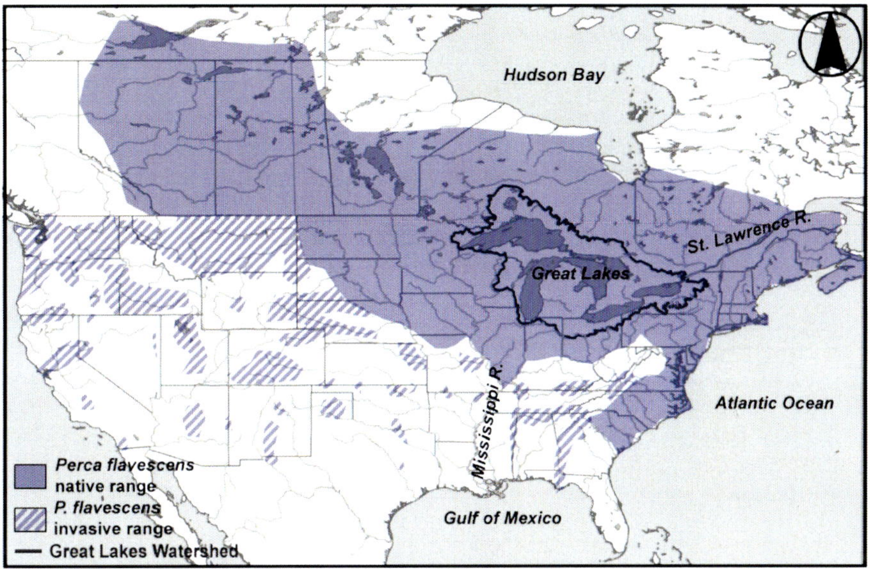

B

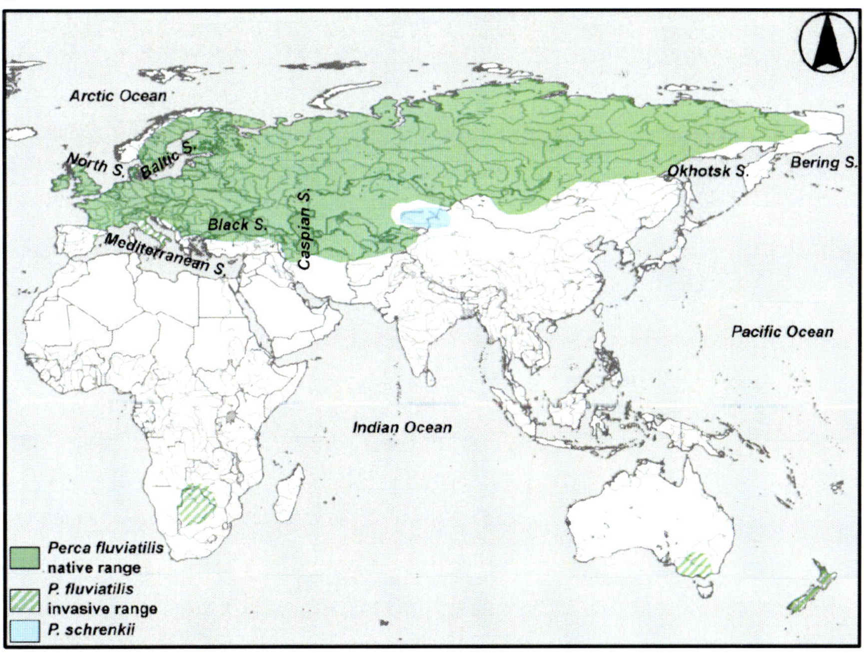

Fig. 1. Maps showing the native and invasive ranges of *Perca* spp. in (A) North America and (B) Eurasia, using information adapted from Collette and Bănărescu (1977), Craig (2000), Page and Burr (2011), and Fuller and Neilson (2012).

A

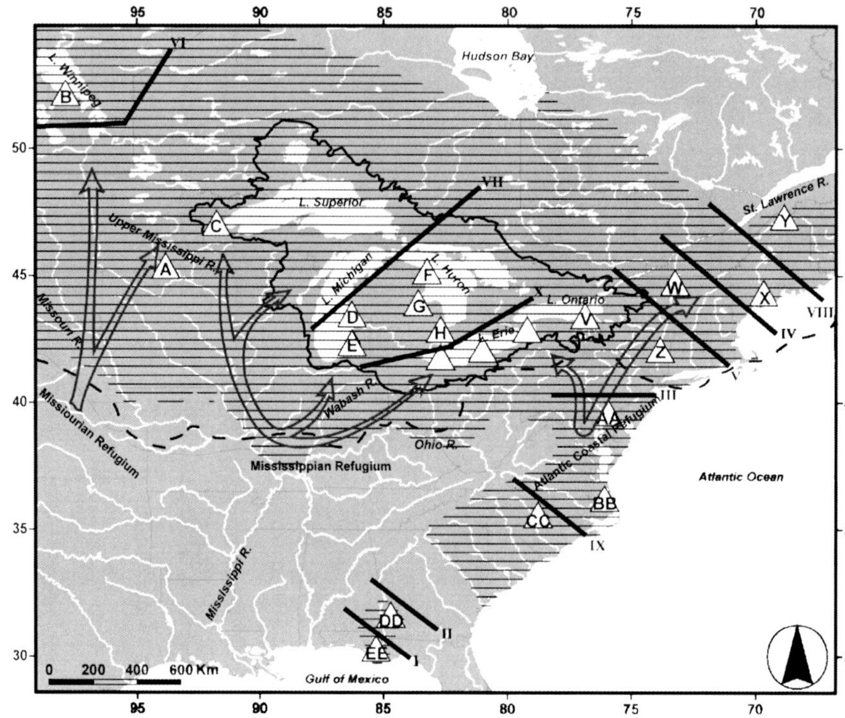

B

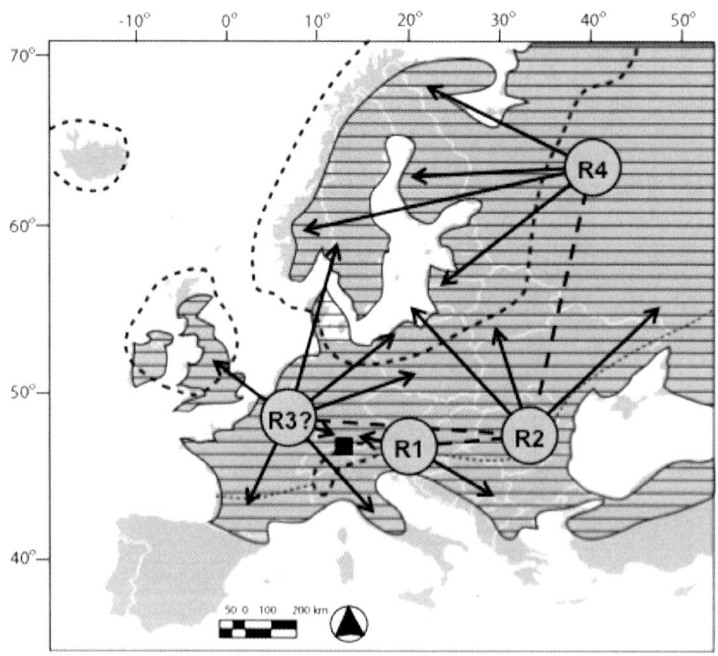

Fig. 2. contd....

and a nuclear DNA gene region, and conducted new phylogenetic analyses. A new phylogeny (Fig. 3), based on concatenated DNA sequence data from the mitochondrial (mt) cytochrome *b* (cyt*b*) and cytochrome *c* oxidase I (COI) genes, and the nuclear recombination-activating gene intron 1 (RAG1), was determined for the present study by the Stepien laboratory using Bayesian analysis in MrBAYES v3.2.1 (Ronquist and Huelsenbeck 2003). We also conducted separate analyses on each gene, whose results were congruent to those obtained from the concatenated gene tree (Fig. 3). Results demonstrate that the genus *Perca* is monophyletic and well-supported (with 1.0 Bayesian posterior probability and 100% maximum likelihood bootstrap support).

Fossil record calibration data indicate that the genus *Perca* originated by an estimated 19.8 million years ago (mya) during the early Miocene Epoch (Fig. 3). It thus shares a similar biogeographic origin and pattern with the percid genus *Sander* (see recent phylogeny by Haponski and Stepien 2013). DNA sequence data indicated that the genus *Perca* is monophyletic and comprises three well-defined species (Fig. 3), which likewise is supported by morphological characters (see Craig et al. 2000). As indicated by the DNA data (Fig. 3), the primary division in the genus separates out the European perch (*P. fluviatilis*) from the Balkhash and yellow perch; the latter shared a more recent common ancestry. The Balkhash and yellow perch are sister species (nearest relatives), but are each highly distinct, having diverged an estimated 13.4 mya during the mid-Miocene Epoch. The common distribution of the ancestral *Perca* might have extended bi-continentally across either the North Atlantic Land Bridge and/or the Beringia Land Bridge, which once linked the continents across the Atlantic and the Pacific oceans, respectively (see Fig. 2). However, given that the native distribution of *P. flavescens* is exclusively east of the Rocky Mountains in North America and extends to the Atlantic Coast (Fig. 1A) and the distribution of *P. fluviatilis* extends northwest to the Atlantic Coast in Eurasia and is absent from far southeastern Asia (Fig. 1B), the distribution of their once-common ancestor most likely extended across the North Atlantic Land Bridge. This is concordant with findings by Wiley (1992)

Fig. 2. Maps showing the distributions of the North American yellow perch *Perca flavescens* (A) and the European perch *P. fluviatilis* (B), with their former areas of glacial refugia noted. A. Sampling locations for yellow perch (triangles) referred to in this study. Thick dashed line indicates the maximum extent of the Wisconsinan glaciations, arrows denote likely routes of post-glacial population colonizations (adapted from Mandrak and Crossman 1992; Stepien et al. 2012; Sepulveda-Villet and Stepien 2012). Grey lines (solid = microsatellite data, dashed = mtDNA control region sequences) denote major barriers to gene flow calculated based on the relationship between geographic coordinates (latitude and longitude) and genetic divergence (F_{ST}) using BARRIER v2.2 analysis (Manni et al. 2004, http://www.mnhn.fr/mnhn/ecoanthropologie/software/ barrier.html). Results are modified from Sepulveda-Villet and Stepien 2012, with genetic barriers ranked I–V for the microsatellite and 1–5 for the mtDNA control region data sets, in order of their decreasing magnitude. B. Map showing the distribution (hatched area) and postulated areas of former glacial refugia (R1–R4) for *P. fluviatilis* in Europe. Hatched lines indicate phylogenetic relationships between mtDNA haplotypes that originated in the refugia, arrows indicate postglacial dispersal pathways. The position of refugium R3 is somewhat unclear, as indicated by a question mark. The area surrounded by a heavy hatched line indicates the extent of the ice during the last glaciation (the Weichselian glaciation) and the glacier covering the Alps. The light hatched line indicates the southern border of the permafrost during the same time. The square indicates the position of the Lake Constance area. This figure was modified from Fig. 6 of Nesbø et al. (1999) and Fig. 1 of Hewitt (1999).

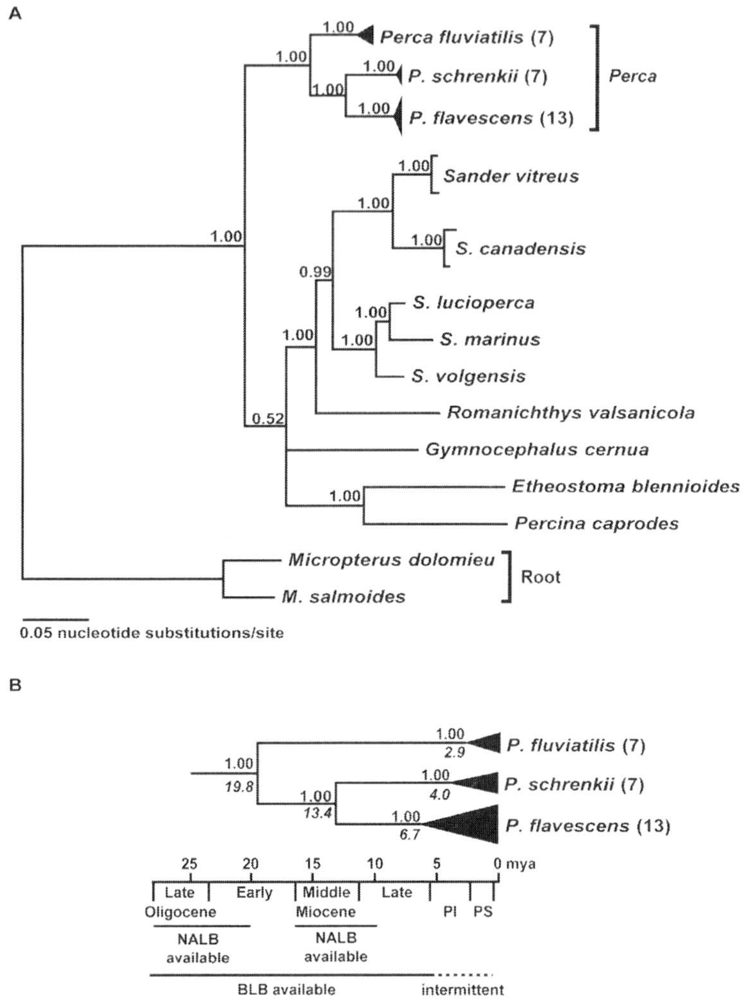

Fig. 3. Phylogeny of *Perca*. (A) Bayesian phylogenetic tree of mtDNA cytochrome *b* sequence haplotypes for the genus *Perca* and members of the family Percidae, determined for this study using concatenated sequence data set from three genes: mitochondrial cytochrome *b* (cyt*b*) and cytochrome *c* oxidase I (COI), and the nuclear recombination-activating gene intron 1 (RAG1), calculated using MrBAYES v3.2.1 (Ronquist and Huelsenbeck 2003, http://mrbayes.sourceforge.net). The program jMODELTEST v2 (Darriba et al. 2012, https://code.google.com/p/jmodeltest2/) was employed to select the best substitution model. The relationships among the *Perca* from the combined concatenated gene data were identical to this from separate analyses of all three genes. Tree is rooted to *Micropterus* based on its hypothesized close relationship to the Percidae, according to Song et al. (1998) and Sloss et al. (2004).

(B) Time-calibrated phylogeny for *Perca* derived using BEAST. Dates for the availability of the North Atlantic Land Bridge (NALB) were determined from Tiffney (1985) and Denk et al. (2011), and for the Bering Land Bridge (BLB) from Gladenkov et al. (2002). PI = Pliocene, PS = Pleistocene.

Above nodes = Bayesian posterior probability (pp) support. Below nodes in italics = estimated divergence times (millions of years) from BEASTv1.71 (Drummond et al. 2012, http://beast.bio.ed.ac.uk/), using the fossil calibration point of 12.0 mya for the genus Micropterus, and three molecular calibration points: 15.4 mya for the origin of the North American Sander, 13.8 mya for the Eurasian Sander, and 9.1 mya for the divergence between S. lucioperca and S. marinus, adapted from Haponski and Stepien (2013).

and Carney and Dick (2000). Moreover, a recent paper for the similarly distributed pikeperch genus *Sander*, likewise pointed to historic connection over the North Atlantic Land Bridge (Haponski and Stepien 2013). The North Atlantic Land Bridge appears to have been disrupted during the estimated interval of 17–20 mya, which matches this estimated time period for *Perca* taxon divergence, determined using BEAST 1.7 analyses (Drummond et al. 2012) and fossil date calibrations shown in Fig. 3. It may be that, given the phylogeny and these dates, this differentiation between *P. fluviatilis* and the *P. flavescens*/*schrenkii* linage began during this disruption in the North Atlantic Land Bridge. The lineages leading to the *P. schrenkii* and *P. flavescens* species later differentiated over the 4.4–6.7 to 13.4 mya time range, as indicated on Fig. 3.

2.2 Broadscale Biogeographic Patterns

2.2.1 Population Relationships Shaped by Glaciations and Colonizations

The yellow perch has a wide native geographic distribution that extends across much of the northeast and northcentral regions of North America, with a few isolated relict populations in the southeast (Figs. 1A and 2A). It inhabits a diversity of lacustrine and fluvial habitats, ranging from large to small in geographic areas, with its most extensive habitats and greatest abundances occurring in the Laurentian Great Lakes—especially in Lake Erie (Scott and Crossman 1973; Hubbs and Lagler 2004) and the St. Lawrence River system (Bernatchez and Giroux 2012).

In the northern regions of the yellow perch's range, the habitats and basins of the Great Lakes region were formed and reshaped by the Laurentian Ice Age glaciations, leading to their present configuration about 4,000–12,000 years ago (ya). Similarly in Europe, the most recent Pleistocene cold stages, especially the Weichselian glaciation 13,000–25,000 ya, shaped the habitats of many freshwater fish species—including the European perch (Nesbø et al. 1999). During the glaciations, perches and other aquatic species migrated to waters south of the ice sheets, where their populations were concentrated in restricted areas, known as glacial refugia (Hocutt and Wiley 1986).

Three primary North American glacial refugia are recognized (marked on Fig. 2A), which are: the Mississippian refugium in the central U.S., the Missourian refugium to the west, and the Atlantic refugium to the east (Bailey and Smith 1981; Crossman and McAllister 1986; Mandrak and Crossman 1992). Following the glacial meltwaters, yellow perch and other aquatic taxa migrated along tributary pathways leading from the refugia into the reformed water bodies of the Great Lakes and other northerly habitats (see Fig. 2A). Today's northern populations of yellow perch and other fishes appear to retain the signatures of their genetic origins from the respective glacial refugia (summarized by Sepulveda-Villet and Stepien 2012 for yellow perch). The yellow perch, although now adapted to the large inland "seas" that comprise the Great Lakes, had its ecological and evolutionary origins in fluvial systems rather than large, lacustrine basins.

A very similar picture occurred in Europe (Fig. 2B). Many European freshwater fish species were driven from their original ranges and their distribution restricted to refugia situated in the three Mediterranean peninsulas: the Iberian (Atlantic-Mediterranean refugium), Italian (Adriatic-Mediterranean refugium), and Balkan peninsulas (Pontic-Mediterranean refugium) (Hewitt 1999; Schmitt 2007), or in eastern continental regions into Asia or in northerly refugia bordering the ice shields (Stewart and Lister 2001). Later, with the glacial retreats, populations dispersed from their respective refugia and re-expanded their ranges. For European perch, a phylogeographic study by Nesbø et al. (1999) indicated the existence of four different refugia (labelled R1–R4 on Fig. 2B), from where European perch commenced their most recent colonisation patterns. The Danubian refugium (R1) presumably served as a founder population that led to few of the present European perch lineages (Nesbø et al. 1999; Behrmann-Godel et al. 2004). Most contemporary European perch populations trace their origins to the other three refugia (R2–R4) (Nesbø et al. 1999).

2.2.2 Phylogeographic Patterns of the European Perch

The European perch possesses a complex evolutionary history that has been delineated through phylogeographic studies. During the last Ice age in Central Europe, a wide plain of permafrost and cold steppe stretched between the northern main ice sheet and the southern mountain ranges—including the Alps—forming a very cold and dry environment (Fig. 2B) (Hewitt 1999). During this time, many freshwater fish species were driven from their original ranges, with their distributions restricted to refugia situated southward and/or eastward of the ice shield (Taberlet et al. 1998; Stewart and Lister 2001; Schmitt 2007).

When the ice later retreated, the populations dispersed again from their respective European refugia and re-expanded their ranges. As supported for other freshwater fish species in Europe, including chub *Leuciscus cephalus* (Durand et al. 1999) and burbot *Lota lota* (Barluenga et al. 2006), the European perch followed a hypothesized two-step expansion model. In the first step, descendants of an eastern lineage (refuge R2 on Fig. 2B), extensively colonized Central Eastern and western Europe likely during the Riss-Würm interglacial (between 115,000–126,000 ya). These survived the next glacial period in various refugia that were located either in Western European rivers, such as Rhine and Rhone (refuge R3 on Fig. 2B), in eastern drainages that enter the Black Sea (refuge R2), and/or in northeastern rivers (refuge R4). During the second step, at the end of the Würm period (~10,000 ya), range expansions took place from the Western stocks into all Atlantic drainages, and from the eastern and northeastern stocks into the rest of Europe. Through these pathways, the lineages reached and met together in the Baltic Sea region, resulting in a high level of heterogeneity (Nesbø et al. 1999).

During this recolonization process in Central Europe, several lineages that diverged in refugial allopatry came into secondary contact (Taberlet et al. 1998; Hewitt 2004a). For European perch, contact between refugial lineages has been indicated in drainages that either enter or postglacially entered the Baltic Sea, the Danube River, and the Lake Constance area (Nesbø et al. 1999; Behrmann et al. 2004). Several studies have identified a contact zone for Atlantic and Danubian fish lineages along the French-German border and in the western Alps, which corresponds to the closest area between the Rhine and Danube rivers (Taberlet et al. 1998; Hewitt 2004b). Lineages in that area possess genetic signatures from both the eastern Danubian and the western cryptic Alpine or Atlantic refugia (see, e.g., Durand et al. 1999; Nesbø et al. 1999; Bernatchez 2001; Volckaert et al. 2002).

Lake Constance is a large pre-alpine lake located in the center of the contact zone between the Rhine and the Danube river systems (Figs. 2B and 4). It was formed after the last glacial period (Würm) with the retreat of the Alpine Rhine glacier, 10,000–15,000 ya (Fig. 4; Keller and Krayss 2000; Behrmann-Godel et al. 2004). It appears that the two neighboring systems in the Rhine and Danube rivers were connected only briefly at the end of the last recessional stage of the glacier (Wagner 1960; Keller and Krayss 2000). With the glacial retreat, melt water streams and proglacial lakes were formed (Fig. 4A–C; Wagner 1960; Keller and Krayss 2000), which provided a potential temporary colonization that could have been used by Danubian fish lineages

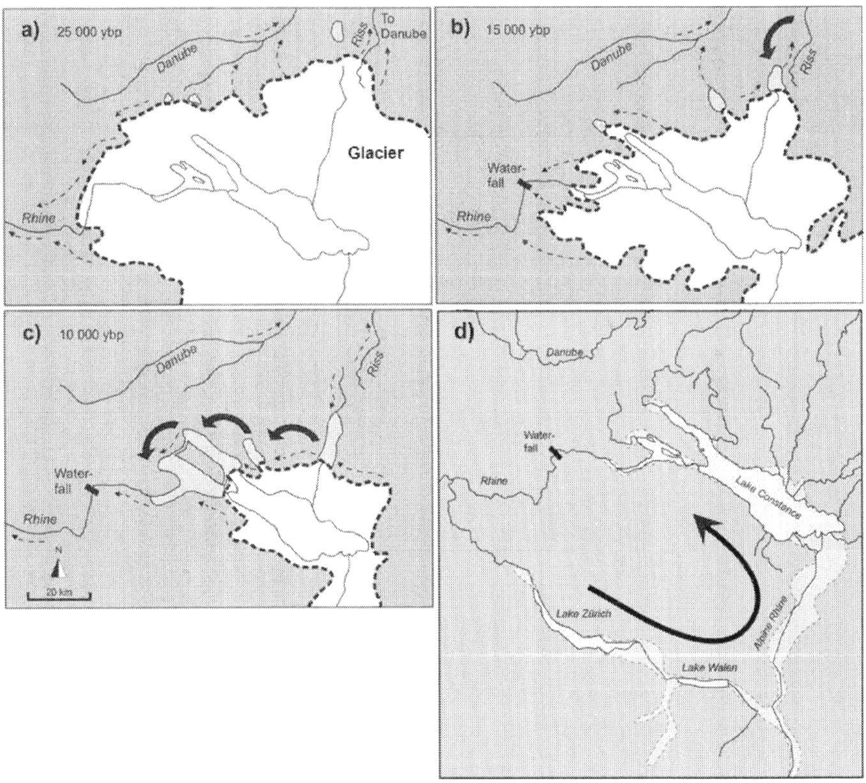

Fig. 4. Colonization route of Danubian (a–c) and Atlantic (d) haplotypes of *P. fluviatilis* into Lake Constance. a) During the last glacial maximum the Alpine glacier extended almost to the Danube system and melt water (hatched arrows) was running into the Danube. b) With the retreating glacier, huge proglacial lakes formed where perch from the Danube system could potentially have entered (black arrow) using the melt water streams as "colonization bridges", whereas colonization directly from the Rhine upstream was prevented by the Rhine waterfall. d) Postglacial connection of the Rhine and the Lake Constance region via Lakes Zürich and Walen after the last glacial period (redrawn after Wagner 1960). The arrow indicates the potential colonization route for Atlantic fish lineages via the connected lake system (area within dashed lines). Figure modified from Behrmann-Godel et al. 2004 and J. Behrmann-Godel, M. Barluenga, A. Meyer, and W. Salzburger, unpublished.

to colonize the lake; this has been shown for European perch, and burbot (Behrmann-Godel et al. 2004; Barluenga et al. 2006). At the end of the glacial period, the Lake Constance region became isolated westward from the upper Rhine River by a 23 meter high Rhine waterfall (Fig. 4c), which cutoff the upstream migration of Atlantic fish lineages (Behrmann-Godel et al. 2004). Directly after the retreat of the Alpine Rhine glacier, the water level of Lake Constance was 410 meters above sea level (15 meters higher than today). Lake Constance then was part of a huge postglacial lake system, which extended along the valley of the Alpine Rhine and connected them via the Linth River and through lakes Walen and Zürich (Fig. 4d). This connection existed until the outflowing Rhine River had excavated its channel, lowering Lake Constance to its present level of 395 meters above sea level. This connection putatively was used by Atlantic lineages of European perch, burbot, bullhead, and the vairone *Telestes muticellus* to colonize the region, which then came into secondary contact with the Danubian lineages within the Lake Constance area (Behrmann-Godel et al. 2004; Barluenga et al. 2006; Behrmann-Godel et al., unpublished).

2.2.3 Contemporary Broadscale Population Relationships of Yellow Perch

Contemporary haplotypes of the yellow perch in North America appear to trace to ~6.0 million years ago (mya; Sepulveda-Villet and Stepien 2012, Fig. 5). Regional distinctiveness of yellow perch metapopulations is very apparent, with those from the south and north, and east versus west, being highly differentiated (Table 2, Figs. 2A, 4–6, also see Stepien et al. 2009; Sepulveda-Villet and Stepien 2012). Divergences of the southern yellow perch haplotypes appear to date to ~2.5–3.6 mya, whereas the northern haplotypes share a common ancestry estimated ~0.6–4.2 mya among them (Fig. 5; Sepulveda-Villet and Stepien 2012).

Pronounced genetic demarcations delineate that the most unique yellow perch populations are located in six major geographic regions: Northwest Lake Plains, Great Lakes watershed, Lake Champlain, North Atlantic coastal, South Atlantic coastal, and Gulf coastal (Figs. 2A and 4–6; Sepulveda-Villet and Stepien 2012). The Atlantic coastal yellow perch populations possess high endemism today (Griffiths 2010), as evidenced by their genetic diversity and unique alleles (Table 1; Sepulveda-Villet and Stepien 2012). Substantial genetic diversity in the southerly, unglaciated populations may be due to their long undisturbed history for evolution and local adaptation (April et al. 2013). The South Atlantic coastal yellow perch populations are adapted to mesohaline conditions, and likely readily migrate from fresh to brackish waters (Grzybowski et al. 2010). The South Atlantic and Gulf coastal haplotypes are more closely related to each other than to those from the North Atlantic region (Sepulveda-Villet and Stepien 2012).

In comparison, the southern Gulf relict population of yellow perch sampled has relatively lower heterozygosity (Table 1), which is characteristic of its small population size, bottlenecks, and genetic drift. Yet, it also possesses a high number and proportion of private alleles, indicative of long-term isolation and distinctiveness (Table 1; Figs. 2A, 4–6). The relict Gulf coastal population appears related to, yet distinct from, populations of the southeast Atlantic seaboard (Sepulveda-Villet et al. 2009; Sepulveda-Villet and Stepien 2012).

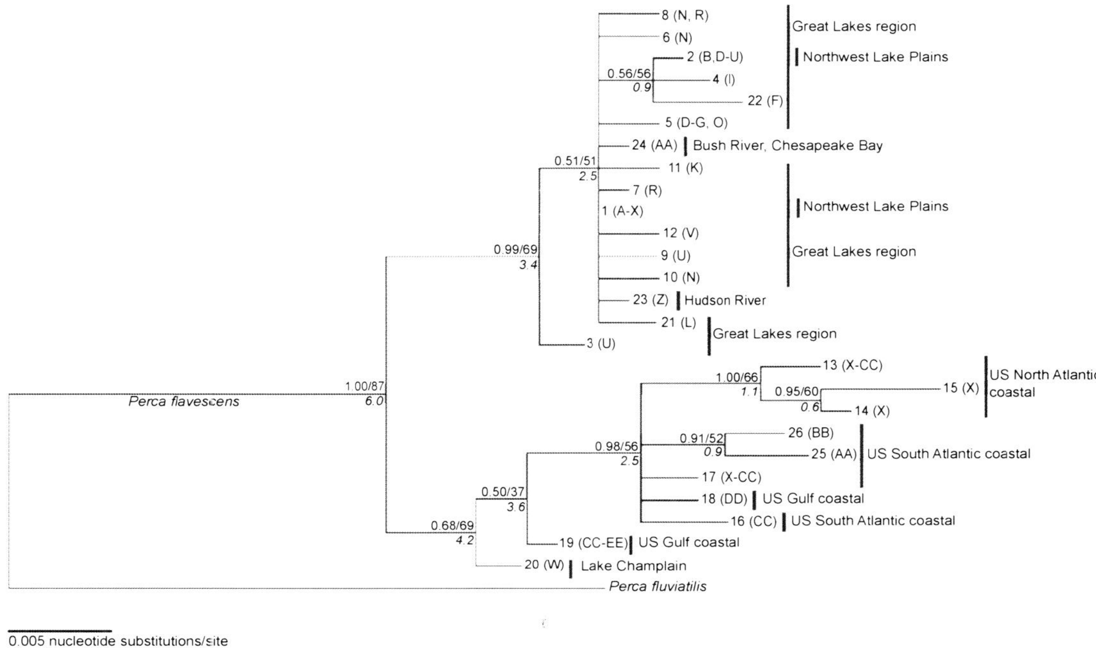

Fig. 5. Bayesian phylogenetic tree of mtDNA control region sequence haplotypes for yellow perch, calculated using MrBAYES v3.2.1 (Ronquist and Huelsenbeck 2003, http://mrbayes.csit.fsu.edu/). Values above nodes = Bayesian posterior probability/percentage support from 2,000 bootstrap pseudo-replications in ML with PHYML v3.0 (Guindon et al. 2010, http://www.atgc-montpellier.fr/ phyml/); those with ≥0.50 pp and ≥50% bootstrap support are reported. jMODELTEST v2 (Darriba et al. 2012, https://code.google.com/p/jmodeltest2/) selected the most likely model of nucleotide substitution for construction of the phylogenetic trees and divergence time estimates. Values below nodes in italics = estimated divergence times (given as millions of years) as determined in r8s v1.71 (Sanderson 2003, http://loco.biosci.arizona.edu/r8s/) and BEAST v1.71 (Drummond et al. 2012, http://beast.bio.ed.ac.uk/Main_Page#Citing_BEAST). Divergence times were calibrated using three fossil and four molecular calibration points following Haponski and Stepien (2013b, 2014). Letters in parentheses denote sampling sites in which haplotypes were recovered (see Fig. 2 map). Vertical bars denote general geographical regions.

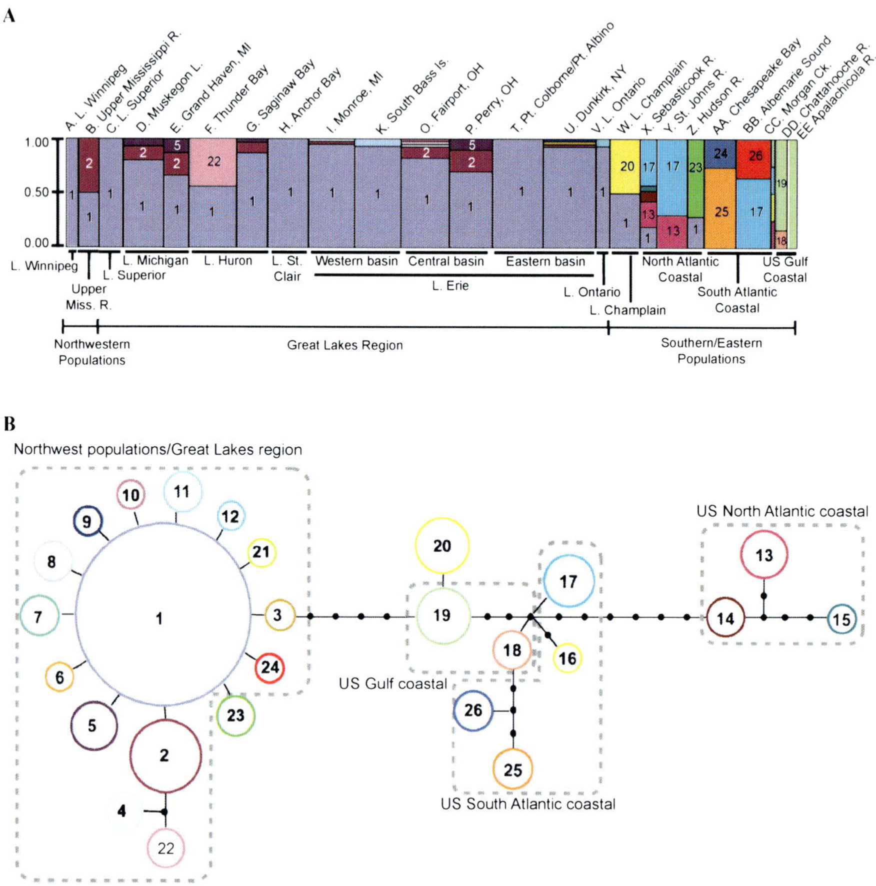

Fig. 6. MtDNA haplotypes of yellow perch across North America (modified from Sepulveda-Villet et al. 2009; Sepulveda-Villet and Stepien 2012). (A) MtDNA control region haplotype frequencies calculated using GENEPOP v4.0 (Rousset 2008, http://kimura.univ-montp2. fr/~rousset/Genepop.htm), and Microsoft Excel 2008 (Redmond, VA). Vertical black lines separate different spawning groups (lettered). Major geographic regions are indicated in the bottom rule. (B) Parsimony network of relationships among yellow perch mtDNA control region haplotypes constructed using TCS v1.21 (Clement et al. 2000, http://darwin. uvigo.es/software/ tcs.html). Circles are sized according to total observed frequency of the haplotype. Lines indicate a single mutational step between the haplotypes. Small, unlabelled circles represent hypothesized unsampled haplotypes. Dashed lines enclosing haplotype groups denote major regional delineations. Circle colors reflect haplotype identities as portrayed in Fig. 6A.

Yellow perch in the Northwest Lake Plains region of North America (Fig. 2A: Lake Winnipeg and the upper Mississippi River) trace their descent to Missourian refugium colonists (Sepulveda-Villet et al. 2009; Sepulveda-Villet and Stepien 2012). Today, the overall differences in these populations from other regions is apparent in Figs. 2A and 6, with the latter depicting a Bayesian STRUCTURE analysis of populations, based on nuclear DNA microsatellite loci (Table 1) which shows that

Table 1. Yellow perch metapopulation regions tested, sample size (N), and mean genetic variability values from (A) 15 microsatellite loci and (B) mtDNA control region sequences. Microsatellite data include: observed (H_O) heterozygosity, inbreeding coefficient (F_{IS}), number of μsat alleles across all loci (N_A), allelic richness (A_R), proportion of private alleles (P_{PA}), and proportion of full siblings (Sib). Values for mtDNA include number of haplotypes (N_H), haplotypic diversity (H_D), and proportion of private haplotypes (P_{PH}). Values were calculated using the programs GENEPOP v4.0 (Rousset 2008, http://kimura.univ-montp2.fr/~rousset/Genepop.htm), FSTAT v2.9.3.2 (Goudet 1995, 2002, http://www2.unil.ch/popgen/softwares/fstat.htm), ARLEQUIN v3.1.5.3 (Excoffier and Lischer 2010, http://cmpg.unibe.ch/software/arlequin35/), and CONVERT v1.31 (Glaubitz 2004, http://www.agriculture.purdue.edu/fnr/html/faculty/rhodes/students%20and%20staff/glaubitz/software.htm). Data are summarized and adapted from Sepulveda-Villet and Stepien (2012).

	A. Nuclear Microsatellite DNA Loci						**B. mtDNA Control Region Haplotypes**			
Locality	N	H_O	F_{IS}	N_A	A_R	P_{PA}	N	H_D	N_H	P_{PH}
Total (or Mean)	892	0.53	0.145	442	8.39	(0.04)	664	0.73	0.029	111
1. Lake Winnipeg	12	0.49	–0.010	68	4.53	0.02	12	0.00	1	0.00
2. Upper Mississippi R. watershed	18	0.52	0.165	112	7.47	0.04	18	0.53	2	0.00
Great Lakes region (3–9):	459	0.55	0.206	363	9.97	0.14	459	0.22	14	0.07
3. Lake Superior	25	0.64	0.080	119	7.93	0.01	25	0.00	1	0.00
4. Lake Michigan	65	0.54	0.174	298	9.93	0.05	65	0.34	3	0.00
5. Lake Huron	80	0.61	0.135	355	11.83	0.02	80	0.40	4	0.15
6. Lake St. Clair	86	0.59	0.098	225	13.22	0.03	39	0.00	1	0.00
Lake Erie:	401	0.55	0.116	313	13.26	0.09	235	0.21	12	0.03
7. Western Basin, L. Erie	189	0.55	0.100	259	12.87	0.05	77	0.27	4	0.03
8. Eastern Basin, L. Erie	212	0.54	0.122	270	13.00	0.06	88	0.07	4	0.03
9. Lake Ontario	62	0.55	0.122	213	13.79	0.04	15	0.13	2	0.07
10. Northeastern populations	60	0.50	0.236	347	7.71	0.05	60	0.48	3	0.29
11. Southeastern populations	68	0.60	0.132	349	7.78	0.06	68	0.63	7	0.62
12. US Gulf coastal region	15	0.39	0.346	108	3.60	0.07	15	0.15	2	0.13

the Lake Winnipeg and Upper Mississippi River populations are different from those in most of the Great Lakes. Yellow perch from western Lake Superior also are very distinctive based on microsatellite DNA data (Fig. 7, Sepulveda-Villet and Stepien 2012). Glacial Lake Agassiz initially occupied much of the Hudson Bay watershed (including Lake Winnipeg), which probably had some southern drainage to Lake Superior (Mandrak and Crossman 1992; Rempel and Smith 1998), facilitating fish movements 8,500–13,000 ya. Ice later blocked this passage (Saarnisto 1974; Teller and Mahnic 1988), isolating the yellow perch populations in our Northwest Lake Plains sites, as is shown by their high divergences from other areas (denoted by distinct colors on Fig. 7, Sepulveda-Villet and Stepien 2012). The Lake Superior region was long covered in ice, except for glacial Lake Duluth in the west until ~8,500–9,000 ya, thus isolating its yellow perch gene pools. Most of the Great Lakes fauna—especially in Lakes Huron, Michigan, St. Clair, and western Lake Erie (Underhill 1986; Mandrak

Table 2. Genetic divergence F_{ST} (Weir and Cockerham 1984) pairwise comparisons among yellow perch regional metapopulations, based on: A. nuclear DNA microsatellite loci (below diagonal) and B. mtDNA control region sequence data (above diagonal), using FSTAT v2.9.3.2 (Goudet 1995, 2002, http://www2.unil.ch/popgen/softwares/fstat.htm) and ARLEQUIN v3.1.5.3 (Excoffier and Lischer 2010, http://cmpg.unibe.ch/software/arlequin35/), with significance tested through 100,000 replicates. Results are congruent to those from exact tests of differentiation comparisons. Note that spawning populations are grouped together in metapopulation regions for purpose of comparison, thus please consult the original papers to examine fine-scale patterns. *= significant with sequential Bonferroni correction (Rice 1989), *italics* = significant at 0.05 prior to Bonferroni correction. Not bold, not * = not significant. Results are modified from Sepulveda-Villet and Stepien (2011, 2012).

Population Region	1.	2.	3.	4.	5.	6.	7.	8.	9.	10.	11.	12.
1. Lake Winnipeg	–	0.552*	0.522*	0.191*	0.081*	0.392*	0.395*	0.490*	0.399*	0.249*	0.501*	0.895*
2. Upper Mississippi R.	0.202*	–	0.000	0.087*	0.107*	0.000	0.012	0.006	0.050	0.282*	0.548*	0.949*
3. Lake Superior	0.116*	0.202*	–	0.077*	0.096	0.000	0.008	0.010	0.036	0.266*	0.531*	0.943*
4. Lake Michigan	0.137*	0.228*	0.142*	–	0.119*	0.027	0.037*	0.059*	0.054	0.322*	0.584*	0.914*
5. Lake Huron	0.133*	0.183*	0.128*	0.030	–	0.045	0.205*	0.177*	0.069	0.189*	0.433*	0.537*
6. Lake St. Clair	0.217*	0.320*	0.226*	0.122*	0.117*	–	0.026	0.044	0.029	0.197*	0.468*	0.909*
7. Lake Erie, Western Basin	0.217*	0.271*	0.198*	0.074*	0.055	0.147*	–	0.002	0.023*	0.455*	0.701*	0.951*
8. Lake Erie, Eastern Basin	0.218*	0.274*	0.202*	0.079*	0.053	0.145*	0.014*	–	0.004	0.404*	0.659*	0.956*
9. Lake Ontario	0.244*	0.353*	0.213*	0.125*	0.111*	0.093*	0.118*	0.111*	–	0.225*	0.492*	0.915*
10. Northeastern populations	0.204*	0.246*	0.133*	0.166*	0.148*	0.222*	0.185*	0.188*	0.179*	–	0.215*	0.240*
11. Southeastern populations	0.250*	0.279*	0.169*	0.222*	0.203*	0.259*	0.273*	0.274*	0.231*	0.117*	–	0.436*
12. Gulf coastal region	0.290*	0.361*	0.237*	0.273*	0.251*	0.320*	0.317*	0.320*	0.294*	0.180*	0.186*	–

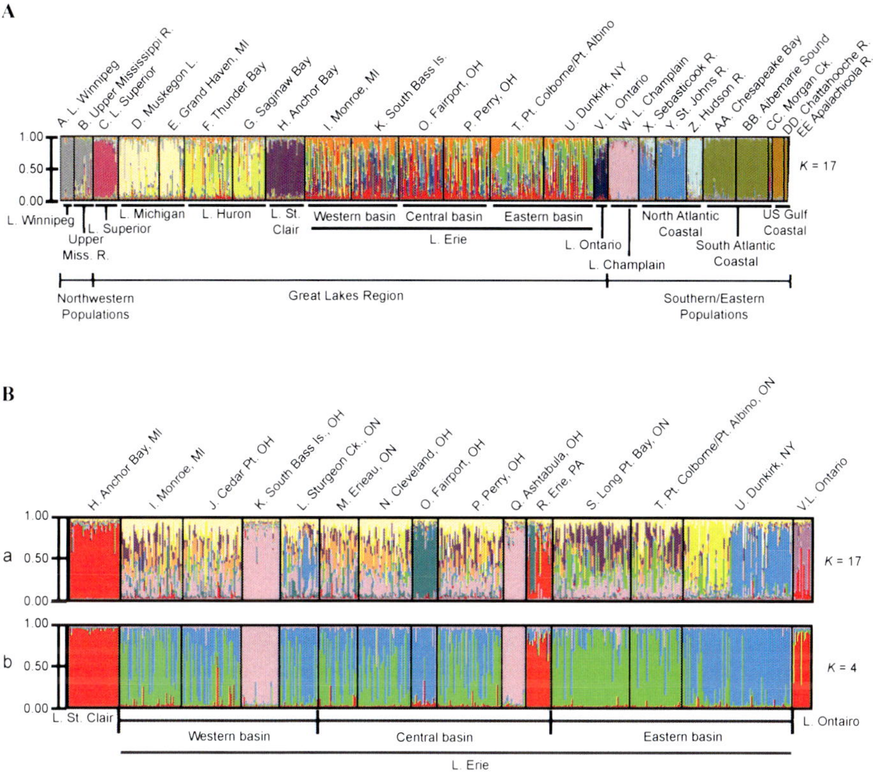

Fig. 7. Estimated comparative population structure for yellow perch from Bayesian STRUCTURE v2.3.3 analyses (Pritchard et al. 2000; Pritchard and Wen 2004, http://pritchardlab.stanford.edu/structure.html) for (A) 24 spawning groups using 15 nuclear DNA microsatellite loci (modified from Sepulveda-Villet and Stepien 2012), for which optimal K = 17; and (B) 15 Lake Erie spawning groups at optimal (a) K = 10 and (b) K = 4 four (adapted from Sepulveda-Villet and Stepien 2011); in reference to outlying groups from Lake St. Clair and L. Ontario. Analyses were run with 100,000 burn-in and 500,000 replicates. Optimal K values were determined by posterior probabilities (Pritchard et al. 2000) and the ΔK method of Evanno et al. (2005). Thin vertical lines represent individuals and thicker bars separate spawning groups at given locations; these are partitioned into K colored segments that represent estimated population group membership. Note that there is no correspondence between the colors of A and B.

and Crossman 1992; Todd and Hatcher 1993; Stepien et al. 2009, 2010)—trace their origins to the Mississippian refugium. In particular, Lake Erie's formation dates to glacial Lake Maumee (~14,000 ya), which then drained west via the Ohio River to the Mississippi, then switched outlets during several lake stages, with its current outlet draining east into Lake Ontario (~10,000 ya) (Underhill 1986; Larson and Schaetzl 2001; Strange and Stepien 2007). As a consequence, Lake Erie yellow perch today are geographically isolated and genetically differentiated from most other Great Lakes populations. Notably, Lake Erie physically is separated from Lake Ontario by Niagara Falls and from the upper Great Lakes by the narrow and short Detroit River, which drains Lake St. Clair. Yellow perch from Lake St. Clair clearly are separated from those spawning in Lake Erie (note the color difference between the red-colored population from Lake St. Clair versus the mixed colors in Lake Erie in Fig. 7B); these appear on opposite sides of a genetic barrier (X), determined from BARRIER v2.2 analysis (Manni et al. 2004), shown in Fig. 2A.

More information about the evolutionary origin of yellow perch stems from mtDNA studies. Thus, yellow perch mtDNA control region haplotype 1 (Figs. 6A and B) likely was widespread pre-glacially and was represented in both the Mississippian and Atlantic refugia populations, but was more common in the west. Today, yellow perch haplotype 1 remains more abundant in the west (see Fig. 6A), apparently reflecting retention of its original predominant proportions in populations colonized from the Mississippian refugium and is only slightly represented in those descendants from the Atlantic refugium to the present day.

The Atlantic coastal refugium (Fig. 2A) formed a warm enclave of diverse habitats in coastal plains and estuaries east of the Appalachian Mountains (Schmidt 1986; Bernatchez 1997); yellow perch from that refugium migrated north to colonize the northeastern and north central regions after the glaciations (Russell et al. 2009; Sepulveda-Villet and Stepien 2012). The northeastern migrating populations split to found the yellow perch populations in Maine (colored blue, sites X and Y, on Fig. 7A) and the Hudson River (colored light blue, site Z); both are very divergent today (also see mtDNA haplotypes on Fig. 6A).

Lake Champlain (site W) drains into the St. Lawrence River and its yellow perch appear to trace to joint origins from the Atlantic and Mississippian refugia, but today has a very divergent genetic composition from other locations (see unique haplotype 20 on Figs. 5 and distinct colors for Lake Champlain yellow perch on Figs. 5A and 6A, denoting different genetic composition). Lake Champlain received meltwaters from glacial Lake St. Lawrence (~11,600 ya), and then Lake Agassiz (~8,000–10,900 ya) and glacial Lake Barlow-Ojibway (~8,000–9,500 ya). This convergence of meltwaters produced an extensive freshwater habitat that replaced the former saline Champlain Sea, which was a temporary inlet of the Atlantic Ocean formed by the retreating glaciers (Rodrigues and Vilks 1994; Sepulveda-Villet and Stepien 2012). Regional flooding presumably led to colonization of Lake Champlain by aquatic taxa from the Atlantic refugium, as suggested by genetic evidence from lake cisco *Coregonus artedi* (Turgeon and Bernatchez 2001) and lake whitefish *C. clupeaformis* (Bernatchez and Dodson 1991); the Lake Champlain yellow perch population appears to reflect joint contributions from the Atlantic and Mississippian refugia (see Fig. 6; Sepulveda-Villet and Stepien 2012).

Variable evolutionary, demographic, and colonization history of the species also impacts the extent of population divergence observed throughout its range of distribution. The greatest inter-population divergences are observed in the upper Mississippi River and the Gulf Coast, reaching a genetic divergence Fixation Index (F_{ST}) value = 0.361, based on microsatellite data (Table 2; Sepulveda-Villet and Stepien 2012). The greatest difference for the mitochondrial DNA data likewise occurs between the same population pair (F_{ST} = 0.949, Table 2). Pronounced divergence also was observed among geographically isolated small populations from northwestern Québec and northeastern Ontario (average F_{ST} = 0.378) (Bourret et al. 2008). Some of the most divergent yellow perch population groups also distinguished the Gulf

Coast (mean F_{ST} = 0.275 among 11 pairwise comparisons), the upper Mississippi River (0.257), the southeast Atlantic Coast (0.224), and then Lake Winnipeg (0.203). Likewise, these population groups appeared highly differentiated according to their mtDNA, in relative order of: the Gulf Coast (mean F_{ST} = 0.786), the southeast Atlantic Coast (0.506), Lake Winnipeg (0.424), and then Lake Erie (0.255).

The relationship between genetic and geographic distances (measured by nearest waterway) is illustrated on Fig. 8A, which indicates broadscale correspondence across

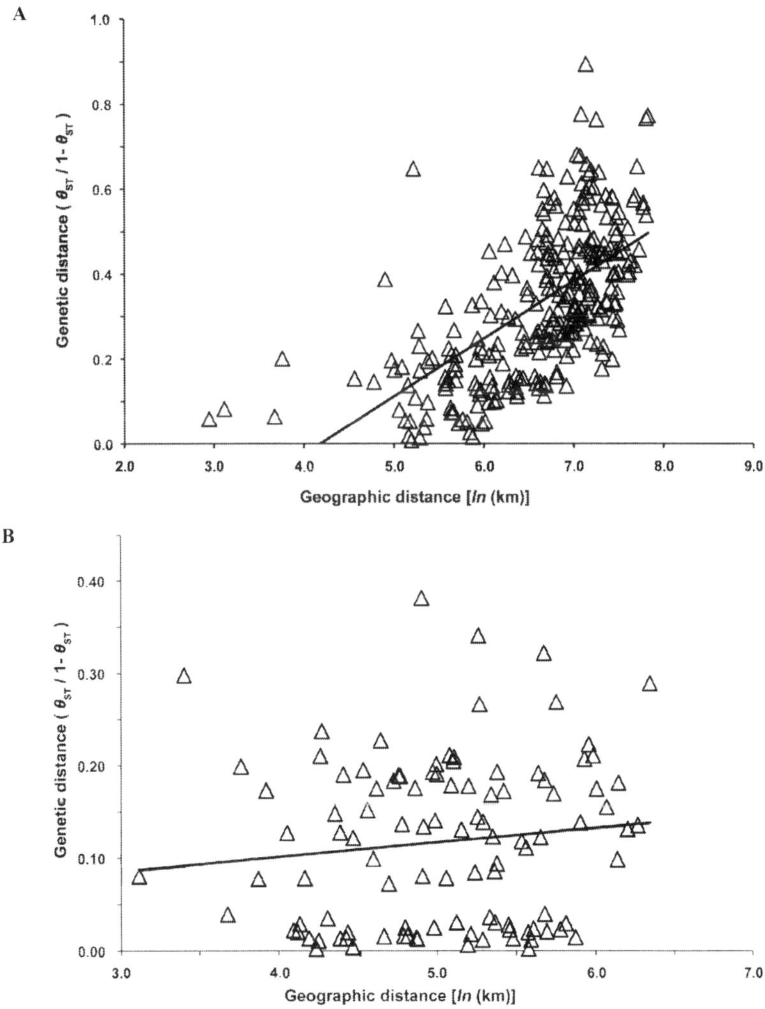

Fig. 8. Mantel (1967) pairwise tests using GENEPOP v4.0 (Rousset 2008, http://kimura.univ-montp2.fr/~rousset/Genepop.htm), with 10,000 permutations, for the relationship between genetic distance (θ_{ST}/1-θ_{ST}) and natural logarithm of geographic distance (kilometers) (A) across the native North American range of yellow perch populations from North America (p < 0.001, R^2 = 0.39, y = 0.14x–0.57; modified from Sepulveda-Villet et al. 2012), and (B) among Lake Erie spawning groups (p = 0.212, R^2 = 0.024, y = 0.016x–0.038; modified from Sepulveda-Villet and Stepien 2011). Note there is no significant relationship in the fine-scale analysis (B).

the range. Thus, there is an overall isolation by genetic distance pattern. However, some populations in relatively close proximity are distinguished by much greater than expected genetic distances, as indicated by the results of the BARRIER analysis (Manni et al. 2004), shown in Fig. 2. The Bayesian STRUCTURE analyses also denote a substantial number of distinctive population groups of yellow perch (K = 17, Fig. 7A). Overall, these results illustrate that pronounced population genetic differentiation in perch may stem from a variety of sources, including: long-term geographic separation, geographic distance separating populations, barriers to dispersal, and genetic drift.

In summary, earlier biogeographic investigations based on fish distribution and knowledge of geological and glacial history led to diverse hypotheses regarding the evolutionary history of the yellow and European perches. Present-day detailed knowledge of the evolutionary history of both species summarized in this chapter stems from recent large-scale phylogeographic and population genetic analyses using mitochondrial DNA and nuclear DNA microsatellite markers, which has led to new insights and consensus. Altogether, these studies have revealed a predominant role of historical to modern biogeography for interpreting contemporary patterns of genetic diversity in both species. In particular, regions that have remained the most stable in time (e.g., were less affected by glaciations) are generally characterized by more pronounced genetic diversity, both at the intra- and inter-specific levels. Higher genetic diversity is also observed in zones of secondary contacts between evolutionary lineages that previously evolved in geographic isolation. Thus, history has played an important role in determining the genetic diversity and its differentiation in perch species across both continents.

2.3 Perch Life Histories and Population Genetic Implications

While historical contingency has deeply influenced broad scale patterns of genetic diversity in perches, it is also clear that the species, and their biological and ecological characteristics also play an important role. In particular, a remarkable similarity in major life history characteristics (Craig 2000) and population genetic implications are shared by both the yellow perch and the European perch across both continents, which invokes their common evolutionary history. Here we introduce some major life history characteristics and population genetic implications for both species and highlight similarities and differences.

2.3.1 Population Genetic Structuring of Reproductive Groups

Cued by gradual changes in water temperature and photoperiod, *Perca* species aggregate to spawn in late spring to early summer on shallow reef complexes in lacustrine systems or slow-moving tributaries (Scott and Crossman 1973; Coles 1981; Wang and Eckmann 1994; Carlander 1997; Jansen et al. 2009; Chapter 3, Chapter 7). Spawning is related to spring water temperatures and starts at >10°C. Because it is mainly dictated by temperature, spawning occurs much earlier in the southern reaches of their ranges, and much later in the extreme north (Thorpe 1977; Carlander 1997; Craig 2000). Spring spawning migrations of yellow perch and European perch are relatively short and it is believed that they return to specific natal sites in shallow waters (Aalto and Newsome 1990; Carlander 1997; Craig 2000; Sepulveda-Villet and Stepien 2012).

Yellow perch captured and tagged during spawning season and released many kilometers distant in the Eastern Basin of Lake Erie were found to return to their tagging locations (MacGregor and Witzel 1987), implicating homing. Separate studies by Clady (1977), Rawson (1980), and Ontario Ministry of Natural Resources (2011) likewise found that most yellow perch tagged during spawning were recovered at or very close to their initial spawning locations in subsequent years. Yellow perch spawning groups located just a few kilometers apart (17 kilometers) in central Lake Erie diverged from one another in genetic and morphological composition (Kocovsky et al. 2013). This genetic divergence suggests that Lake Erie yellow perch populations are highly structured and likely congregate in natal groups at specific spawning locations (Sepulveda-Villet and Stepien 2012; Sullivan and Stepien 2014, 2015). Aalto and Newsome (1990) removed yellow perch egg masses from given spawning sites, which led to fewer fish returning to that location in subsequent years than in control sites, suggesting that they returned to the same spawning areas year after year. It is hypothesized that imprinting occurs during the early life history of yellow perch and European perch, with their highly developed olfactory systems used to detect natal spawning sites and/or the pheromones of neighbors and relatives (see Horal 1981; Gerlach et al. 2001). As observed in Lake Erie, most recaptures of thousands of tagged yellow perch from Lake Saint-Louis in the St. Lawrence River, Québec, were made within 10 kilometers from tagging sites (Leclerc et al. 2008). Moreover, movements between the two areas were limited, with only 4% of recaptured yellow perch found in alternate locations over a 3-year period (Dumont 1996). As a consequence, significant genetic differences were observed between yellow perch sampled on the north vs. south shores of the lake (Leclerc et al. 2008).

Genetic composition of yellow perch spawning groups differs significantly from location to location across broad and fine geographic scales (Sepulveda-Villet and Stepien 2011, 2012; Sullivan and Stepien 2014). However, some significant differences have been found at some of the same locations from year to year (Sullivan and Stepien 2015). This local genetic variability suggests that although yellow perch may spawn together with a specific group (believed to be their natal group), specific spawning locations may vary from year to year. A similar pattern was found for European perch in Lake Erken, Sweden by Bergek and Olsson (2009). When comparing perch aggregations caught at four different locations in the lake, consistent genetic differentiation was found among these locations over time. However, local European perch groups were genetically differentiated when comparing perch from the same location from different years (Bergek and Olsson 2009).

The spawning process is similar in both species. During the spawning season, males move into the spawning areas first, arriving before females by a few weeks and lingering longer at the sites (Scott and Crossman 1973; Craig 2000; Simon and Wallus 2006). The female perch lays a long gelatinous egg strand (up to 2.1 meters long), which contains 10,000 to 40,000 eggs, over submerged vegetation or other structures at night or in early morning. As the egg mass is released by the female, it is externally fertilized by a cluster of 2–25 males, who closely follow the female and often are in close proximity to other spawning clusters (Scott and Crossman 1973; Mangan 2004; Simon and Wallus 2006).

2.3.2 Early Life History and Population Patterns

Soon after hatching in littoral areas, the young-of-the-year move into deeper water as juveniles in late spring, where they occupy a wide variety of habitats (Craig 2000). As a consequence, Parker et al. (2009) found that age one juvenile yellow perch differed in morphology and genetically at 12 nuclear DNA microsatellite loci between populations in Lake Huron and Lake Michigan, with those from Lake Huron having deeper, longer bodies and larger dorsal fins. The researchers also discerned morphological and genetic differences between juveniles living in nearshore versus wetland habitats in Lake Michigan. Juveniles inhabiting nearshore areas from both lakes had deeper, longer bodies and larger dorsal fins than did those occupying wetlands, which might reflect an adaptive response to predators and open-water cruise swimming. Although phenotypic differences between habitats across the lakes were hypothesized to reflect plasticity between phenotypic and genetic divergence (Parker et al. 2009), a genetic basis for such variation cannot be ruled out, given the evidence for some reproductive isolation based on genetic data. Bergek and Björklund (2009) discerned a very similar genetic and morphological divergence pattern for European perch from the east coast of Sweden in the Baltic Sea between local populations sampled at two different spatial scales, a near scale (300 m–2 kilometers) and a far scale (2–13 kilometers). The morphological differences between them were hypothesized to reflect phenotypic plasticity, but again, a genetic basis could not be ruled out.

2.3.3 Adult Perch Movements and Migrations: Implications for Genetics

After the spawning season, adult perch movements largely are determined by habitat complexity, food availability, and foraging capacity (Radabaugh et al. 2010). Likewise, juvenile and adult yellow perch and European perch typically occur in shoals, which may facilitate foraging and predator avoidance (Helfman 1984; Craig 2000). Interestingly, shoals of European perch have been shown to contain large numbers of related individuals; they are believed to recognize each other from chemical and physical cues (Gerlach et al. 2001; Behrmann-Godel et al. 2006). However, whether the shoals of yellow perch likewise are structured based on kinship has not yet been evaluated (see Sullivan and Stepien 2015).

A study of yellow perch tag returns by Haas et al. (1985) determined that post-spawning movements are moderate; individuals tagged at Lake Erie spawning sites did not move upstream through the corridor connecting Lake Huron to Lake Erie, which is termed the Huron-Erie Corridor (HEC), connecting the two Lakes via the St. Clair River, Lake St. Clair, and the Detroit River. Some yellow perch that were tagged in Lake St. Clair migrated to nearby tributaries (Haas et al. 1985). Note that although individuals may move among water bodies to feed, their reproductive groups determine their overall population genetic structures.

A tagging study of European perch in prealpine Lake Constance, Germany discerned moderate movements of adults beyond the spawning season (Godel 1999). Tagged perch occupied home ranges of approximately 500 meters, located along distinct depth contours (ca. 10 meter water depth in summer, Fig. 9), within which they spent several days to weeks. Typically, their home ranges were located near structures that resembled fish nurseries (submerged wooden structures built by fishers in Lake Constance) or near harbors. During this stationary time in the home range, their typical behavioral patterns revealed highest activity levels during or close to dusk and dawn, low activity during the day, and no activity at night (Fig. 10). After a period ranging from two days to three weeks, most perch individuals became migratory and began to move straightforward along the shoreline between 1 and 20 kilometers longitudinally for a few hours to a few days before they again settled to occupy a new home range (Fig. 9). Congruent with movement patterns depicted in previous studies, evidence for yellow perch metapopulations in Lake Ontario embayments were discerned by otolith microchemistry (Murphy et al. 2012), indicating discrete assemblages in connected bays and impoundments, as were found among spawning groups along Lake Erie coastal sites (Fig. 7B, Fig. 11; Sepulveda-Villet and Stepien 2011; Sullivan and Stepien 2015). These populations likely display seasonal mixing, as described by Parker et al. (2009).

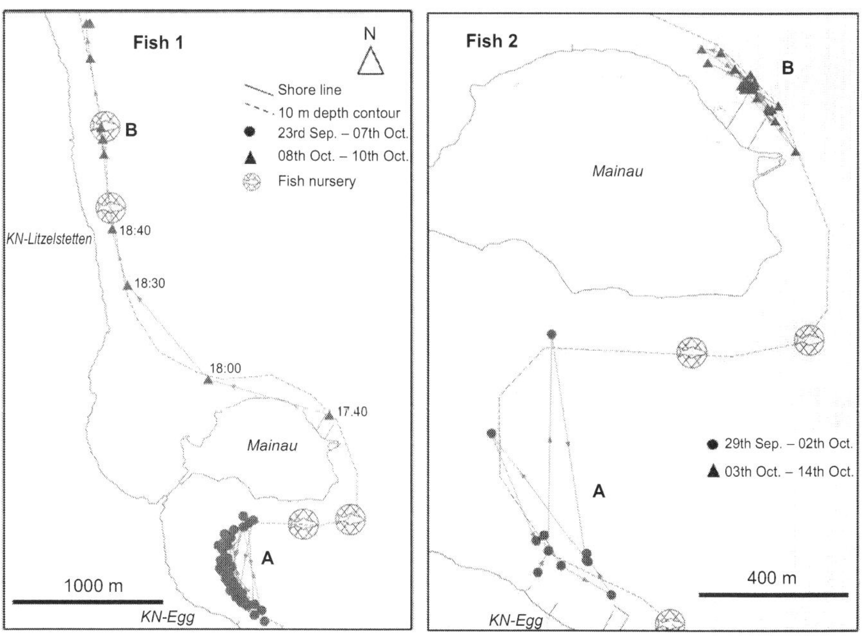

Fig. 9. Swimming paths of two individual European perch during a tagging study using ultrasonic transmitters in Lake Constance near the isle of Mainau. Dots and triangles are fish positionings, arrows indicate swimming direction. For Fish 1, time (CET) of positionings is given for a migratory movement in the evening of the 08th October, where the fish moved from home range A to home range B.

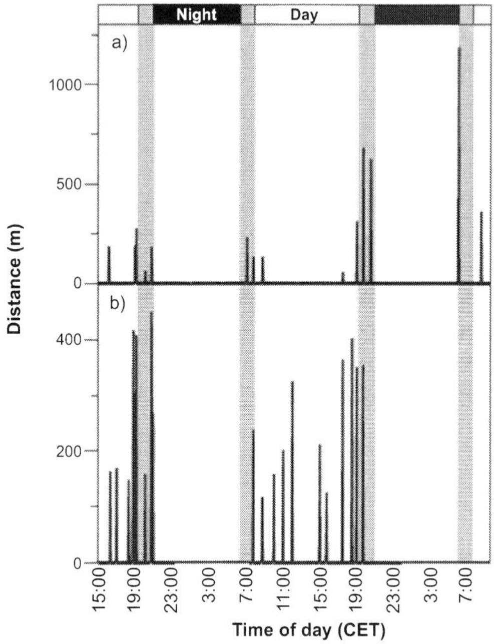

Fig. 10. Swimming activity for two European perch (a, b) during 48 h observation periods in Lake Constance. Fish were tagged with ultrasonic transmitters. Activity is given as swimming distance between two positionings. Perch typically showed highest activity during or close to twilight or during the day and no activity at night.

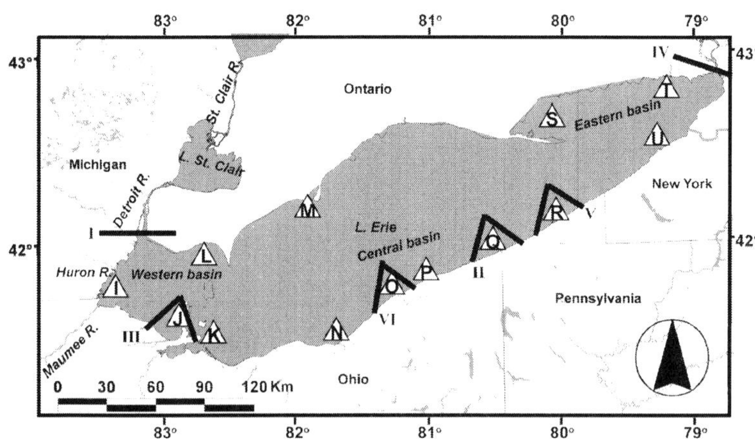

Fig. 11. Fine-scale map of Lake Erie in the North American Laurentian Great Lakes showing locations of yellow perch spawning groups (triangles) tested and genetic barriers delineating specific groups from 15 nuclear microsatellite loci and BARRIER v2.2 analysis (Manni et al. 2004, http://www.mnhn.fr/mnhn/ecoanthropologie/software/ barrier.html), which are ranked I–VI, in order of their decreasing magnitude. Sites of spawning groups are lettered, as follows: I. Monroe, MI; J. Cedar Pt., OH; K. S. Bass Isl., OH; L. Sturgeon Ck., ON; M. Erieau, ON; N. Cleveland, OH; O. Fairport, OH; P. Perry, OH; Q. Ashtabula, OH; R. Erie, PA; S. Long Pt. Bay, ON; T. Pt. Colborne, ON; U. Dunkirk, NY. Figure was modified from results of Sepulveda-Villet and Stepien (2011, 2012).

2.4 Influence of Habitat Connectivity, Isolation, and Dispersal on Population Genetic Patterns

In addition to the role of intrinsic species characteristics described above, patterns of population structure in both yellow and European perch are influenced by habitat characteristics and landscape, which directly impact their dispersal capability and connectivity. Large connected spans of suitable habitats, as found in the Great Lakes basins and tributaries for yellow perch, offer a variety of environmental resources for robust and diverse populations, reflecting the interplay between migration opportunity and localized adaptation (see Lindsay et al. 2008; Vandewoestijne et al. 2008; Kunin et al. 2009).

Aquatic habitats frequently are connected by narrow and relatively ephemeral connections that link populations during migration and dispersal, but whose habitats may pose distinct biological challenges. For example, small connecting channels may offer limited food and shelter, and extensively differ in size and habitat complexity, which then influence the distribution of population genetic variability. In other areas, such as in the St. Lawrence River, patches of suitable habitat for reproduction and nursing for yellow perch may be separated by extensive stretches of unsuitable habitats, which also may result in meta-population dynamics (e.g., Mingelbier et al. 2008) that will influence the distribution of genetic diversity. In contrast, isolated relict populations having little connectivity may possess lower overall genetic diversity due to the influences of genetic drift, bottlenecks, and selection (Moran and Hopper 1983; Petit et al. 2003; Coulon et al. 2012; Sepulveda-Villet and Stepien 2012).

2.4.1 Population Genetic Diversity Comparisons

The overall genetic diversity of yellow perch is relatively low compared to freshwater fishes in general (DeWoody and Avise 2000). For instance, it is much lower than in walleye for both nuclear DNA (mean heterozygosity is 0.53 for yellow perch versus 0.73 for walleye) and mtDNA sequence variability (mean haplotypic diversity is 0.31 for yellow perch versus 0.77 for walleye) (Stepien et al. 2012; Sepulveda-Villet and Stepien 2012; Haponski and Stepien 2014). The relatively low diversity for yellow perch likewise was revealed by other genetic data sets, including allozymes (Leary and Booke 1982; Todd and Hatcher 1993; Moyer and Billington 2004), mtDNA restriction fragment length polymorphisms (RFLPs) (Billington 1993; Moyer and Billington 2004), mtDNA sequences (Sepulveda-Villet et al. 2009; Sepulveda-Villet and Stepien 2012), as well as nuclear microsatellites (Miller 2003; Bourret et al. 2008; Leclerc et al. 2008; Sepulveda-Villet and Stepien 2011, 2012; Sullivan and Stepien 2014, 2015), and single nucleotide polymorphisms (SNPs; Bélanger-Deschênes et al. 2013).

Overall mtDNA genetic diversity of yellow perch roughly matches that of the European perch (Refseth et al. 1998; Nesbø et al. 1998, 1999), which also has relatively low allozymic genetic diversity (Gyllensten et al. 1985; Bodaly et al. 1989). In fact, relatively low genetic diversity in both mtDNA and nuclear DNA appears characteristic of the genus *Perca*. Some individual yellow perch spawning groups possess a relatively high degree of kin relationship (see Table 1), which may result in lower variability within samples (Sullivan and Stepien 2015). Genetically similar individuals of European perch have been found to aggregate with one another (Gerlach et al. 2001), recognizing their relatives via olfactory cues at the fry life stage and beyond (Behrmann-Godel et al. 2006). Kin recognition by olfactory cues has not yet been studied in yellow perch, but might yield important insights on their fine-scale population structure and the distribution of their respective diversities.

Across most of their North American ranges, yellow perch populations exhibit relatively consistent levels of genetic variability for the nuclear microsatellite data (Table 1; Sepulveda-Villet and Stepien 2012). Less genetic diversity is found in some of the northwest populations and the Gulf coastal region; both areas house small, isolated populations that likely experienced bottlenecks. In contrast, population genetic diversities are relatively high across the Great Lakes and in the southeastern populations (Sepulveda-Villet and Stepien 2012). The southeastern populations were never glaciated, and thus maintained relatively stable population sizes. In contrast, the Great Lakes populations were founded from variable combinations of the Missourian, Mississippian, and Atlantic glacial refugia (Stepien et al. 2009; Sepulveda-Villet and Stepien 2012). The Great Lakes populations today exhibit relatively high genetic diversities as a result of this admixture of founding from multiple refugia, in combination with their large population sizes and abundant habitat expanses. These factors appear to have superseded any effects of bottlenecks stemming from their former restriction to glacial refugia during the Pleistocene and any founder effects during the subsequent post-glacial resurgence to new habitats opened in the modern-day Great Lakes.

Levels of mitochondrial DNA diversity, measured as haplotypic diversity (Table 1), are comparatively quite low for yellow perch. MtDNA diversity in populations is much more influenced by bottlenecks as its theoretical effective population size is just ¼ that of nuclear DNA (see Avise 2004). As a consequence, values of haplotypic diversity of yellow perch are much lower than their nuclear diversity levels. This may reflect a history of bottlenecks for yellow perch, with most of the Great Lakes and the northwestern populations dominated by a single yellow perch haplotype (haplotype 1 of Fig. 6). Nuclear DNA diversity levels for yellow perch populations overall also are relatively modest (Sepulveda-Villet and Stepien 2011) in comparison to walleye from the same regions (Stepien et al. 2009, 2012; Haponski and Stepien 2014).

Private alleles are those that are found only in a specific population or set of populations. In the Great Lakes overall, 14% of nuclear microsatellite alleles and 7% of the mitochondrial haplotypes were private in yellow perch (Table 1). In Lake Erie, 9% of the microsatellite alleles and 3% of the mitochondrial haplotypes were private. In the Gulf Coastal population, 7% of yellow perch microsatellite alleles and 13% of the mitochondrial haplotypes were private (Table 1, Fig. 6). Thus, these distributions reflect appreciable differentiation among metapopulations.

2.4.2 Fine-scale Population Genetic Structure in Yellow Perch and European Perch

Although relationships among yellow perch populations typically follow a broad-scale pattern of genetic isolation by geographic distance (Fig. 8A), relationships among spawning groups within individual lakes do not reflect geographic distance (Figs. 4 and 8B). Some closely situated spawning groups are markedly different, whereas others are more closely related. Fine-scale relationships among yellow perch spawning groups appear to be driven by spawning aggregations, natal homing behavior, and localized adaptations, rather than due to simple geographic connectivity (see Sepulveda-Villet and Stepien 2011, 2012; Kocovsky et al. 2013; Sullivan and Stepien 2014, 2015).

Sepulveda-Villet and Stepien (2011) found significant differences at 15 microsatellite loci among Lake Erie yellow perch at spawning sites (shown in Fig. 7B), discerning no relationship between genetic distance and geographic distance between sampling locations. Kocovsky and Knight (2012) reported similar trends using morphometric data from yellow perch sampled from many of the same spawning locations used by Sepulveda-Villet and Stepien (2011). Yellow perch spawning groups in the Central Basin of Lake Erie that are separated by 17–94 kilometers were distinguished by significant genetic divergences of $F_{ST} = 0.016–0.056$ using the same 15 loci, and also displayed significant morphological differences (Kocovsky et al. 2013). Grzybowski et al. (2010) described fine-scale genetic structure between yellow perch spawning in Lake Michigan open water versus those in Green Bay, also from microsatellite data ($F_{ST} = 0.126$).

Such differentiation among spawning groups within a system appears to result from spawning site philopatry to specific natal locations, maintained from generation to generation. European perch form long-term population groups of related individuals, according to microsatellite data (Behrmann-Godel et al. 2006; Bergek and Björklund 2007, 2009). Reproductive success was significantly lower in breeding experiments when two subpopulations were hybridized, with reduced pre-zygotic and post-zygotic fitness manifested by lower fertilization rates and less hatching success (Behrmann-Godel and Gerlach 2008). One of the likely barriers to gene flow for European perch thus is reproductive isolation, either via kin recognition using olfactory cues (Gerlach et al. 2001; Behrmann et al. 2006) or due to reduced hybrid fitness between sympatric but divergent cohorts (Behrmann-Godel and Gerlach 2008). Likewise, it is possible that yellow perch returning to natal locations are guided by olfactory information imprinted during early stages of their life history. If so, it may be the primary mechanism for maintaining divergence among spawning aggregations, but this hypothesis remains to be tested.

There apparently is no effect of gender in the establishment of these fine-scale genetic structure trends, as both male and female yellow perch have analogous genetic patterns, and thus appear to have similar site fidelity (Sepulveda-Villet and Stepien 2011, 2012; Sullivan and Stepien 2015). Eight spawning groups from Lake Erie locations (sites on Fig. 11) were all genetically distinguishable from one other (mean $F_{ST} = 0.068 \pm 0.008$, range = 0.002–0.168), but some also differed in allelic composition between two sampling time periods (2001–2004 versus 2009), at ~1/4 the magnitude of the difference among locations. Sullivan and Stepien (2015) found significant differences among yellow perch spawning groups and between different sampling years at some of these sites. An example of annual variation within the yellow perch spawning group sampled at Van Buren Bay in eastern Lake Erie is given in Table 3. A study by Demandt (2010) likewise found significant variations in microsatellite allelic

Table 3. Fine-scale pairwise genetic divergences for yellow perch spawning at Dunkirk NY in eastern Lake Erie sampled in six different collection years: 1985, 2001, 2004, 2008, 2009, and 2010, based on 15 nuclear DNA microsatellite loci (modified from Sullivan and Stepien 2015). Sample sizes are in parentheses. Calculations used *F*STAT v2.9.3.2 (Goudet 1995, 2002, http://www2.unil.ch/popgen/softwares/fstat.htm) and ARLEQUIN v3.1.5.3 (Excoffier and Lischer 2010, http://cmpg.unibe.ch/software/arlequin35/), with 100,000 replicates to test for significance. * = All were significantly different following sequential Bonferroni correction (Rice 1989).

Year (*N*)	1985 (34)	2001 (37)	2004 (48)	2008 (30)	2009 (30)
2001 (37)	0.037*	–			
2004 (48)	0.056*	0.055*	–		
2008 (30)	0.125*	0.138*	0.141*	–	
2009 (30)	0.041*	0.072*	0.088*	0.105*	–
2010 (36)	0.011*	0.041*	0.057*	0.113*	0.014*

frequencies of European perch among sampling years for a population in Sweden. This observation suggests a similar trend for annual variability at spawning sites for both yellow and European perch.

Yellow perch spawning groups varied among individual sampling years and age cohorts, with the 2003 cohort being the most distinctive of those sampled (Sepulveda-Villet and Stepien 2011, 2012; Sullivan and Stepien 2014, 2015). This 2003 cohort was an especially large and successful group for yellow perch recruitment in Lake Erie (YPTG 2013). Spawning groups of yellow perch contained high numbers of full siblings (mean = 18.5%, ranging to 75% for the 2001 age cohort spawning at Van Buren Bay in eastern Lake Erie; Sullivan and Stepien 2015). Temporal genetic divergence at spawning locations was not explained by genetic isolation over time, but appeared to be due to yellow perch spawning in kin-related groups, which varied slightly from year to year. Spatial patterns were attributed to limited migration and natal homing, whereas temporal patterns may reflect kin group structuring and differential reproductive success (Sullivan and Stepien 2015).

Although there is ample evidence for fine scale population structuring in both yellow and European perch, there may be some exceptions to this, as exemplified by the population structure revealed in a landscape genetics study of yellow perch from the St. Lawrence River. Leclerc et al. (2008) employed a landscape genetics approach to document the population genetic structure of yellow perch using microsatellite markers, assess to what extent the structure was explained by landscape heterogeneity, and interpret the relevance of interactions between genetics and landscape for management and conservation. Genetic analysis of 1715 individuals from 16 localities, distributed across 310 kilometers in the freshwater section of the Saint Lawrence River, revealed a modest level of genetic structuring (F_{ST} = 0.039). BARRIER analysis (Manni et al. 2004), which combined geographical and genetic information, identified three zones of restricted gene flow. These delineated just four distinct populations over a large geographic distance (Fig. 12). Results showed that physical barriers (e.g., occurrence of dams) played a more important role on gene flow and genetic structure than waterway geographical distance. The authors also found correlations between genetic differentiation and the presence of distinct water masses, and with fragmentation of

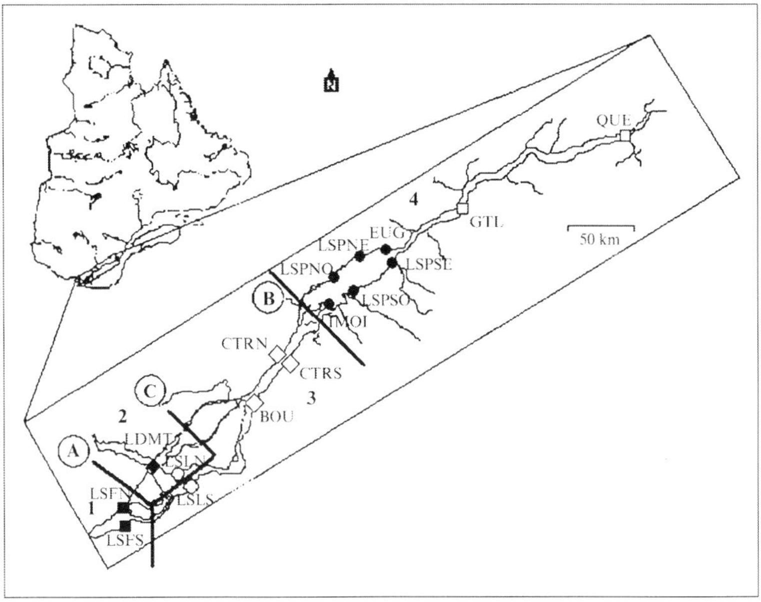

Fig. 12. Areas of population genetic breaks among yellow perch along the St. Lawrence River system identified by BARRIER 2.2 analysis (Manni et al. 2004) using Monmonier's algorithm in the study by Leclerc et al. (2008). Genetic barriers that were retained under the majority-rule criterion are identified by order of importance (A, B, and C). These barriers separated the system into four genetically distinct populations: (1) Lake Saint-François; (2) North of Lake Saint-Louis and Lake des Deux-Montagnes; (3) South of Lake Saint-Louis downstream to Contrecoeur; (4) and Lake Saint-Pierre downstream to Quebec City. Symbols denote sampling locations, from west to east, dark squares, Lake Saint-François; black diamonds, Lake des Deux-Montagnes; white circles, Lake Saint-Louis; white diamonds, fluvial section from Boucherville to Contrecoeur; black circles, Lake Saint-Pierre; white squares, fluvial section from Gentilly to Quebec City.

spawning habitats. The study also showed that landscape genetics is a powerful means to identify environmental barriers to gene flow, which create genetic discontinuities in apparently highly connected aquatic landscapes.

2.5 Applications to Perch Population Maintenance and Restoration

As molecular-based population dynamics and structure analyses increasingly provide ways to better assess past and present levels of diversity in fish populations, a need for greater use of these techniques has been proposed in concert with traditional management approaches. For instance, dating to the mid-1800s, yellow perch was stocked (artificially introduced to new areas) to many areas of North America to support recreational fishery and to provide fishing opportunity (USFWS/GLFC 2010). These introductions sometimes mixed nonindigenous hatchery broodstock with local genotypes. Molecular analyses may help to discern whether stocking may have partly blurred the evolutionary history of the species and may have impacted the genetic integrity of some indigenous populations. Today's fishery managers increasingly recognize the importance of preserving local population variability, and it is advisable to perform any supplementation solely with native genotypes specific to that particular locale. The most prudent action still remains to protect the habitats of locally adapted populations and avoid negative effects of overexploitation, thereby circumventing any "need" to stock.

Understanding and maintaining yellow perch population structure are critically important fisheries-management goals designated by the Great Lakes Fishery Commission (Ryan et al. 2003; GLFC 2011). For example, the findings of just four biological units in the study by Leclerc et al. (2008), were in contrast with the current basis for yellow perch management and thus called for a re-evaluation of management strategy of the species in this system. This also has been true for Lake Erie yellow perch, for which genetic data reveal many more population subunits than the number that are managed for (Fig. 7B, 11; Sepulveda-Villet and Stepien 2011; Kocovsky et al. 2013; Sullivan and Stepien 2014, 2015). When demographic data on exploited fisheries are collected on a larger scale than population subunits, valuable data may be lacking for management decisions to conserve local genetic and morphological diversity and adaptedness.

Fish habitats in the Great Lakes, including the St. Lawrence River, Lake Erie, Lake St. Clair, and connecting tributaries were subject to extensive and deleterious changes in the 20th and 21st centuries, marked by loss of wetlands, channelization of major streams, construction of dams, oxygen depletion, shoreline modification, siltation of spawning areas, nutrient enrichment, water-quality deterioration, sand and gravel extraction, and invasive species introductions (Trautman 1981; Bolsenga and Herdendorf 1993; Fielder 2002; Hoff 2002; Ryan et al. 2003; Mailhot et al., Chapter 5 of this volume). Discerning whether and how perch adapted and coped with such pronounced environmental disturbance is crucial, and may provide significant insights into how populations will respond in the future.

Molecular tools that inform about adaptive differences between populations as well as population response to environmental stressors are readily available, yet these have been largely untapped in perch studies. As one example of informative applications of recent genomic tools, Bélanger-Deschênes et al. (2013) documented functional polymorphisms of chronically metal-contaminated wild yellow perch. Based on a *de novo* transcriptome scan; they first contrasted subsets of individuals from clean and contaminated lakes to identify 87 candidate annotated coding SNPs. Candidate genotypes and liver metal concentration were obtained in 10 populations (N = 1,052) and a genome scan distinguished outliers between polluted and unpolluted sites: one nuclear (cyclin G1 gene) and two mitochondrial (cytochrome *b* and NADH dehydrogenase subunit 2) genes also displayed allelic correlation to mean liver cadmium concentration. Based on associated functions and inter-population differentiation, the authors proposed that contaminated perch may have been selected for fast life cycle completion involving the p53 pathway and memorization impairment mitigation through this long-term potentiation pathway. In accordance with predicted evolutionary trajectory for stressed and energy deprived organisms, adapted perch would not compensate for repair mechanism inhibition, instead reallocating energy towards growth and favoring inexpensive impairment mitigation adaptations over costly detoxification. Overall, this study showed that a few dozen generations of selection apparently drove rapid, potentially adaptive evolution by selecting for alleles that increased perch fitness in polluted environments. This result is in line with the growing evidence that human-driven environmental change may cause rapid evolutionary change that must be taken into consideration for sound management and conservation strategies (Smith and Bernatchez 2008).

More recently, somatic and genetic markers were employed to evaluate the reproductive health of yellow perch populations for which fisheries monitoring revealed reduced recruitment, in urbanized and developed streams of the Chesapeake Bay watershed (Blazer et al. 2013). Results showed gonadal anomalies and changes in DNA integrity in those yellow perch population samples. These findings suggest that pollution can significantly impact reproduction and recruitment, the effects of which can be detected with molecular markers.

Genetic findings to date, as illustrated in this chapter, reveal that most perch populations have appreciable genetic diversity and significantly differ from other populations, both nearby and distant, despite and sometimes because of anthropogenic influences. These diversity and divergence patterns may translate to localized adaptations, which merit preservation. Accordingly, we recommend conserving their genetic composition and differentiation patterns by maintaining and restoring spawning habitats, and continued careful management of fisheries.

2.6 Conclusions and Perspectives for Future Research

Temperatures are predicted to increase over the next 50 years, with those in the North American Great Lakes region predicted to increase by 5–5.5ºC to become more like today's Gulf of Mexico Coast (Hayhoe et al. 2010). Climate change may disproportionally increase or decrease genetic variability across a taxon's range due to shifts in physical conditions or biological resources (Hewitt 1999; Petit et al. 2003; Hampe and Jump 2011), as occurred during Pleistocene glaciations (Oberdorff et al. 1997; Davis and Shaw 2001; Soltis et al. 2006) and is ongoing today (Araújo and Rahbek 2006; Harris and Taylor 2010).

Bergek et al. (2010) suggested that environmental factors other than geographic distance distinguished European perch spawning groups, implicating water temperature differences among groups spawning in various habitats during the spring. The onset of perch spawning is highly controlled by spring season water temperature, in combination with day length (see Chapter 7 for an extensive review on the topic). Water temperatures need to be below 10°C during the winter to ensure gonad maturation and reproductive success (Hokanson 1977; Dabrowski et al. 1996). Spring spawning of perches is initiated by a rise in water temperature above 10ºC along with increasing day length; spawning occurs over a very short period of time (approximately 14 days). By manipulating water temperature and day length, Dabrowski et al. (1996) delayed yellow perch spawning for several months. Bergek et al. (2010) tested different environmental parameters including: salinity, turbidity, surface temperature in August, mean temperature in April (the spawning time of European perch at the location analyzed), and mean water depth for their correlation with genetic differentiation of spawning groups analyzed from spawning places along an environmental gradient.

Of all parameters tested, only mean temperature in April correlated significantly with genetic isolation of various spawning groups. Similarly, yellow perch populations were significantly affected by water level fluctuations of glacial lakes in North Dakota, with their greatest recorded abundances and body weights occurring during high water periods (Dembkowski et al. 2013), underscoring potential deleterious effects of increased evaporation and water losses linked to climate change.

These findings highlight the importance of spawning sites and localized variations governing relationships among the associated reproductive groups for yellow perch and European perch. It appears likely that genetic structure among spawning localities will continue to reflect a product of the interplay between ancestral lineages and environmental variation among spawning areas, rather than simple isolation by distance. If this concept holds true, then we should expect effects on genetic diversity and composition from the increasing pace of climate change and higher surface water temperatures with shifting population distributions.

In addition, there might be a number of indirect effects affecting the genetic structure of perch populations, such as an alteration of the exposure to parasites and pathogens (Poulin 2006; Chapter 8 in this volume). It is well known that the interaction between hosts and parasites is controlled by the environment (Wolinska and King 2009). For example, the period of parasite transmission can be prolonged, and the abundance and virulence of distinct pathogens and parasites may be increased by rising temperatures (e.g., Poulin 2006). Alien parasites and diseases may be favored and cause epizootic outbreaks in naïve host populations that either lack the genetic adaptations to reduce pathogenicity or to defend the invaders (Marcogliese 2001; Britton et al. 2011; Behrmann-Godel et al. 2013). For example, the recent outbreaks and deleterious fish die-offs of the viral hemorrhagic septicemia virus (VHS) in the Great Lakes, which first appeared ~2005, may be related to climate change (see Pierce and Stepien 2012). Additionally the cumulative effects that multiple stressors exert on native species, including anthropogenic pollutants in combination with parasites or pathogens currently are raising a major concern (see section 10.3 in Chapter 10 for further discussion on this topic). For example, Vidal-Martínez et al. (2010) and Marcogliese and Pietrock (2011) discussed their negative effects on immune function and animal health. The exact details of the effect of a changing environment on host-parasite interactions are hard to predict due to reciprocal evolutionary effects among multiple factors. This coevolution first will affect the genes that play major roles in infection and susceptibility of parasites and hosts, respectively (Woolhouse et al. 2002; Stepien et al. 2015). However they may cause local coadaptation, which is a prerequisite for further population sub structuring. The antagonistic co-evolution is believed to be the major driver of the extraordinarily high polymorphism usually found for infection and resistance alleles in parasites and their hosts, as evidenced by the major histocompatibility complex (MHC) genes in vertebrate hosts. It has already been shown that MHC classII receptor genes in European perch are highly variable (Michel et al. 2009; Oppelt et al. 2012) and it was recently found that a long-term rise in temperature (over 35 years) strongly affected those genes resulting in a massive change in MHC variability and an alteration of the cycling pattern of several MHC alleles in a European perch population enclosed in an artificially heated Swedish lake (Björklund et al. 2015). This observation indicated strong selection on the MHC classII receptor genes of perch via an alteration of their parasitic community. Future studies will reveal whether such long-term changes in temperature will indeed result in stable changes of the genetic composition of perch populations.

Evaluating diversity and divergence patterns resulting from post-glacial dispersal and adaptation in new environments, and the genetic reservoirs comprising isolated relict groups, may help us to predict the challenges faced by taxa during this era of rapid climate and habitat alterations. In effect, climate change patterns rapidly are extending the northward post-glacial expansion trajectory of many taxa in north temperate regions; meanwhile their southerly rear-edge groups may experience greater isolation, habitat reduction, and bottlenecks. The southern genotypes may move northward, given connection or transport opportunity, and likely house valuable genetic adaptations to warmer climates (Hampe and Petit 2005). For example, the diverse Atlantic coastal yellow perch populations in the southeastern U.S. may prove especially well-adapted to tolerating salinity fluctuations and increasing water temperatures, facilitating their northward coastal migration, if sea levels rise to eventually connect low-lying estuaries, which are currently isolated by barrier island and sandbar systems. Distributional changes in populations are significant in the context that they may interbreed with long-term native populations in the north. It is possible that the adaptive potential of native populations may be either positively or negatively influenced by these changes.

Due to the uncertain nature and amplitude of climate change it is difficult to predict how genetic diversity would be expected to respond to climate change or how one would use such information to interpret climate shifts. Nevertheless, it may prove useful to employ functional genomics (e.g., genome-wide genotyping at coding gene regions) monitoring in other to temporally track what biological functions are most associated (and therefore most affected) by climate change.

Warming temperatures and increases in storm events may influence fish population structure and overall productivity via biological and climate-related effects as outlined by Newbry and Ashworth (2004). For example, Hill and Magnuson (1990) suggested that changes in bioenergetics accompanying climate change might modify growth and prey consumption, thereby affecting food-web dynamics. Shuter and Post (1990) suggested that an increase of 4°C may move the distributional limit of yellow perch northward and, depending on lake morphometry and productivity, might also greatly affect survival, relative year-class strength, and ecosystem carrying capacity. Moreover, climate change may affect various regions of the distribution range of perches and other taxa in unpredictable ways, resulting in habitat fragmentation and leading to genetic diversification. To date, yellow and European perch populations possess relatively consistent levels of genetic diversity and high local distinctiveness. These appear to have been maintained despite anthropogenic habitat loss, degradation, fragmentation, and exploitation, likely offset by their large population sizes and the relative abundance of habitats.

Genetic structure of today's perch populations reflects interplay among climatic events, ephemeral waterway connections, population sizes, and likely spawning group philopatry. Delineation of the genomic adaptations that underlie the patterns of genetic diversity and diversity described here will aid predictions of likely response to changing environments, new habitat areas, and exploitation pressures (see Allendorf et al. 2010; Avise 2010). The study of Bélanger-Deschênes (2013) described above, as well as the use of transcriptomics to investigate population response to pollutants (Pierron et al. 2011; Bougas et al. 2013; Couture et al. Chapter 10 this volume), clearly illustrate the benefits of taking advantage of the modern genomics toolbox for perch management and conservation. A combined fisheries management and genetics/genomic approach will provide a bridge for understanding the unique challenges faced by aquatic taxa due to their constrained dispersal and gene flow via habitat connectivity. Understanding the historical and present day factors that shaped today's populations may aid their continued conservation in the face of future challenges.

2.7 Acknowlegements

We thank Patrice Couture and Gregory Pyle for inviting us to contribute to this volume, and for valuable suggestions made by Joseph Rasmussen in review. This is contribution #2015-001 from the University of Toledo's Lake Erie Research Center. We thank Timothy Sullivan, Amanda Haponski, and Jhonatan Sepulveda-Villet for supplying results from their graduate degree work in CAS's Great Lakes Genetics/Genomics Lab at the University of Toledo. Amanda Haponski helped by running phylogenetic analyses, making the maps, and formatting the figures. Shane Yerga-Woolwine helped with DNA sequencing, checking and formatting the references, and running the phylogenetic analyses. Funding for research reported here by CAS came from the National Science Foundation NSF GK-12 DGE#0742395, USEPA #CR-83281401-0, NOAA Ohio Sea Grant R/LR-013, and USDA ARS 3655-31000-020-00D. LB has been funded by a Canadian Research Chair in Genomics and Conservation of Aquatic Resources as well as by the Natural Sciences and Engineering research Council of Canada. JBG is thankful to Sara Bergek, Mats Björklund and Reiner Eckmann for their kind support. Funding for JBG's work came from the German Science Foundation (DFG) within the CRC 454 "littoral of Lake Constance" as well as from the Konrad-Adenauer-Stiftung and the University of Konstanz.

2.8 References

Aalto, S.K. and G.E. Newsome. 1990. Additional evidence supporting demic behavior of a yellow perch (*Perca flavescens*) population. Can. J. Fish. Aquat. Sci. 47: 1959–1962.

Aldenhoven, J.T., M.A. Miller, P.S. Corneli and M.D. Shapiro. 2010. Phylogeography of nine spine sticklebacks (*Pungitius pungitius*) in North America: glacial refugia and the origins of adaptive traits. Mol. Ecol. 19: 4061–4076.

Allendorf, F.W., P.A. Hohenlohe and G. Luikart. 2010. Genomics and the future of conservation genetics. Nature Rev. Genet. 11: 697–710.

April, J., R. Hanner, R. Mayden and L. Bernatchez. 2013. Metabolic rate and climatic fluctuations shape continental wide pattern of genetic divergence and biodiversity in fishes. Plos One 8: e70296.

Araújo, M.B. and C. Rahbek. 2006. How does climate change affect biodiversity? Science 313: 1396–1397.

Avise, J.C. 2004. Molecular Markers, Natural History, and Evolution, 2nd edn. Sinauer Associates, Sunderland, Massachusetts.

Avise, J. 2010. Conservation genetics enters the genomics era. Conserv. Genet. 11: 665–669.

Azizishirazi, A., W.A. Dew, H.L. Forsyth and G.G. Pyle. 2013. Olfactory recovery of wild yellow perch from metal contaminated lakes. Ecotoxicol. Environ. Safety 88: 42–47.

Bailey, R.M. and G.R. Smith. 1981. Origin and geography of the fish fauna of the Laurentian Great Lakes basin. Can. J. Fish. Aquat. Sci. 38: 1539–1561.

Barluenga, M., M. Sanetra and A. Meyer. 2006. Genetic admixture of burbot (Teleostei: *Lota lota*) in Lake Constance from two European glacial refugia. Mol. Ecol. 15: 3583–600.

Barton, B.A. and T.P. Barry. 2011. Reproduction and environmental biology. pp. 199–231. *In*: B.A. Barton (ed.). Biology, Management, and Culture of Walleye and Sauger. American Fisheries Society, Bethesda, Maryland.

Behrmann-Godel, J., G. Gerlach and R. Eckmann. 2004. Postglacial colonization shows evidence for sympatric population splitting of European perch (*Perca fluviatilis* L.) in Lake Constance. Mol. Ecol. 13: 491–497.

Behrmann-Godel, J., G. Gerlach and R. Eckmann. 2006. Kin and population recognition in sympatric Lake Constance perch (*Perca fluviatilis* L.): Can assortative shoaling drive population divergence? Behav. Ecol. Sociobiol. 59: 461–468.

Behrmann-Godel, J. and G. Gerlach. 2008. First evidence for postzygotic reproductive isolation between two populations of European perch (*Perca fluviatilis* L.) within Lake Constance. Front. Zool. 5: 1–7.

Behrmann-Godel, J., S. Roch and A. Brinker. 2014. Gill worm *Ancyrocephalus percae* (Ergens 1966) outbreak negatively impacts the Eurasian perch *Perca fluviatilis* L. stock of Lake Constance, Germany. J. Fish Dis. 37: 925–930.

Bélanger-Deschênes, S., P. Couture, P.G. Campbell and L. Bernatchez. 2013. Evolutionary change driven by metal exposure as revealed by coding SNP genome scan in wild yellow perch (*Perca flavescens*). Ecotoxicology 22: 938–957.

Beletsky, D., D.M. Mason, D.J. Schwab, E.S. Rutherford, J. Janssen, D.F. Clapp and J.M. Dettmers. 2007. Biophysical model of larval yellow perch advection and settlement in Lake Michigan. J. Great Lakes Res. 33: 842–866.

Bergek, S. and M. Björklund. 2007. Cryptic barriers to dispersal within a lake allow genetic differentiation of European perch. Evolution. 61: 2035–2041.

Bergek, S. and M. Björklund. 2009. Genetic and morphological divergence reveals local subdivision of perch (*Perca fluviatilis* L.). Biol. J. Linn. Soc. 96: 746–758.

Bergek, S. and J. Olsson. 2009. Spatiotemporal analysis shows stable genetic differentiation and barriers to dispersal in the European perch (*Perca fluviatilis* L.) Evol. Ecol. Res. 11: 827–840.

Bergek, S., G. Sundblad and M. Björklund. 2010. Population differentiation in perch *Perca fluviatilis*: environmental effects on gene flow? J. Fish. Biol. 76: 1159–1172.

Bernatchez, L. and J.J. Dodson. 1991. Phylogeographic structure in mitochondrial DNA of the Lake Whitefish (*Coregonusclupeaformis*) in North America and its relationships to Pleistocene glaciations. Evolution 45: 1016–1035.

Bernatchez, L. 1997. Mitochondrial DNA analysis confirms the existence of two glacial races of rainbow smelt *Osmerusmordax* and their reproductive isolation in the St. Lawrence River estuary (Quebec, Canada). Mol. Ecol. 6: 73–83.

Bernatchez, L. 2001. The evolutionary history of brown trout (*Salmo trutta* L.) inferred from phylogeographic, nested clade, and mismatch analyses of mitochondrial DNA variation. Evolution 55: 351–379.

Bernatchez, L. and M. Giroux. 2012. Les Poissons d'eau douce du Québec et leur répartition dans l'est du Canada. Editions Broquet 348 p.

Billington, N. 1993. Genetic variation in Lake Erie yellow perch (*Perca flavescens*) demonstrated by mitochondrial DNA analysis. J. Fish. Biol. 43: 941–943.

Björklund, M., T. Aho and J. Behrmann-Godel. 2015. Isolation over 35 years in a heated biotest basin causes selection on MHC class IIß genes in the European perch (*Perca fluviatilis* L.) Ecol. Evol. 5(7): 1440–1455. 5: DOI: 10.1002/ece3.1426.

Blazer, V.S., A.E. Pinkney, J.A. Jenkins, L.R. Iwanowicz, S. Minkkinen, R.O. Draugelis-Dale and J.H. Uphoff . 2013. Reproductive health of yellow perch *Perca flavescens* in selected tributaries of the Chesapeake Bay. Sci. Total Environ. 447: 198–209.

Bodaly, R.A., R.D. Ward and C.A. Mills. 1989. A genetic stock study of perch, *Perca fluviatilis* L., in Windermere. J. Fish. Biol. 34: 965–967.

Bolsenga, S.J. and C.E. Herdendorf. 1993. Lake Erie and Lake St. Clair Handbook. Wayne State University Press, Detroit, Michigan.

Bougas, B., E. Normandeau, F. Pierron, P.G.C. Campbell, L. Bernatchez and P. Couture. 2013. How does exposure to nickel and cadmium affect the transcriptome of yellow perch (*Perca flavescens*)— Results from a 1000 candidate-gene microarray. Aquat. Toxic. 142-143C: 355–364.

Bourret, V., P.P. Couture, P.C.G. Campbell and L. Bernatchez. 2008. Evolutionary toxicology of wild yellow perch (*Perca flavescens*) populations chronically exposed to a polymetallic gradient. Aquatic Toxicology 86: 76–90.

Bozek, M.A., D.A. Baccante and N.P. Lester. 2011. Walleye and sauger life history. pp. 233–301. *In*: B.A. Barton (ed.). Biology, Management, and Culture of Walleye and Sauger. American Fisheries Society. Bethesda, Maryland.

Britton, J.R., J. Pegg and C.F. Williams. 2011. Pathological and ecological host consequences of infection by an introduced fish parasite. PLoS ONE 6 (10): e26365.

Carlander, K.D. 1997. Handbook of Freshwater Fishery Biology, Vol. 3: Life History Data on Ichthyopercid and Percid Fishes of the United States and Canada. Iowa State University Press. Ames, Iowa.

Carney, J.P. and T.A. Dick. 2000. The historical ecology of yellow perch (*Perca flavescens* [Mitchill]) and their parasites. Journal of Biogeography 27: 1337–1347.

Clady, M.D. 1977. Distribution and relative exploitation of yellow perch tagged on spawning grounds in Oneida Lake. NY Fish. Game J. 24: 46–52.

Clement, M., D. Posada and K.A. Crandall. 2000. TCS: A computer program to estimate gene genealogies. Mol. Ecol. 9: 1657–1660. Available at http://darwin.uvigo.es/software/tcs.html

Coles, T. 1981. The distribution of perch, *Perca fluviatilis* L. throughout their first year of life in LlynTegid, North Wales. Arch. Fisch. 15: 193–204.

Collette, B.B., M.A. Ali, K.E.F. Hokanson, M. Nagiec, S.A. Smirnov, J.E. Thorpe, A.H. Weatherly and J. Willemsen. 1977. Biology of the percids. J. Fish. Res. Board Can. 34: 1891–1899.

Coulon, A., J.W. Fitzpatrick, R. Bowman and I.J. Lovette. 2012. Mind the gap: Genetic distance increases with habitat gap size in Florida scrub jays. Biol. Letters 8: 582–585.

Craig, J.F. 2000. Percid Fishes Systematics, Ecology, and Exploitation. Blackwell Science Ltd., Oxford, England.

Crossman, E.J. and D.E. McAllister. 1986. Zoogeography of freshwater fishes of the Hudson Bay Drainage, Ungava Bay and the Arctic Archipelago. pp. 53–104. *In*: C.H. Hocutt and E.O. Wiley (eds.). The Zoogeography of North American Freshwater Fishes. John Wiley and Sons, New York.

Dabrowski, K., R.E. Ciereszko, A. Ciereszko, G.P. Toth, S.A. Christ, D. El-Saidy and J.S. Ottobre. 1996. Reproductive physiology of yellow perch (*Perca flavescens*): environmental and endocrinological cues. J. Appl. Ichthyol. 12: 139–148.

Darriba, D., G.L. Taboada, R. Doallom and D. Posada. 2012. jModelTest 2: More models, new heuristics and parallel computing. Nat. Methods 9: 772. Available at http://code.google.com/p/jmodeltest2/.

Davis, M.B. and R.G. Shaw. 2001. Range shifts and adaptive responses to quaternary climate change. Science 292: 673–679.

Demamdt, M.H. 2010. Temporal changes in genetic diversity of isolated populations of perch and roach. Conserv. Genet. 11: 249–255.

Dembkowski, D.J., S.R. Chipps and B.G. Blackwell. 2013. Response of walleye and yellow perch to water-level fluctuations in glacial lakes. Fish. Man. and Ecol. 21: 89–95.DOI: 10.1111/fme.12047.

DeWoody, J.A. and J.C. Avise. 2000. Microsatellite variation in marine, freshwater and anadromous fishes compared with other animals. J. Fish. Biol. 56: 461–473.

Diekmann, O.E. and E.A. Serrão. 2012. Range-edge genetic diversity: locally poor extant southern patches maintain a regionally diverse hotspot in the seagrass *Zostera marina*. Mol. Ecol. 7: 1647–1657.

Drummond, A.J., M.A. Suchard, D. Xie and A. Rambaut. 2012. Bayesian phylogenetics with BEAUti and the BEAST 1.7. Mol. Biol. Evol. 29: 1969–1973. Available at http://beast.bio.ed.ac.uk/Main_Page#Citing_BEAST.

Dumont, P. 1996. Comparaison de la dynamique des populations de perchaudes (*Perca flavescens*) soumises a des niveaux differents de stress anthropique. Ministere de l'Environmement et de la Faune, Service de l'amenagemetn et de l'exploitation de la faune. Repport technique 06–46. Montreal, Canada.

Durand, J.D., H. Persat and Y. Bouvet. 1999. Phylogeography and postglacial dispersion of the chub (*Leuciscus cephalus*) in Europe. Mol. Ecol. 8: 989–97.

Evanno, G., S. Regnaut and J. Goudet. 2005. Detecting the number of clusters of individuals using the software STRUCTURE: a simulation study. Mol. Ecol. 14: 2611–2620.

Excoffier, L. and H.E. Lischer. 2010. ARLEQUIN suite ver. 3.5: A new series of programs to perform population genetics analyses under Linux and Windows. Mol. Ecol. Resour. 10: 564–567. Available at http://cmpg.unibe.ch/software/arlequin35/.

Faber, J.E. and C.A. Stepien. 1997. The utility of mitochondrial DNA control region sequences for analyzing phylogenetic relationships among populations, species, and genera of the Percidae. pp. 129–144. *In*:

T.D. Kocher and C.A. Stepien (eds.). Molecular Systematics of Fishes. Academic Press, London, England.

Ferguson, M.M. and G.A. Duckworth. 1997. The status and distribution of lake sturgeon, *Ascipenser fulvescens*, in the Canadian provinces of Manitoba, Ontario and Quebec: A genetic perspective. Environ. Biol. Fishes. 48: 299–309.

Fielder, D.G. 2002. Sources of walleye recruitment in Saginaw Bay, Lake Huron. N. Am. J. Fish. Manage. 22: 1032–1040.

Fulford, R.S., J.A. Rice, T.J. Miller, F.P. Binkowski, J.M. Dettmers and B. Belonger. 2006. Foraging selectivity by larval yellow perch (*Perca flavescens*): implications for understanding recruitment in small and large lakes. Can. J. Fish. Aquat. Sci. 63: 28–42.

Gerlach, G., U. Schardt, R. Eckmann and A. Meyer. 2001. Kin–structured subpopulations in European perch (*Perca fluviatilis* L.). Heredity 86: 213–221.

Glaubitz, J.C. 2004. CONVERT: A user–friendly program to reformat diploid genotypic data for commonly used population genetic software packages. Mol. Ecol. Notes 4: 309–310. http://www.agriculture.purdue.edu/fnr/html/faculty/rhodes/students%20and%20staff/glaubitz/software.htm.

GLFC (Great Lakes Fishery Commission). 2011. Strategic vision of the Great Lakes Fishery Commission 2011–2020. Great Lake Fish. Comm. Spec. Pub. Ann Arbor, Michigan. Available at http://www.glfc.org/pubs/SpecialPubs/StrategicVision2012.pdf.

Godel. 1999. Seasonal migrations of perch *Perca fluviatilis* L. in Lake Constance (in German) Master thesis, University of Konstanz.

Goudet, J. 1995. Fstat version 1.2: a computer program to calculate *F* statistics. J. Hered. 86: 485–486.

Goudet, J. 2002. Fstat version 2.9.3.2.Available at http://www2.unil.ch/popgen/softwares/fstat.htm.

Griffiths, D. 2010. Pattern and process in the distribution of North American freshwater fish. Biol. J. Linn. Soc. 100: 46–61.

Grzybowski, M., O.J. Sepulveda-Villet, C.A. Stepien, D. Rosauer, F. Binkowski, R. Klaper, B.S. Shepherd and F. Goetz. 2010. Genetic variation of 17 wild yellow perch populations from the Midwest and east coast analyzed via microsatellites. Trans. Am. Fish. Soc. 139: 270–287.

Guindon, S., J.F. Dufayard, V. Lefort, M. Anisimova, W. Hordijk and O. Gascuel. 2010. New Algorithms and methods to estimate maximum–likelihood phylogenies: assessing the performance of PhyML 3.0. Systematic Biol. 59: 307–321. Available at http://www.atgc–montpellier.fr/phyml/.

Gyllensten, U.L.F., N. Ryman and G. Ståhl. 1985. Monomorphism of allozymes in perch (*Perca fluviatilis* L.). Hereditas 102: 57–61.

Haas, R.C., W.C. Bryant, K.D. Smith and A.J. Nuhfer. 1985. Movement and harvest of fish in Lake St. Clair, St. Clair River, and Detroit River. Final Report Winter Navigation Study U.S. Army Corps of Engineers.

Hampe, A. and A.S. Jump. 2011. Climate relicts: Past, present, future. Ann. Rev. Ecol. Evol. Syst. 42: 313–333.

Hampe, A. and R. Petit. 2005. Conserving biodiversity under climate change: The rear edge matters. Ecol. Lett. 8: 461–467.

Haponski, A.E. and C.A. Stepien. 2013a. Genetic connectivity and diversity of walleye (*Sander vitreus*) spawning groups in the Huron-Erie Corridor. J. Great Lakes Res. 49: 89–100.

Haponski, A.E. and C.A. Stepien. 2013b. Phylogenetic and biogeographic relationships of the *Sander* pikeperches (Perciformes: Percidae): Patterns across North America and Eurasia. Biol. J. Linn. Soc. 110: 156–179.

Haponski, A.E. and C.A. Stepien. 2014. A population genetic window into the past and future of the walleye *Sander vitreus*: Relation to historic walleye and the extinct blue pike variant. BMC Evol. Biol. 14: 133.

Harris, L.N. and E.B. Taylor. 2010. Pleistocene glaciations and contemporary genetic diversity in a Beringian fish, the broad whitefish, *Coregonusnasus* (Pallas): Inferences from microsatellite DNA variation. J. Evol. Biol. 23: 72–86.

Hayhoe, K., J. VanDorn, T. Croley, N. Schlegal and D. Wuebbles. 2010. Regional climate change projections for Chicago and the US Great Lakes. J. Great Lakes Res. 36: 7–21.

Helfman, G. 1984. School fidelity in fishes: the yellow perch pattern. Anim. Behav. 32: 663–672.

Hewitt, G.M. 1999. Post–glacial re-colonization of European biota. Biol. J. Linn. Soc. 68: 87–112.

Hewitt, G.M. 2004a. The structure of biodiversity—insights from molecular phylogeography. Front. Zool. 1: 4.

Hewitt, G.M. 2004b. Genetic consequences of climatic oscillations in the Quaternary. Philos. T. R. Soc. B. 359: 183–195.

Hill, D.K. and J.J. Magnuson. 1990. Potential effects of global climate warming on the growth and prey consumption of Great Lakes fish. Trans. Am. Fish. Soc. 119: 265–275.

Hoagstrom, C.W. and C.R. Berry. 2010. The native range of walleyes in the Missouri River drainage. N. Am. J. Fish. Manage. 30: 642–654.

Hocutt, C.H. and E.O. Wiley. 1986. The Zoogeography of the North American Freshwater Fishes. Wiley, New York 880 pp.

Hoff, M.H. 2002. A rehabilitation plan for walleye populations and habitats in Lake Superior. Great Lakes Fish. Comm. Misc. Pub. 2003-01. Ann. Arbor, Michigan.

Hokanson, K.E.F. 1977. Temperature requirements of some percids and adaptations to the seasonal temperature cycle. J. Fish. Res. Brd. Can. 34: 1524–1550.

Horrall, R.M. 1981. Behavioral stock-isolating mechanisms in Great Lakes fishes with special reference to homing and site imprinting. Can. J. Fish. Aquat. Sci. 38: 1481–1496.

Hubbs, C.L. and K.F. Lagler. 2004. Fishes of the Great Lakes Region (G.R. Smith, revised). University of Michigan. Ann. Arbor, Michigan.

Jansen, A.C., B.D.S. Graeb and D.W. Willis. 2009. Effect of a simulated cold-front on hatching success of yellow perch eggs. J. Freshwater Ecol. 24: 651–655.

Jones, M.L., J.K. Netto, J.D. Stockwell and J.B. Mion. 2003. Does the value of newly accessible spawning habitat for walleye (*Stizostedionvitreum*) depend on its location relative to nursery habitats? Can. J. Fish. Aquat. Sci. 60: 1527–1538.

Kamilov, G. 1966. New data on the composition of ichthyofauna of reservoirs in Uzbekistan. pp. 110–111 (in Russian). *In*: Biological Background of Commercial Fishery in Reservoirs of Middle Asia and Kazakhstan. Alma-Ata, Kainar.

Keller, O. and E. Krayss. 2000. Die Hydrographie des Bodenseeraums in Vergangenheit und Gegenwart. Ber. St. Gall. Naturw. Ges. 89: 39–56.

Kerr, S.J., B.W. Corbett, N.J. Hutchinson, D. Kinsman, J.H. Leach, D. Puddister, L. Stanfield and N. Ward. 1997. Walleye habitat: A synthesis of current knowledge with guidelines for conservation. Percid Community Synthesis, Walleye Habitat Working Group, Ontario Ministry of Natural Resources, Peterborough.

Kocovsky, P.M. and C.T. Knight. 2012. Morphological evidence for discrete stocks of yellow perch in Lake Erie. J. Great Lakes Res. 38: 534–539.

Kocovsky, P.M., T.J. Sullivan, C.T. Knight and C.A. Stepien. 2013. Genetic and morphometric differences demonstrate fine–scale population substructure of the yellow perch *Perca flavescens*: Need for redefined management units. J. Fish. Biol. 82: 2015–2030.

Kornis, M.S., N. Mercado-Silva and M.J. Vander Zanden. 2012. Twenty years of invasion: a review of *Neogobiusmelanostomus* biology, spread, and ecological implications. J. Fish. Biol. 80: 235–285.

Kunin, W.E., P. Vergeer, T. Kenta, M.P. Davey, T. Burke, F.I. Woodward, P. Quick, M-E. Mannarelli, N.S. Watson-Haigh and R. Butlin. 2009. Variation at range margins across multiple spatialscales: Environmental temperature, population genetics and metabolomic phenotype. P. R. Soc. B. 276: 1495–1506.

Larson, G. and R. Schaetzl. 2001. Origin and evolution of the Great Lakes. J. Great Lakes Res. 27: 518–546.

Leary, R. and H.E. Booke. 1982. Genetic stock analysis of yellow perch from Green Bay and Lake Michigan. Trans. Am. Fish. Soc. 111: 52–57.

Leclerc, E., Y. Mailhot, M. Mingelbier and L. Bernatchez. 2008. The landscape genetics of yellow perch (*Perca flavescens*) in a large fluvial ecosystem. Mol. Ecol. 17: 1702–1717.

Li, S. and J.A. Mathias. 1982. Causes of high mortality among cultured larval walleyes. Trans. Am. Fish. Soc. 111: 710–721.

Lindsay, D.L., K.R. Barr, R.F. Lance, S.A. Tweddale, T.J Hayden and P.L. Leberg. 2008. Habitat fragmentation and genetic diversity of an endangered, migratory songbird, the golden-cheeked warbler (*Dendroicachrysoparia*). Mol. Ecol. 17: 2122–2133.

MacGregor, R.B. and L.D. Witzel. 1987. A twelve year study of the fish community in the Nanticoke Region of Long Point Bay, Lake Erie. Lake Erie Fisheries Assessment Unit Report 1987–3. Ontario Ministry of Natural Resources, Port Dover.

Mandrak, N.E. and E.J. Crossman. 1992. Postglacial dispersal of freshwater fishes into Ontario. Can. J. Zool. 70: 2247–2259.

Mangan, M.T. 2004. Yellow perch production and harvest strategies for semi-permanent wetlands in Eastern South Dakota. M.S. Thesis, Wildlife and Fisheries Sciences, South Dakota State University, South Dakota.

Manni, F., E. Guérard and E. Heyer. 2004. Geographic patterns of (genetic, morphologic, linguistic) variation: How barriers can be detected by using Monmonier's algorithm. Hum. Biol. 76: 173–190. Available at http://www.mnhn.fr/mnhn/ecoanthropologie/software/barrier.html.

Manning, N.F., C.M. Mayer, J.M. Bossenbroek and J.T. Tyson. 2013. Effects of water clarity on the length and abundance of age-0 yellow perch in the Western Basin of Lake Erie. J. Great Lakes Res. 39: 295–302.

Mantel, N. 1967. The detection of disease clustering and a generalized regression approach. Cancer Res. 27: 209–220.

Marcogliese, D.J. 2001. Implications of climate change for parasitism of animals in the aquatic environment. Can. J. Zool. 79: 1331–1351.

Marcogliese, D.J. and M. Pietrock. 2011. Combined effects of parasites and contaminants on animal health: parasites do matter. Trends Parasitol. 27: 123–130.

Michel, C., L. Bernatchez and J. Behrmann-Godel. 2009. Diversity and evolution of MHII beta genes in a non-model percid species—the Eurasian perch (*Perca fluviatilis* L.). Mol. Immunol. 46: 3399–3410.

Miller, L.M. 2003. Microsatellite DNA loci reveal genetic structure of yellow perch in Lake Michigan. Trans. Am. Fish Soc. 132: 503–513.

Mingelbier, M., P. Brodeur and J. Morin. 2008. Spatially explicit model predicting the spawning 578 habitat and early stage mortality of Northern pike (*Esoxlucius*) in a large system: the 579 St. Lawrence River between 1960 and 2000. Hydrobiologia 601: 55–69.

Moran, G.F. and S.D. Hopper. 1983. Genetic diversity and the insular population structure of the rare granite rock species, *Eucalyptus caesia* Benth. Aust. J. Bot. 31: 161–172.

Moyer, G.R. and N. Billington. 2004. Stock structure among yellow perch populations throughout North America determined from allozyme and mitochondrial DNA analysis. pp. 96–97. *In*: T.P. Barry and J.A. Malison (eds.). Proceedings of Percis III: The Third International Percid Fish Symposium. University of Wisconsin Sea Grant Institute, Madison, Wisconsin.

Murdoch, M.H. and P.D. Hebert. 1997. Mitochondrial DNA evidence of distinct glacial refugia brown bullhead (*Ameiurusnebulosus*). Can. J. Fish. Aquat. Sci. 54: 1450–1460.

Murphy, S., N. Collins, S. Doka and B. Fryer. 2012. Evidence of yellow perch, largemouth bass and pumpkinseed metapopulations in coastal embayments of Lake Ontario. Environ. Biol. Fish. 95: 213–226

Nesbø, C.L., C. Magnhagen and K.S. Jakobsen. 1998. Genetic differentiation among stationary and anadromous perch (*Perca fluviatilis*) in the Baltic Sea. Hereditas 129: 241–249.

Nesbø, C.L., T. Fossheim, L.A. Vollestad and K.S. Jakobsen. 1999. Genetic divergence and phylogeographic relationships among European perch (*Perca fluviatilis*) populations reflect glacial refugia and postglacial colonization. Mol. Ecol. 8: 1387–1404.

Newbrey, M.G. and A.C. Ashworth. 2004. A fossil record of colonization and response of lacustrine fish populations to climate change. Can. J. Fish. Aquat. Sci. 61: 1807–1816.

Nuriyev, H. 1967. Occurrence of Balkhash perch in Kattakurgan reservoir (pool of Zeravshan River). pp. 208–209 (in Russian). *In*: Biological background of commercial fishery in reservoirs of Middle Asia and Kazakhstan (summeries of reports). Balkhash, Kazakhstan.

Oberdorff, T., B. Hugueny and J.-F. Guégan. 1997. Is there an influence of historical events on contemporary fish species richness in rivers? Comparisons between western Europe and North America. J. Biogeogr. 24: 461–467.

OMNR (Ontario Ministry of Natural Resources). 2011. 2006–2009 Annual Report. Lake Erie MU.

Oppelt, C. and J. Behrmann-Godel. 2012. Genotyping MHC classIIB in non-model species by reference strand-mediated conformational analysis (RSCA). Con. Gen. Res. 4: 841–844.

Parker, A.D., C.A. Stepien, O.J. Sepulveda-Villet, C.B. Ruehl and D.G. Uzarski. 2009. The interplay of morphology, habitat, resource use, and genetic relationships in young yellow perch. Trans. Am. Fish Soc. 138: 899–914.

Petit, R.J., I. Aguinaglade, J.-L. de Beaulieu, C. Bittkau, S. Brewer, R. Cheddadi, R. Ennos, S. Fineschi, D. Grivet, M. Lascoux, A. Mohanty, G. Müller-Starck, B. Demesure-Musch, A. Palmé, J.P. Martín, S. Rendell and G.G. Vendramin. 2003. Glacial refugia: Hotspots but not melting pots of genetic diversity. Science 300: 1563–1565.

Pierce, L.R. and C.A. Stepien. 2014. Evolution and biogeography of an emerging quasi species: Diversity patterns of the fish viral hemorrhagic septicemia virus (VHSv). Mol. Phyl. Evol. 63: 327–341.

Pierron, F., E. Normandeau, M.A. Defo, P.G.C. Campbell, L. Bernatchez and P. Couture. 2011. Effects of chronic metal exposure on wild fish populations revealed by high-throughput cDNA sequencing. Ecotoxicology 6: 1388–1399.

Pivnev, I. 1985. Fish species in pool of Chu and Talas rivers. pp. 189 (in Russian). Frunze: Ilim.

Poulin, R. 2006. Global warming and temperature-mediated increases in cercarial emergence in trematode parasites. Parasitology 132: 143–151.

Pritchard, J.K. and W. Wen. 2004. Documentation for STRUCTURE software: Ver. 2.3.3. Stanford University. Available at http://pritchardlab.stanford.edu/software.html.

Pritchard, J.K., M. Stephens and P. Donnelly. 2000. Inference of population structure using multilocus genotype data. Genetics 155: 945–959.

Provan, J. and K.D. Bennett. 2008. Phylogeographic insights into cryptic glacial refugia. Trends Ecol. Evol. 23: 564–571.

Radabaugh, N.B., W.F. Bauer and M.L. Brown. 2010. A comparison of seasonal movement patterns of yellow perch in simple and complex lake basins. N. Am. J. Fish. Manage. 30: 179–190.

Rawson, M.R. 1980. Yellow perch movements. Ohio Department of Natural Resources Job Program Report, Dingell–Johnson Project Number F–35–R–18, Study Number. 4, November 1 1979–June 30, 1980.

Redman, R.A., S.J. Czesny and J.M. Dettmers. 2013. Yellow Perch Population Assessment in Southwestern Lake Michigan. INHS Technical Report 2013 (25., Division of Fisheries, Illinois Department of Natural Resources, Champaign, IL.

Refseth, U.H., C.L. Nesbø, J.E. Stacy, L.A. Vøllestad, E. Fjeld and K.S. Jakobsen. 1998. Genetic evidence for different migration routes of freshwater fish into Norway revealed by analysis of current perch (*Perca fluviatilis*) populations in Scandinavia. Mol. Ecol. 7: 1015–1027.

Rempel, L.L. and D.G. Smith. 1998. Postglacial fish dispersal from the Mississippi refuge to the Mackenzie River basin. Can. J. Fish. Aquat. Sci. 55: 893–899.

Rice, R.M. 1989. Analyzing tables of statistical tests. Evolution. 43: 223–225.

Rodrigues, C.G. and G. Vilks. 1994. The impact of glacial lake runoff on the Goldthwait and Champlain Seas: The relationship between glacial Lake Agassiz runoff and the younger dryas. Quat. Sci. Rev. 13: 923–944.

Ronquist, F. and J.P. Huelsenbeck. 2003. MrBayes 3: Bayesian phylogenetic inference under mixed models. Bioinformatics. 19: 1572–1574. http://mrbayes.csit.fsu.edu/. (v3.1.2 2005).

Roseman, E.F., W.W. Taylor, D.B. Hayes, J.T. Tyson and R.C. Haas. 2005. Spatial patterns emphasize the importance of coastal zones as nursery areas for larval walleye in western Lake Erie. J. Great Lakes Res. 31: 28–44.

Rousset, F. 2008. Genepop'008: A complete re–implementation of the Genepop software for Windows and Linux. Mol. Ecol. Resour. 8: 103–106. Available at http://kimura.univmontp2.fr/ ~rousset/ Genepop.htm.

Russell, D.A., F.J. Rich, V. Schneider and J. Lynch-Stieglitz. 2009. A warm thermal enclave in the Late Pleistocene of the south-eastern United States. Biol. Rev. 84: 173–202.

Ryan, P.A., R. Knight, R. MacGregor, G. Towns, R. Hoopes and W. Culligan. 2003. Fish-community goals and objectives for Lake Erie. Great Lakes Fishery Commission Special Publication 03–02.

Saarnisto, M. 1974. The deglaciation history of the Lake Superior region and its climatic implications. Quatern. Res. 4: 316–339.

Sanderson, M.J. 2003. r8s: inferring absolute rates of molecular evolution and divergence times in the absence of a molecular clock. Bioinformatics 19: 301–302. Available at http://loco.biosci.arizona.edu/r8s/.

Schmidt, R.E. 1986. Zoogeography of the northern Appalachians. pp. 137–159. *In*: C.H. Hocutt and E.O. Wiley (eds.). The Zoogeography of North American Freshwater Fishes. John Wiley and Sons, New York.

Schmitt, T. 2007. Molecular biogeography of Europe: Pleistocene cycles and postglacial trends. Front. Zool. 4: 11.

Scott, W.B. and E.J. Crossman. 1973. Freshwater fishes of Canada. J. Fish. Res. Board Can. 184: 1–196.

Sepulveda-Villet, O.J., A.M. Ford, J.D. Williams and C.A. Stepien. 2009. Population genetic diversity and phylogeographic divergence patterns of the yellow perch (*Perca flavescens*). J. Great Lakes Res. 35: 107–119.

Sepulveda-Villet, O.J. and C.A. Stepien. 2011. Fine-scale population genetic structure of the yellow perch *Perca flavescens* in Lake Erie. Can. J. Fish. Aquat. Sci. 68: 1435–1453.

Sepulveda–Villet, O.J. and C.A. Stepien. 2012. Waterscape genetics of the yellow perch (*Perca flavescens*): Patterns across large connected ecosystems and isolated relict populations. Mol. Ecol. 21: 5795–5826.

Shuter, B.J. and J.R. Post. 1990. Climate, population viability, and the zoogeography of temperate fishes. Trans. Am. Fish Soc. 119: 314–336.

Simon, T.P. and R. Wallus. 2006. Reproductive Biology and Early Life History of Fishes in the Ohio River Drainage, Vol. 4: Percidae–Perch, Pikeperch, and Darters. CRC Taylor and Francis, Boca Raton, Florida.

Sloss, B.L., N. Billington and B.M. Burr. 2004. A molecular phylogeny of the Percidae (Teleostei, Perciformes) based on mitochondrial DNA sequence. Mol. Phylogenet. Evol. 32: 545–562.

Smith, T.B. and L. Bernatchez (eds.). 2008. Evolutionary change in human-altered environments. Mol. Ecol. 17: 1–8.

Sokolovsky, V., S. Galouschak and V. Skakun. 2000. Current condition of Balkhash perch *Perca schrenki* (Percidae) in lakes of Alakol system. Problems of Ichthyology 40(2): 228–234.

Soltis, D.E., A.B. Morris, J.S. McLachlan, P.S. Manos and P.S. Soltis. 2006. Comparative phylogeography of unglaciated eastern North America. Mol. Ecol. 15: 4261–4293.

Song, C.B., T.J. Near and L.M. Page. 1998. Phylogenetic relations among percid fishes as inferred from mitochondrial cytochrome *b* DNA sequence data. Mol. Phylogenet. Evol. 10: 343–353.

Stepien, C.A. and J.E. Faber. 1998. Population genetic structure, phylogeography, and spawning philopatry in walleye (*Stizostedionvitreum*) from mtDNA control region sequences. Mol. Ecol. 7: 1757–1769.

Stepien, C.A., A.K. Dillon and M.D. Chandler. 1998. Genetic identity, phylogeography, and systematics of ruffe *Gymnocephalus* in the North American Great Lakes and Eurasia. J. Great Lakes Res. 24: 361–378.

Stepien, C.A., J.E. Brown, M.E. Neilson and M.A. Tumeo. 2005. Genetic diversity of invasive species in the Great Lakes versus their European source populations: Insights for risk analysis. Risk Anal. 25: 1043–1060.

Stepien, C.A., D.J. Murphy and R.M. Strange. 2007. Broad- to fine-scale population genetic patterning in the smallmouth bass *Micropterus dolomieu* across the Laurentian Great Lakes and beyond: An interplay of behaviour and geography. Mol. Ecol. 16: 1605–1624.

Stepien, C.A., D.J. Murphy, R.N. Lohner, O.J. Sepulveda–Villet and A.E. Haponski. 2009. Signatures of vicariance, postglacial dispersal, and spawning philopatry: Population genetics and biogeography of the walleye *Sander vitreus*. Mol. Ecol. 18: 3411–3428.

Stepien, C.A., D.J. Murphy, R.N. Lohner, A.E. Haponski and O.J. Sepulveda–Villet. 2010. Status and delineation of walleye (*Sander vitreus*) genetic stock structure across the Great Lakes. pp. 189–223. *In*: E. Roseman, P. Kocovsky and C. Vandergoot (eds.). Status of Walleye in the Great Lakes: Proceedings of the 2006 Symposium. Great Lakes Fishery Commission Technical Report 69, Ann. Arbor.

Stepien, C.A., J.A. Banda, D.J. Murphy and A.E. Haponski. 2012. Temporal and spatial genetic consistency of walleye (*Sander vitreus*) spawning groups. Trans. Am. Fish Soc. 141: 660–672.

Stepien, C.A., L.R. Pierce and D. Leaman, M. Niner and B. Shepherd. 2015. Genetic diversification of an emerging pathogen: A decade of mutation by the fish Viral Hemorrhagic Septicemia (VHS) virus in the Laurentian Great Lakes. In review.

Stewart, J.R. and A.M. Lister. 2001. Cryptic northern refugia and the origins of the modern biota. Trends. Ecol. Evol. 16: 608–613.

Strange, R.M. and C.A. Stepien. 2007. Genetic divergence and connectivity among river and reef spawning groups of walleye (*Sander vitreus*) in Lake Erie. Can. J. Fish. Aquat. Sci. 64: 437–448.

Sullivan, T.J. and C.A. Stepien. 2014. Genetic diversity and divergence of yellow perch spawning populations across the Huron–Erie Corridor, from Lake Huron through western Lake Erie. J. Great Lakes Res. 40. 101–109.

Sullivan, T.J. and C.A. Stepien. 2015. Temporal population genetic structure of yellow perch (*Perca flavescens*) spawning groups in the lower Great Lakes. Trans. Am. Fish Soc. 144: 211–226. DOI: 10.1080/00028487.2014.982260.

Taberlet, P., L. Fumagalli, A.-G. Wust-Saucy and J.F. Cosson. 1998. Comparative phylogeography and postglacial colonization routes in Europe. Mol. Ecol. 7: 453–464.

Teller, J.T. and P. Mahnic. 1988. History of sedimentation in the northwestern Lake Superior basin and its relation to Lake Agassiz overflow. Can. J. Earth Sci. 25: 1660–1673.

Thorpe, J.E. 1977. Morphology, physiology, behaviour and ecology of *Perca fluviatilis* L. and *Perca flavescens* Mitchill J. Fish. Res. Board Can. 34: 1504–1514.

Timmerman, A.J. 1995. Walleye assessment and enhancement projects in the middle Grand River watershed 1987–1995. Ontario Ministry of Natural Resources, Cambridge District, Guelph.

Todd, T.N. and C.O. Hatcher. 1993. Genetic variability and glacial origins of yellow perch (*Perca flavescens*) in North America. Can. J. Fish. Aquat. Sci. 50: 1828–1834.

Trautman, M.B. 1981. The Fishes of Ohio. Ohio State University Press, Columbus, Ohio.

Truemper, H.A. and T.E. Lauer. 2005. Gape limitation and piscine prey size-selection by yellow perch in the extreme southern area of Lake Michigan, with emphasis on two exotic prey items. J. Fish. Biol. 66: 135–149.

Turgeon, J. and L. Bernatchez. 2001. Mitochondrial DNA phylogeography of lake cisco (*Coregonusartedi*): evidence supporting extensive secondary contacts between two glacial races. Mol. Ecol. 10: 987–1001.

Underhill, J.C. 1986. The fish fauna of the Laurentian Great Lakes, the St. Lawrence lowlands, Newfoundland, and Labrador. pp. 105–136. *In*: C.H. Hocutt and E.O. Wiley (eds.). The Zoogeography of North American Freshwater Fishes. John Wiley and Sons, New York.

USFWS/GLFC (United States Fish and Wildlife Service/Great Lakes Fishery Commission). 2010. Great Lakes Fish Stocking database. U.S. Fish and Wildlife Service, Region 3 Fisheries Program, and Great Lakes Fishery Commission. Available at ttp://www.glfc.org/fishstocking/.

Vandewoestijne, S., N. Schtickzelle and M. Baguette. 2008. Positive correlation between genetic diversity and fitness in a large, well-connected metapopulation. BMC Biol. 6: 46.

Vidal-Martínez, V.M., D. Pech, B. Sures, S.T. Purucker and R. Poulin. 2010. Can parasites really reveal environmental impact? Trends Parasitol. 26: 44.

Volckaert, F.A.M., B.H. Hänfling, B. Hellmans and G.R. Carvalho. 2002. Timing of the population dynamics of bullhead *Cottusgobio* (Teleostei: Cottidae) during the Pleistocene. J. Evol. Ecol. 15: 930–944.

Wagner, G. 1960. Einführung in die Erd- und Landschaftsgeschichte mit besonderer Berücksichtigung Süddeutschlands, Verlag der Hohenlohe´schen Buchhandlung F. Rau, Öhringen.

Wang, N. and R. Eckmann. 1994. Distribution of perch (*Perca fluviatilis* L.) during their first year of life in Lake Constance. Hydrobiologia 277: 135–143.

Weir, B.S. and C.C. Cockerham. 1984. Estimating *F*–statistics for the analysis of population structure. Evolution 38: 1358–1370.

Wiley, E.O. 1992. Phylogenetic relationships of the Percidae (Teleostei: Perciformes): A preliminary hypothesis. pp. 247–267. *In*: R.L. Mayden (ed.). Systematics, Historical Ecology, and North American Freshwater Fishes. Stanford University Press, Stanford (CA).

Wilson, C.C. and P.D. Hebert. 1996. Phylogeographic origins of lake trout (*Salvelinus namaycush*) in eastern North America. Can. J. Fish. Aquat. Sci. 53: 2764–2775.

Wolinska, J. and K.C. King. 2009. Environment can alter selection in host–parasite interactions. Trends Parasitol. 25: 236–244.

Woolhouse, M.E.J., J.P. Webster, E. Domingo, B. Charlesworth and B.R. Levin. 2002. Biological and biomedical implications of the co-evolution of pathogens and their hosts. Nature Genetics 32: 569.

YPTG (Yellow Perch Task Group of the Lake Erie Committee, Great Lakes Fishery Commission). 2013. Report of the Lake Erie yellow perch task group. Great Lakes Fishery Commission Ann Arbor. Available at http://www.glfc.org/lakecom/lec/YPTG_docs/annual_reports/YPTG_report_2013.pdf.

Zhao, Y., M.L. Jones, B.J. Shuter and E.F. Roseman. 2009. A biophysical model of Lake Erie walleye (*Sander vitreus*) explains interannual variations in recruitment. Can. J. Fish. Aquat. Sci. 66: 114–125.

3

Biology of Balkhash Perch (*Perca schrenkii* Kessler, 1874)

Nadir Shamilevich Mamilov

ABSTRACT

The Balkhash perch *Perca schrenkii* is one of the indigenous and endemic fish species of the Balkhash Lake watershed, including the Alakol Lakes system. There are two opinions about its origin: It either is a relict of the ancient fish fauna, or is a postglacial intruder. A chapter in this volume (Stepien et al., Chapter 2) provides strong evidence from DNA sequencing data that it is a true species, and is the sister species of *P. flavescens* from North America. The Balkhash perch once constituted one of the most important traditional local fisheries, however, in the second part of the 20th century other fish species were introduced into its native range, which became preferred as food. At the same time, the Balkhash perch was introduced to the Nura River and the Chu River basins. The native range of the Balkhash perch has become significantly reduced and it is now rare in most of the Balkhash Lake basin. Introduced populations of this species to other areas were not successful. Although the external morphology and life strategy of the Balkhash perch appear rather adaptable, this species' future is uncertain.

Keywords: Balkhash perch, endemic, morphology, threatened species

Balkhash Perch—*Perca schrenkii* Kessler, 1874

The native area of the Balkhash perch *Perca schrenkii* (Fig. 1) is restricted exclusively to water bodies of the Balkhash watershed (Fig. 2). The hydrological system of the

PhD, Director of Laboratory of Zoology, Al-Farabi Kazakh National University, Institute of Biology and Biotechnology, Al-Farabi av., 71. 050038 Almaty, Republic of Kazakhstan.
E-mail: mamilov@gmail.com

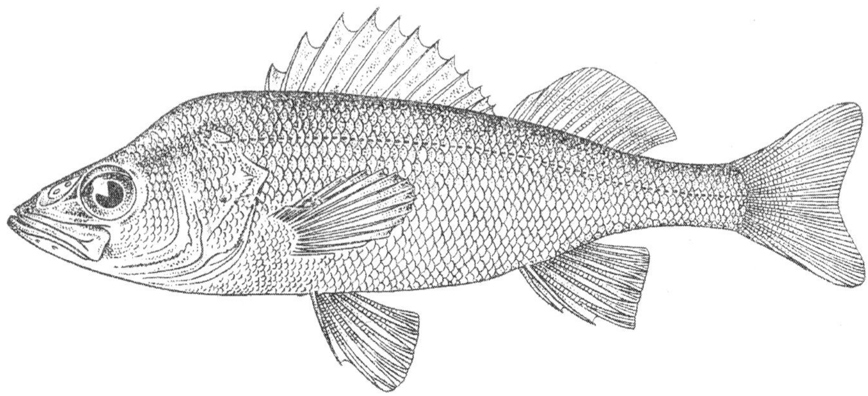

Fig. 1. Balkhash perch (from Berg 1949).

Fig. 2. Contemporary distribution of Balkhash perch in the Balkhash watershed. Numbered arrows point to isolated populations: 1—Lake Alakol, 2—Shynzhaly River and 2'—Tentek River, 3—Bakanas River, 4—Lake Big Altay, 5—Kurty Reservoir, 6—Sorbulak sewage pond, 7–9—ponds at the inflows of the Ili River, 10—Lake Issyk. Arrows without numbers point to water bodies where Balkhash perch was not found in 1993–2013.

watershed geologically developed over several million to 10,000 years ago. Therefore, the first ichthyological fauna comprised fish species that originated from rivers of the Tian Shan mountain range and rivers to the north. Following the isolation of the

Balkhash watershed, a stable complex of several fish species developed (Mitrofanov 1986), including (scientific fish names are given after Froese and Pauly 2014): representatives of the Cyprinidae family like the Balkhash marinka *Schizothorax argentatus* Kessler, 1874, Ili marinka *Schizothorax pseudaksaiensis* Herzenstein, 1889, scaled (or scaly) osman *Diptychus maculates* Steindachner, 1866, naked osman *Gymnodiptychus dybowskii* (Kessler 1874), Eurasian minnow *Phoxinus phoxinus* (Linnaeus 1758), Seven River's minnow *Ph. Brachyurus* Berg, 1912, Balkhash minnow *Rhynchocypris poljakowii* (Kessler 1879), representatives of the Balitoridae family like spotted stone loach *Triplophysa strauchii* (Kessler 1874), plain stone loach *T. labiata* (Kessler 1874), Tibet stone loach *T. stoliczkai* (Steindachner 1866), grey stone loach *T. dorsalis* (Kessler 1872), Severtzov's stone loach *T. sewerzowi* (Nikolskii 1938), and the Balkhash perch *Perca schrenkii* Kessler, 1874. It is assumed that Siberian dace *Leuciscus leuciscus baicalensis* (Dybowski 1874) also is an indigenous fish species, but its origin in the Balkhash watershed is disputable (Mitrofanov 1986). The Balkhash perch was the sole predatory fish among them.

3.1 Description (External Morphology)

The discovery of Balkhash perch has a long history. Some were caught in 1842 during the expedition of Alexander von Schrenk to the Semiretchie (Seven Rivers) in eastern Turkestan. Specimens were conveyed to the depository of the Russian Academy of Sciences, where they were later rediscovered by academic Alexander Strauch. Professor Strauch forwarded these specimens to Karl Ferdinand Kessler, who described them as a new fish species (Kessler 1874). Today a single specimen of the Balkhash perch from Balkhash Lake (holotype, No2326 ZIN RAN) and four specimens from Alakol Lake (paratypes No2327 ZIN RAN) reside in the collection of the Institute of Zoology of Russian Academy of Sciences in St. Petersburg.

The Balkhash perch is one of the endemic fish species in the Balkhash Lake watershed (including the Alakol Lakes system) situated in the centre of Eurasia (Fig. 2). Its general profile is very similar to that of the European perch (*P. fluviatilis*) and the yellow perch (*P. flavescens*). Like the other species of *Perca*, the Balkhash perch (1) is small to moderate-sized with spines in its dorsal, anal, and pelvic (ventral, abdominal) fins, as well as on the opercular tip, (2) its mouth contains many fine teeth, (3) the body is oval shaped and laterally compressed, (4) its general dorsal body coloration is black, deep grey, or sometimes olive-grey, and brightens ventrally, (5) possesses ctenoid scales, which give the Balkhash perch a rough feel. It is distinguished from other *Perca* by: (1) its more elongated body, (2) relatively larger scales (numbering 44–54 scales on the lateral line), (3) a lesser height of the first dorsal fin, and (4) a greater number of gill rakers on the first gill arch (most often from 27 to 33) (Berg 1949a). Individuals with a protruded lower jaw and a sub-superior mouth are frequently observed, in contrast to other representatives of the genus. The body coloration is a trait that distinguishes the Balkhash perch from the yellow and European perch, which have dark stripes on a green background. The dorsal and caudal fins are usually grey coloured. Coloration of the paired and anal fins varies from the usual light grey to light yellow, reddish, or orangey. A black spot on the posterior and basal end of the first dorsal fin is occasionally found, but is atypical. Large individuals of the Balkhash perch are distinguished from the other *Perca* species by the light, almost white, body coloration and vertical stripes that are either absent or very faint (Fig. 3).

Balkhash perch from the small Bakanas River display some different morphological features. Their body coloration is drab, with 6–8 dark vertical stripes, and the paired fins (pectoral and pelvic) are yellow. Melanophores on the head form complex patterns, appearing as oval, ribbon-like, and irregular spots. Stripes on the body and yellow coloration—especially on the paired fins—are characteristic of the other two *Perca* species, but are absent in most other Balkhash perch populations.

Zhadin (1949) revealed some skeletal features of the Balkhash perch that differ from the European perch *P. fluviatilis*. The cranium of the former is characterized by a larger mandible that extends forward, and its mouth thus appears semi superior. Its eye sockets are relatively small and dorsally located, with the quadratoarticular joint situated lower than the line of the closed mouth. This position of the mouth and eyes suggests that the Balkhash perch spends much of its time near the bottom and catches its prey from below. The form and position of the preoperculum is another distinctive trait of its cranium; the preoperculum is very large, particularly its lower part, and the distance between the preoperculum and the back corner of the eye is appreciably longer than for the European perch. The frontal bones (frontale) of the Balkhash perch are longer, narrower, and situated further back in comparison with the European perch. The sensory canals on the head of the Balkhash perch are more pronounced, especially the supraorbital and preoperculomandibular canals, which protrude over frontal and preoperculum bones; Zhadin (1949) considered these characters as primitive (pleisiomorphic). The vertebral column of Balkhash perch has two distinctive features:

Fig. 3. Photograph of a Balkhash perch captured in the Tekes River, a tributary of the Ili River (N. Mamilov).

(1) a fewer number of vertebrae (average 38) than in European perch (average 41) and (2) a greater number of caudal vertebrae (average 20) (Table 1), as in percoids (thus this also is pleisiomorphic). Svetovidov and Dorofeeva (1963) found that the shape of the skull of yellow and European perches were similar to each other and well distinguished from the skull of Balkhash perch.

Table 1. Number of corporeal and caudal vertebrae of *Perca* and *Sander* (by Zhadin 1949).

Species	Average number of vertebrae		
	Abdominal	Caudal	Total
Perca fluviatilis	22	19	41
Perca flavescens	22	19	41
Perca schrenkii	18	20	38
Sander lucioperca	25	21	46
Sander volgensis	22	21	43
Sander vitreum	25	21	46

Many scientists who studied Balkhash perch noted its significant morphological and biological variability. This intrapopulational variability led to the description of different forms (Amirgaliev et al. 2006), including pelagic, coastal, dwarf, and riverine variants. For example, Golodov and Mitrofanov (1968) noted local stocks in lakes of the Ili river watershed, and Strelnikov (1970) mentioned two forms that differed in growth rates in Alakol Lake, but they did not detail their characteristics. Goryunova (1950) separated "reed" and "lake" forms from the delta of the Ili River and Balkhash Lake, respectively, with the former having deep grey coloration and the latter a lighter colour and deeper body. Unfortunately, since many populations of the Balkhash perch have now vanished, it is impossible to evaluate their earlier diversity.

Presently, the most abundant population of Balkhash perch inhabits Alakol Lake, where early ichthyologists described the existence of two distinguishable forms: (1) normal growing or pelagic and (2) slow growing or reed form (Nekrashevich 1965b; Tsyba 1965; Strelnikov 1974). Unfortunately, no detailed description was performed, thus here we investigated morphological diversity, including fish age and biology, in order to evaluate modern intrapopulation variability (Mamilov and Mitrofanov 2002). All fish analysed were divided into two groups by age and length (Fig. 4). Since the minimal body lengths of the normal growing form and the maximal length for slowly growing ones overlapped for perch aged 1, 3 and 4 years, this discrimination was subjective. We grouped 92 individuals as being slowly growing, and 59 as normal.

Comparing our data on age specific lengths with those from Tsyba (1965) and Strelnikov (1974) it appears that the average body lengths of both forms in 1988 and 1996 were higher than during 1960–1970. General characteristics of the two forms, presented in Table 2, show no differences between them in condition factors despite their differences in body lengths and weights. Individuals with more oblong-shaped or deeper bodies occurred within each form, and a number of intermediate forms were present. Three types of body colorations were observed: (1) head and back light-grey or grey, sides and belly silvery-white, dorsal and caudal fins grey in color and binate and anal fins pale-grey with an occasional light yellow hue. This coloration was observed

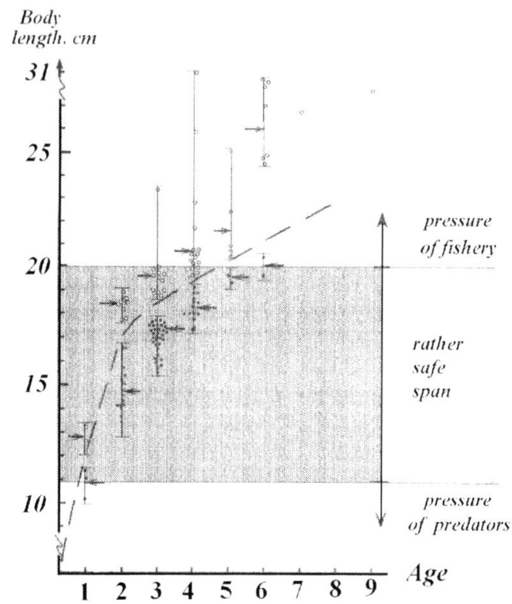

Fig. 4. An analysis of growth patterns of Balkhash perch in Lake Alakol. The "slowly growing" form is shown with solid circles and the "normal growing" form has empty circles. Horizontal arrows show average length for each age and form (reprinted with permission from Mamilov and Mitrofanov 2002).

Table 2. General characteristics of the two forms of Balkhash perch from Alakol Lake.

Characteristics	Normal growing form (n = 59)		Slowly growing form (n = 92)	
	min–max	**M ± m**	**min–max**	**M ± m**
Total body length, cm	20.9–34.5	22.3 ± 1.70	11.8–23.5	18.4 ± 2.85
Standard body length, cm	17.8–31.0	19.3 ± 1.7	10.1–20.4	16.2 ± 2.6
Total body weight, g	58–330	108 ± 3	12–94	52.1 ± 3.6
Carcass weight (gutted), g	50–320	92 ± 3.2	10–71	46.3 ± 4.0
Fulton's Condition factor	0.98–1.54	1.31 ± 0.04	1.08–1.49	1.32 ± 0.03
Clark's Condition factor	0.85–1.26	1.12±0.05	0.83–1.29	1.14 ± 0.03

M-mean, ± m – mean deviation
Fulton's Condition factor = Q x $100/l^3$
Clark's Condition factor = q x $100/l^3$
where
Q = Total body weight, g
q = Carcass weight (gutted), g
l = standard body length, cm

only in relatively large fish >18 cm total length (TL); (2) head, back and sides grey or dark-grey, with a tendency to be lighter ventrally, with the ventral surface light-grey and never white, dorsal and lateral surfaces occasionally a pale brown or cream hue, 6–8 vertical dark bands with diffuse edges sometimes present, all fins dark-grey or grey, and the edges of the paired and anal fins pale yellow in some cases; and

(3) primary coloration of the head, dorsal, and lateral surfaces usually grey, with a composite pattern of black or dark-grey rounded, oval, or ribbon-shaped spots, ventral surface always uniformly grey, dorsal, caudal and pectoral fins black-grey, as well as the abdominal and anal fins, with occasional yellow edges. Such colored individuals were rare in both forms.

Morphological indices for the two putative forms of Balkhash perch from Alakol Lake are shown in Table 3. Differentiation indexes (CD) were calculated according to Mayr (1969); those above 1.28 may be considered appreciable.

Table 3. Variability of morphological indices of Balkhash perch from Alakol Lake. CD = Differentiation Index.

Indices	Normal growing form (n = 59)		Slowly growing form (n = 92)		CD
	min–max	M ± m	min–max	M ± m	
Relative measurement estimates (indices in % of standard length):					
aD	31.5–36.5	33.4 ± 0.1	29.8–38.3	33.4 ± 0.2	0
pD	16.8–22.0	19.5 ± 0.3	15.9–21.8	19.2 ± 0.2	0.10
aA	66.2–72.8	70.0 ± 0.3	61.3–72.5	69.5 ± 0.2	0.14
aV	34.6–40.1	36.6 ± 0.2	32.4–41.5	36.6 ± 0.2	0
aP	29.3–34.0	31.4 ± 0.2	28.6–34.7	31.3 ± 0.2	0.04
VA	31.4–35.4	33.3 ± 0.2	29.1–36.8	32.5 ± 0.2	0.27
ca	18.9–23.2	20.8 ± 0.2	19.1–22.5	20.6 ± 0.2	0.06
c	28.7–33.3	31.1 ± 0.2	28.7–43.9	31.1 ± 0.3	0
hc	15.3–18.5	16.8 ± 0.2	14.9–20.2	17.3 ± 0.1	0.26
ao	6.5–8.2	7.5 ± 0.1	6.7–9.0	7.5 ± 0.1	0.03
o	4.9–6.7	5.8 ± 0.1	5.0–7.1	6.0 ± 0.1	0.22
l_{mx}	9.8–12.5	11.4 ± 0.1	9.0–12.8	11.1 ± 0.1	0.23
h_{mx}	2.3–3.7	3.1 ± 0.0	2.7–3.8	3.3 ± 0.1	0.36
l_{md}	14.6–16.8	15.7 ± 0.1	13.4–17.3	15.3 ± 0.1	0.31
op	15.4–18.3	16.6 ± 0.1	14.5–19.3	16.5 ± 0.1	0.07
io	4.2–5.6	5.1 ± 0.1	4.6–6.6	5.4 ± 0.1	0.42
H	21.3–27.2	24.7 ± 0.2	22.8–27.4	25.1 ± 0.1	0.18
h	7.1–9.2	8.0 ± 0.1	6.4–9.5	8.3 ± 0.1	0.30
lD1	26.4–32.1	30.0 ± 0.2	27.6–35.4	30.6 ± 0.2	0.22
lD2	13.5–18.2	16.1 ± 0.2	14.3–18.8	16.6 ± 0.1	0.24
hD1	12.6–18.0	14.6 ± 0.2	13.2–18.8	15.9 ± 0.1	0.53
hD2	7.7–14.4	11.7 ± 0.2	10.7–14.6	12.3 ± 0.1	0.28
lA	7.5–10.8	9.0 ± 0.2	7.1–11.6	9.2 ± 0.1	0.09
hA	10.3–13.5	12.0 ± 0.1	10.4–15.5	12.9 ± 0.1	0.49
lP	14.9–19.6	17.1 ± 0.2	15.4–18.9	17.5 ± 0.1	0.19
lV	14.1–17.9	15.9 ± 0.2	11.9–19.8	16.9 ± 0.1	0.49
lC_s	14.2–19.5	16.7 ± 0.2	14.7–19.6	17.3 ± 0.1	0.28
lC_i	15.1–19.6	16.8 ± 0.2	14.7–19.4	17.0 ± 0.1	0.09

Table 3. contd....

Table 3. contd.

Indices	Normal growing form (n = 59)		Slowly growing form (n = 92)		CD
	min–max	M ± m	min–max	M ± m	
Relative measurements (indices in % of head length):					
hc	48.3–57.9	54.2 ± 0.5	50.0–67.3	56.4 ± 0.5	0.33
ao	19.7–26.2	24.1 ± 0.3	21.2–29.6	24.5 ± 0.2	0.13
o	15.4–21.4	18.6 ± 0.3	14.9–24.2	19.2 ± 0.2	0.17
mx	31.0–40.7	36.8 ± 0.4	28.6–40.7	36.1 ± 0.3	0.15
hmx	8.6–12.3	10.6 ± 0.2	7.6–12.0	9.9 ± 0.1	0.38
md	45.3–53.7	50.2 ± 0.3	45.5–54.0	49.9 ± 0.3	0.08
op	47.0–59.3	53.4 ± 0.5	46.8–63.6	53.8 ± 0.3	0.08
io	13.2–19.2	16.2 ± 0.2	14.6–21.2	17.6 ± 0.2	0.55
Counted characters:					
l.l.	47–57	51.9 ± 0.4	45–58	51.2 ± 0.3	0.13
$l.l_s$	6–8	7.1 ± 0.1	6–8	7.0 ± 0.1	0.11
$l.l_i$	13–16	14.3 ± 0.1	12–16	14.1 ± 0.1	0.12
D_1	12–14	13.2 ± 0.1	12–14	13.3 ± 0.1	0.08
D_2r	1–3	2.0 ± 0.1	1–3	2.0 ± 0.1	0
D_2s	11–14	12.5 ± 0.1	11–14	12.6 ± 0.1	0.07
As	7–9	7.7 ± 0.1	7–10	7.8 ± 0.1	0.08
Sp.br.	27–35	30.8 ± 0.3	26–35	30.2 ± 0.2	0.16
Vert.	36–39	37.6 ± 0.1	36–40	37.5 ± 0.1	0.05

Measurements: aD—predorsal length; pD—postdorsal length; aA—preanal length; aV—preventral length; aP—prepectoral length; VA—pelvic base to anal origin length; ca—length of the caudal peduncle; c—head length; hc—head depth at the level of occiput; ao—snout length; o—horizontal length of orbit; mx—upper jaw length; hmx—upper jaw width; md—lower jaw length; op—postorbital head length; io—interorbital length; H—maximum body depth; h—the least depth of the caudal peduncle; lD1—length of the base of the first dorsal fin; lD2—length of the base of the second dorsal fin; hD1—height of the first dorsal fin; hD2—height of the second dorsal fin; lA—length of the base of the anal fin; hA—height of the anal fin; lP—length of the pectoral fin; lV—length of the pelvic fin; lC_s—length of upper part of the caudal fin; lC_i—length of lower part of the caudal fin.

Counted: l.l. —scales in the lateral line; *l.l.s*—scale rows above the lateral line; *l.l.i*—scale rows below the lateral line; D1, D2r—spines (sharpened rays in the first and the second dorsal fins); D2s and As—branched rays in the second dorsal and anal fins; Sp.br. —the number of gill rakers on the first gill arch; Vert. —total number of vertebrae (all vertebrae were counted including the last one bearing the urostyle).

CD is subspecies differentiation indexes calculated according to Mayr (1969):

$CD = (M_1-M_2)/(s_1+s_2)$, where *M* is mean for samples 1 and 2, and *s* is standard error

Presently, the Balkhash perch population in the Alakol Lakes system is experiencing strong impacts from introduced fish species and human activities. Predatory fishes like *Sander* and the Balkhash perch itself are intensively feeding on young Balkhash perch, and thus fast growing young individuals likely avoid predation more efficiently than slower growing individuals. However, individuals >20 cm experience intensive fishing pressure. These two population influences impact Balkhash perch populations, at early and later life history stages (Fig. 4). Despite these pressures, morphometric data (Table 3) suggest that Balkhash perch from Lake Alakol represent a single population owing to no significant differences between normal-growing and slow-growing forms.

Form and number of vertical bands on the body were used as characters to describe European perch population variability (Yakovlev et al. 1988; Shaykin 1989). Development and coloration of bands depend on many factors, including the size of individuals (Dukravets and Mitrofanov 1989), and thus cannot be used as a reliable feature. We observed these bands only on Balkhash perch from the Sorbulak and Issyk lakes and from the pond on the Kurlep River (an inflow of the Ili River). Bands in the form of the letter "W" (the penultimate band on Fig. 5) also were observed in some specimens from Sorbulak Lake.

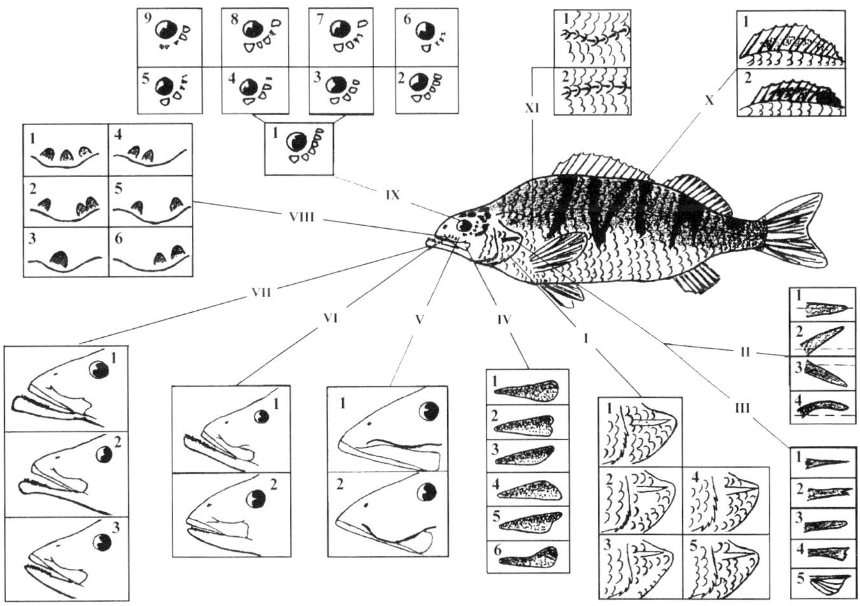

Fig. 5. Variability of some anatomical characters of Balkhash perch. Roman numerals designate characters and Arabic numerals label their states: I presence and position of scales on the operculum; II direction of spine on the operculum; III form of the spine; IV form of dental bone; V position of lacrimal bone relative to dental bone; VI position of the lower jaw; VII form of the lower jaw; VIII number and disposition of mucus hollows over the lacrimal bone; IX number and disposition of mucus hollows around the eye; X absence or presence of the black spot at the end of the first dorsal fin; XI form of the lateral line (reprinted with permission from Mamilov 2000).

Since estimation of morphological diversity in body coloration of Balkhash perch is quite subjective, we evaluated other morphological features, shown in Fig. 5, whose frequencies are in Table 4. Such morphological variation data may provide a basis for long-term monitoring of these populations (indicated with bold font in Table 4). A better knowledge of their relation with genetic diversity would be valuable.

Table 4. Frequency of different states of some features in populations of Balkhash perch.

Feature	State	Water bodies*									
		A	B	C	D	E	F	G	H	I	J
I	1	0.556	0	0.100	0.625	0.071	0.265	0.125	0	0	0
	2	0.444	0	0.667	0.250	0.571	0.706	0.500	1.000	0.583	0.333
	3	0	0	0.233	0	0.214	0.029	0	0	0	0
	4	0	1.000	0	0.125	0.143	0	0.250	0	0.417	0.667
	5	0	**0**	**0**	**0**	**0**	**0**	**0.125**	**0**	**0**	**0**
II	1	0.167	0.250	0.033	0.375	0.214	0.676	0.250	0.346	0.636	0.167
	2	0.778	0.750	0.967	0.500	0.714	0.206	0	0.654	0.364	0.833
	3	0	**0**	**0**	**0**	**0**	**0.118**	**0.750**	**0**	**0**	**0**
	4	0.056	**0**	**0**	**0.125**	**0.072**	**0**	**0**	**0**	**0**	**0**
III	1	0.944	0.500	0.100	0.500	0.643	0.676	1.000	0.654	1.000	1.000
	2	0.056	0.125	0	0	0.071	0.118	0	0.346	0	0
	3	0	0	0.167	0	0.071	0	0	0	0	0
	4	0	0.375	0	0	0.214	0.206	0	0	0	0
	5	0	**0**	**0.833**	**0.500**	**0**	**0**	**0**	**0**	**0**	**0**
IV	1	0.167	0.750	0.733	0.500	0.688	0.735	1.000	0.654	0.167	0.167
	2	0.778	0.250	0.067	0.375	0.188	0.265	0	0.308	0.833	0
	3	0	0	0.167	0	0.125	0	0	0	0	0.667
	4	0.056	0	0.033	0	0	0	0	0.038	0	0
	5	0	0	0	0	0	0	0	0	0	0.167
	6	**0**	**0**	**0**	**0.125**	**0**	**0**	**0**	**0**	**0**	**0**
V	1	0.167	0	0.433	0.250	0.333	0.118	0	0.385	0.083	0.167
	2	0.833	1.000	0.567	0.750	0.667	0.882	1.000	0.615	0.917	0.833
VI	1	0.944	0.875	0.800	0.625	0.714	0.471	0.375	0.538	0.500	0.500
	2	0.056	0.125	0.200	0.375	0.286	0.529	0.625	0.462	0.500	0.500
VII	1	0.056	0	0.267	0.125	0.071	0	0	0	0.042	0
	2	0.111	0	0.133	0	0	0	0.125	0	0.083	0
	3	0.833	1.000	0.600	0.875	0.929	1.000	0.875	1.000	0.875	1.000
VIII	1	0.611	0.625	0.200	0	0.500	0.727	0.750	0.308	0.533	0.500
	2	0	0	0	0	0	0	0	0	0	0.333
	3	0	0	0	0.071	0	0	0	0	0	0
	4	0.389	0.250	0.800	0.875	0.357	0.212	0.250	0.692	0.467	0
	5	0	0.125	0	0	0.071	0.061	0	0	0	0
	6	0	**0**	**0**	**0.125**	**0**	**0**	**0**	**0**	**0**	**0.167**
IX	1	0	0	0	0	0	0.059	0	0	0.067	1.000
	2	0.444	0.125	0.100	0.500	0.571	0.529	0.750	0.231	0.800	0
	3	0	0	0	0.125	0	0	0	0	0	0
	4	0.278	0.250	0.533	0.125	0.071	0.235	0.250	0	0	0
	5	0.222	0	0.266	0	0.143	0.118	0	0	0.133	0
	6	0.056	0.625	0	0	0.143	0.059	0	0	0	0
	7	0	0	0	0.071	0.071	0	0	0	0	0
	8	0	0	0	0.125	0	0	0	0	0	0
	9	0	**0**	**0**	**0**	**0**	**0**	**0**	**0.769**	**0**	**0**
X	1	1.000	1.000	1.000	1.000	1.000	0.882	0.885	1.000	0.958	1.000
	2	0	0	0	0	0	0.118	0.115	0	0.042	0
XI	1	1.000	1.000	1.000	1.000	1.000	0.971	0.875	1.000	1.000	0.833
	2	0	0	0	0	0	0.029	0.125	0	0	0.167

*A: Alakol Lake; B: Shynzhaly River; C: Bakanas River; D: Big Altay Lake; E: Kurty water reservoir (for 1993 only); F: Sorbulak Lake (sewage pond); G, H and I: ponds on the Ily River inflows; J: Yssik Lake. Boldface font shows specific forms.

The presence of a black spot on the posterior end of the first dorsal fin was found in two young of the year individuals from the Tentek River and in several adults from Sorbulak Lake and from two ponds on inflows of the Ili River. This spot is typical of yellow and European perch, as well as in hybrids between European and Balkhash perch (Mina 1974; Dukravets and Birukov 1976). It is a well-known fact (Turkiya 1997) that fish breeders in China introduced fish into water bodies of Northern Tien Shan and in the upper reach of the Ili River several times. Hence, the presence of the black spot on Balkhash perch could indicate hybrid origins. However, despite intensive commercial fishing, the European perch has not been detected in catches.

Several external morphological abnormalities were found, including (Table 5): (1) distortion of one or more spines or rays in the fins (Lake Big Altay, Sorbulak sewage pond, two ponds on the Ili River inflows), (2) reduction of the last rays in the first dorsal fin (Bakanas River), (3) forked rays in the first dorsal fin (Sorbulak sewage pond), (4) abnormalities in the lines of scales (one pond on the Ili River inflow), (5) truncated pelvic fins (Shynzhaly River), (6) incisure of the upper lip (one pond on the Ili River inflow), and (7) an eyelid-like overgrown epithelium around the eyes (Sorbulak sewage pond). External morphological abnormalities occurred in 4–13% of the individuals sampled from the Shymzhaly and Bakanas rivers, Big Altay Lake, and ponds adjacent to the Ili River, and in ~50% of the samples from Sorbulak Lake. No abnormalities were detected in perch from the Alakol and Yssik lakes and Kurty water reservoir.

Fluctuating asymmetry refers to differences between the two sides of a bilateral character that may reflect instability during an individual's development, and possibly lower fitness (Zakharov and Ruban 1985). Values of fluctuating asymmetry lower than 0.30 correspond to stable living conditions (Zakharov et al. 2000) and characterized most investigated Balkhash perch populations (Table 5). That index was higher in Lake Issik (0.38) and higher in the 1996 sample from Lake Alakol (0.47), likely indicating

Table 5. Asymmetry and incidence of abnormalities in some samples of Balkhash perch (reprinted with permission from Mamilov 2000).

Index	**Water bodies***									
	A	B	C	D	E	F	G	H	I	J
Fish number	18	8	30	8	14	34	8	26	24	6
Asymmetry	0.47	0.19	0.18	0.29	0.24	0.81	0.16	0.29	0.20	0.38
External abnormalities	0	0.125	0.033	0.125	0	0.529	0.250	0.154	0.292	0
Internal abnormalities:										
Heart	0.063	0	0	0	0	0.250	0	0	0	0
Liver	1.000	0.875	0	0	1.000	1.000	1.000	1.000	1.000	1.000
Bowels	0	0	0	0	0	0.531	0	0	0	0
Fat	0	1.000	0	0.750	0	1.000	0	0	0	0
Kidneys	0	0	0	0	0	0.313	0.125	0.083	0.091	0
Gonads	0	0	0	0	0	0.094	0	0	0	0.167

*A: Alakol Lake; B: Shynzhaly River; C: Bakanas River; D: Big Altay Lake; E: Kurty water reservoir (for 1993 only); F: Sorbulak Lake (sewage pond); G, H and I: ponds on the Ily River inflows; J: Yssik Lake.

poor environmental quality of the habitat. The highest value of the index (0.81) was recorded from the Sorbulak sewage pond, denoting the poorest environmental conditions.

Incidences of external and internal abnormalities were investigated as well, following the methods of Reshetnikov et al. (1999) and Chebotareva et al. (1999). Internal organs appeared normal solely in individuals from Big Altay Lake and Bakanas River. In other water bodies, irregular liver coloration was observed in Alakol Lake individuals, whereas all those from Shynzhaly River had pale-grey livers with greenish or brown spots.

Introduction of sander and roach to the Kurty water reservoir led to a food deficit for the Balkhash perch. Perch from the last sampling effort, in 1993, had drawn-in bellies with reduced liver lobes and an absence of internal fat. Perch from ponds situated close to the city of Almaty had abnormalities including irregular coloration of the liver and kidneys and constrictions of the kidneys, which may have been a result of water contamination with fertilizers and pesticides from surrounding agricultural areas (Mitrofanov 1991).

3.2 Origin, Distribution and Habitat Characteristics

In Chapter 2 (this volume), analysis of mitochondrial DNA cytochrome *b* sequence data has clearly shown that all three species of *Perca* each are monophyletic, comprising well-defined species. The primary division in the genus separates out the western European taxon (the European perch *P. fluviatilis*) from the Balkhash and yellow perch; the latter share a more recent common ancestry. The Balkhash and yellow perch are sister species (nearest relatives), but are each highly distinct, having diverged an estimated 15 Mya during the mid-Miocene epoch. This separation is confirmed by large fossil remains of representatives of the *Perca* genus from the Zaissan depression in the eastern Kazakhstan (Lebedev 1959, 1960; Yakovlev 1964).

The Balkhash depression was separated early from other water systems in Kazakhstan, with Kassin (1947) hypothesizing that the watershed had been isolated since the Paleozoic period. Mitrofanov (1986) demonstrated that the segregation of the Balkhash depression occurred before the appearance of many modern families of freshwater fishes. In any case, significant fluctuations of the water level in the Balkhash Lake from a huge lake to almost total desiccation have been shown (Sinitsyn 1962). The ancient pre-Balkhash Lake included the system of modern Alakol Lakes (Sassykkol, Alakol, Koshkrkol and Zhalanashkol Lakes), Ebi-Nur and Telli-Nur lakes during the Tertiary period (Obruchev 1949), which is supported by paleontological confirmation. Palaeontologists have found fossilized remains of ancient Amiidae (bowfin) fish dating to the mid-late Eocene simultaneously with a tapir on the slopes of the Dzhungar Alatau in the Alakol Lake basin. It is assumed that Balkhash Lake dried out completely in recent times, and its present basin formed after the last glacial period (Berg 1958).

Zhadin (1949) hypothesized that an ancestor of the Balkhash perch may have, along with other northern fishes, such as the pike *Esox lucius*, penetrated into the Balkhash watershed from a water body in the Zaysan depression, which was probably the nearest glacial refugium at the end of Pleistocene. A progressive drying of the lake began in the early Holocene. A number of lakes formed, of which the largest is Balkhash Lake. Most northern fish species have become extirpated due to climate and loss of riparian spawning areas, but the perch survived due to its adaptive plasticity and cannibalistic tendency. Later, in the Quaternary period, mountain Asian fish species like *Schizothorax, Gymnodipthychus,* and *Triplophysa* penetrated into the Balkhash watershed founding the modern fish fauna. At that time, Balkhash perch could have migrated into the Balkhash watershed from the Sherubay-Nura River (an inflow of the Nura River, the Irtysh basin) to the Tokraun River (a northern inflow of the Balkhash Lake) (Mitrofanov 1986) during post-glacial recolonization.

The native area of the Balkhash perch remains in water bodies adjacent to the system of Alakol Lakes and the Balkhash Lake watershed. The species was accidentally introduced in the 1960s into some other water bodies of Kazakhstan and Central Asia during transfers of young carp from fish farms located in the Balkhash watershed. Such introduced populations have been described for the following rivers: Olenty (Mina 1974), Nura (Dukravets and Birukov 1976), Chu (Pivnev 1985; Dukravets and Mamilov 1992), some lakes in Central Kazakhstan—Tleuberdy, Sryoba, Kurbet, Maybalyk (Gayduchenko 1986) and in the watershed of the Zeravshan River in Uzbekistan (Nuriev 1985). If European perch already existed in those water bodies, hybridization with the introduced Balkhash perch might have occurred (Mina 1974; Dukravets and Biryukov 1976; Dukravets and Mamilov 1992).

Some indigenous fishes in the Balkhash watershed, including the Balkhash perch, Balkhash marinka, and Ili marinka are considered by local people as delicacies despite their small size. However, their commercial exploitation was solely feasible during the winter, due to issues with storage and shipping. For this reason, fishery authorities in the former U.S.S.R. recommended that the government change the fish fauna in the Balkhash watershed. More than 20 alien fish species from different ecological niches were intentionally or accidentally introduced and became established in the second half of the 20th century. Intentionally introduced species included: common carp *Cyprinus carpio,* crucian carp *Carassius gibelio,* roach *Rutilus rutilus,* bream *Abramis brama,* asp *Aspius aspius,* wells *Silurus glanis*, and sander *Sander lucioperca.* Accidentally introduced species which have been spreading widely include: stone moroco (topmouth gudgeon) *Pseudorasbora parva*, sawbelly *Hemiculter leucisculus,* abbottina (false gudgeon) *Abbottina rivularis*, eleotris *Hypseleotris (Micropercops) cintus,* and Amur goby *Rhinogobius* spp. New alien fish species have become established in the watershed of the Ili River despite cessation of governmental support for introductions since the early 1990s (Karpov and Kaldaev 2005; Ismukhanov and Skakun 2008). The rapid spread of the new and illegally introduced predatory fish Amur snakehead *Channa argus* threatens indigenous fish populations (Mamilov et al. 2010). The introduction of alien species resulted in assemblages of exclusively alien species in Lake Balkhash and in the Ili River and in the displacement of indigenous species to secondary and smaller water bodies (Mitrofanov and Dukravets 1992).

Introduced sander in the Balkhash watershed has experienced substantially increased fishery demand since the beginning of the 1990s. This demand has led to an intentional increase in sander's natural range to the detriment of the indigenous fish fauna. *Sander* completely eliminated indigenous fish species in the Kurty water reservoir and Saz-Talgar ponds systems during the first 4–6 years after its introduction, and then suffered from lack of food resources.

We examined the dynamics of Balkhash perch and sander abundance in the Saz-Talgar ponds system for 20 years (Table 6). In 1989, that fish fauna consisted of the indigenous fishes Balkhash perch, spotted stone loach, plain stone loach and the introduced common carp, crucian carp, stone moroco, Amur goby and eleotris. Sander first was discovered there in 1993, with three individuals that were about 40 cm TL. By 1996, sander had become so numerous that it exterminated the loach and had significantly diminished all other fishes. Sander then became prey-limited; about 50% of adults had empty stomachs, about 25% were cannibalistic on their young, and the remaining 25% consumed young crayfish. These ponds lost their fisheries as a consequence of the introduced sander and the subsequent introduction of the exotic bream (*Abramis brama*); these new populations took about a decade to become stabilized (Fig. 6), during which time the Balkhash perch survived but lost its predominance in this fish community.

The native area of the Balkhash perch has become progressively reduced since the beginning of the 1960s, as a result of the introduction of alien fish species, notably sander and bream (Dukravets and Mitrofanov 1989). Sander consumes a large

Table 6. Average percentage of total number and biomass (in kg) of fish per landing in the Saz-Talgar ponds system (4 gill nets with 25 m length and 2 m depth with mesh size 30, 40, 50 and 60 mm during 24 hours).

Fish species	**Year**					
	1989 (n = 105)		**1996 (n = 19)**		**2008 (n = 37)**	
	% of number	**biomass**	**% of number**	**biomass**	**% of number**	**biomass**
Balkhash perch	34.3	1.80	5.2	0.08	2.7	0.06
Sander	0	0	68.4	2.57	48.6	3.65
Common carp	4.8	4.52	0	0	0	0
Crucian carp	61.0	5.12	26.3	0.38	21.6	1.07
Bream	0	0	0	0	27.0	0.81
Total	100	11.44	100	3.03	100	5.59

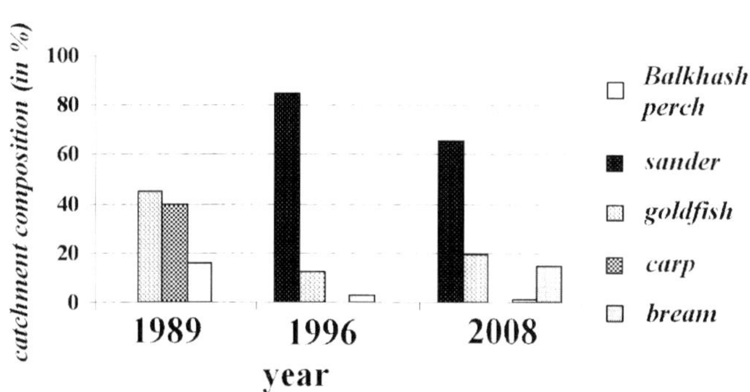

Fig. 6. Changes in the total catch proportions of Balkhash perch from the Saz-Talgar pond system in the period 1989–2008.

proportion of small Balkhash perch and simultaneously decreases the food base for the larger ones. Adult bream prefer benthic organisms, including Balkhash perch, and also consume zooplankton and submerged vegetation (Mamilova 1982; Baimbetov et al. 1988). Occasionally, introduced stone moroco (topmouth gudgeon), false gudgeon, Amur goby, and eleotris spread widely in lakes and rivers of the Balkhash watershed, which preyed upon considerable quantities of Balkhash perch eggs and competed with young Balkhash perch (Mitrofanov and Dukravets 1992).

Today, the Balkhash perch has practically vanished not only from Balkhash Lake, Sassykkol Lake, and substantial portions of the Ili River, but also from main inflows like Lepsy, Karatal, Ayaguz, Kaskelen, as well as from lower reaches of the Issyk, Turgen, and Chilik rivers (arrows without numbers in Fig. 2). The largest contemporary population of Balkhash perch inhabits Alakol Lake, which has saline water (from 1.2 to 11.6 ppt), and abundant areas with dense reeds and cattails that spread for more than 100 m from the shore. Its primary fish species include the Balkhash perch, bream, common carp, crucian carp, and sander. Fishing pressure is quite intensive in this lake, and the number and total mass of Balkhash perch has become significantly reduced compared to the period before the introduction of sander and bream (Mitrofanov et al. 1992).

There is no commercial fishing in many other water bodies that remain inhabited by Balkhash perch. Those water bodies (numbered arrows in Fig. 2) fall into the following groups:

1. The Bakanas River and the lower reaches of the Tentek River are mostly shallow and are well connected to terminal lakes under favourable hydrological conditions. The flow is slow and there are a number of reaches with dense reeds and submerged vegetation.
2. Big Altay Lake and ponds near the city of Almaty, which are quite shallow (1–2 m depth with maximum up to 4 m), with dense reeds and submerged vegetation. The Kuraksuk lakes system also belongs to this group, where Balkhash perch were found by Timirkhanov and Merkulov (1998).
3. The Shynzhaly and Tekes rivers are typical mountain rivers with fast flow and numerous sand bars, having a few small adjacent ponds with riparian and submerged vegetation. Yssyk Lake and the Tekes water reservoir situated in the mountains of Zailiskiy Alatau (Northern Tian Shan) also belong to this group. The Balkhash perch likely was introduced here at the end of 1990s.
4. Sorbulak Lake is a wastewater (sewage) pond for the city of Almaty and has a total area of about 58 km^2 with an average depth of 14–15 m (Baimbetov et al. 1995). The water is polluted with phenols, detergents, metals, oil products and contains excesses of biogenic elements (Korobkin 1990). Riparian vegetation is poor and submerged vegetation is abundant in some areas where discharges of sewage water began around 1996.

Minimal changes to indigenous fish complexes have been observed in the Bakanas, Tekes and Shynzhaly rivers. For the Tekes and Shynzhaly rivers, the dominant species are the Balkhash perch, spotted stone loach, naked osman, and Balkhash marinka. The sole alien fish species are common carp and crucian carp. The Balkhash perch is the single representative of the indigenous fish fauna in common to all of the above-listed water bodies.

An analysis of Balkhash perch and associated fish complexes indicates that it has significant adaptive capabilities. In apparent response to the establishment of alien species, the Balkhash perch migrates against the current in fast mountain rivers and reaches much higher elevations from its previously reported range. Dukravets and Mitrofanov (1989) noted that Balkhash perch had never been found in areas above 600 m (sea level), but now stable populations exist in Yssik Lake and Tekes River at elevations >1500 m above sea level. Given that the average temperature of these mountain rivers has not changed in recent history, remaining very cold given their glacial origins, this upstream migration is likely a response to alien fish introductions in the original low altitude habitats of the Balkhash perch.

Balkhash perch preferentially inhabits small and shallow ponds and rivers having dense submerged and riparian vegetation that lack regular connection with the main water bodies. Vegetation provides perch shelter from sander. In addition, shallow water levels and low or absent flow create favorable oxygen regimes, since sander is very sensitive to hypoxia. Conditions for sander spawning also are often absent in such water bodies.

Balkhash perch is quite resistant to water pollution and can survive even in ponds receiving sewage water (Mamilov and Kostyuk 2000). Thus, Balkhash perch can be described as a highly adaptable species that can occupy a wide range of different habitats. Many other percid species similarly are generally hardy with broad environmental tolerances (Craig 1987).

During recent years the number of trout breeding farms has increased rapidly in the upper reaches of many mountain rivers in the Ili River basin. Trout is mostly a predatory fish, and some have escaped from farms and threaten Balkhash perch (Mitrofanov and Dukravets 1992). Measures to protect the Balkhash perch include the addition of the Balkhash-Ili population into the Red Book of the Republic of Kazakhstan (1996, 2008) and closing fishing.

3.3 Food Preferences and Nutritional Status

Balkhash perch consume a wide variety of prey, similar to the two other perch species (Table 7), which vary with age and body size. Larval and young perch tend to consume zooplankton, (including *Cladocera* crustaceans and some *Copepoda*), which may shift to benthic larvae of midges and dipterans. Larger perch are predominantly predatory (piscivorous) as well as cannibalistic. According to Berg (1949a), food of adults consisted of 90% of young fish from their own species as well as young carp, marinka, loach and fish eggs. The most intensive feeding occurred during the colder seasons, from autumn to spring. Balkhash perch almost stopped feeding during the warm period from the second part of June to the end of summer.

The present study found that prey fish species vary among water bodies, and include cannibalism (Alakol Lake and Kurty water reservoir.) No cannibalism was observed in the other water bodies investigated here. Cannibalism was reported as important in perch diets prior to the 1970s, due to the introduction of alien competitors.

Table 7. Summer diet of adult Balkhash perch from different water bodies (in % from all investigated stomachs) (Mamilov 2000).

Food items	Water bodies*									
	A	B	C	D	E	F	G	H	I	J
Fish	6.0	37.5	47.0	37.5	21.4	8.8	37.5	3.8	8.3	0
Eggs	0	12.5	0	0	0	0	25.0	0	0	0
Larvae of midges and dipterans	11.1	0	0	25.0	0	0	0	57.7	0	50.0
Zooplankton	0	0	2.0	0	0	2.9	0	0	0	16.7
Imago of insects	0	0	8.0	25.0	0	0	0	15.4	0	0
Other prey	0	25.0	2.0	0	0	0	0	0	0	0
Chyme (not identified)	11.1	0	6.7	0	0	38.2	25.0	15.4	8.3	0
Empty	83.3	25.0	42.0	25.0	78.6	55.9	25.0	11.5	83.3	33.3

*A: Alakol Lake; B: Shynzhaly River; C: Bakanas River; D: Big Altay Lake; E: Kurty water reservoir (for 1993 only); F: Sorbulak Lake (sewage pond); G, H and I: ponds on the Ily River inflows; J: Yssik Lake.

In the present study, loaches, predominantly the spotted stone loach, formed the main prey for Balkhash perch in Big Altay Lake and Bakanas River. Small sander, common carp, crucian carp, topmouth gudgeon and grass carp (*Ctenopharyngodon idella*) were part of the diet of perch in the other water bodies. Other dietary items of Balkash perch varied depending on sampling location and included: chironomid larvae (*Chironomidae*), planktonic crustaceans including, freshwater shrimps, beetles, dipterans imago-stage mayflies (*Ephemeroptera*), fish eggs, leeches, macrovegetation and detritus. Until its extirpation from the plains section of the Ili River, Balkhash perch consumed there loaches, young marinka and perch, and in the floodplain lakes they fed on carp and on young of their own species (Nikilsky and Evtukhov 1940). Like for other perch species, the diet of Balkhash perch varies seasonally (Dukravets and Mitrofanov 1989).

According to Malinovskaya's data (1959), young Balkhash perch of the 'reed form' in Alakol Lake consumed zooplankton and when 3–4 years old ate benthic organisms, including larval insects (up to 97% of total food mass) and 4–5 year-old fish consumed their own young. Balkhash perch was the first fish in Alakol lakes that began to eat introduced opossum shrimps (*Mysidacea*), which reached up to 80% of the diet of individuals 80–100 mm long (Loginovskikh and Strelnikov 1973). Dukravets and Mitrofanov (1989) concluded that in the past, Balkhash perch displayed a wide range of food strategies from predation and cannibalism to benthivory and planktivory.

Thus, the general dietary pattern and feeding strategy of Balkhash perch is very similar to European perch and yellow perch. During the first year, they feed mainly on plankton and benthic organisms. Some populations become cannibalistic during the second year of life. A mass transition to predation occurs in the 4–5th year, with about 80% piscivory, including fish of their own species, together with young marinka, common carp and crucian carp.

Condition factors of Balkhash perch have varied widely, from 0.51 to 4.33 (Fulton's Condition Factor) and from 0.41 to 3.36 (Clark's Condition Factor) (Dukravets and Mitrofanov 1989). Their average condition at the beginning of 21st century was slightly higher than reported earlier—from 1.7 to 1.9 and from 1.3 to 1.6 for Fulton's and Clark's condition factors, respectively. Only perch from the Alakol Lake had slightly lower condition indices—about 1.62 and 1.34 for Fulton's and Clark's condition factors, respectively. Similar values were recorded prior to the introduction of sander and bream, with an observed average Clark's condition factor of 1.4 in 1962 (Tsyba 1965).

Only the population of Balkhash perch from the Bakanas River had significant fat reserves in 1995–1997. Perch from Alakol Lake, Kurty water reservoir and some ponds on the Ili River inflows had no or little fat. Perch from the other waterbodies had variable amounts of visceral lipids (Mamilov 2000; Dukravets 2005; Mamilov unpublished data 2000–2011).

3.4 Reproduction

Reproduction of Balkhash perch is very similar to European perch (Craig 1987) and yellow perch (Brown et al. 2009). Male Balkhash perch become sexually mature at 1–3 years and females at 2–4 years, depending on living conditions and probably on intrapopulation genetic differentiation (Dukravets and Mitrofanov 1989). According to Zhadin (1948), males of the pelagic form in Balkash Lake matured at 2–3 years with a mean standard length (SL) of 9.2 cm while females matured at 2–4 years with a mean SL of 10.3 cm. Coastal-form males matured at 1–3 years with a mean of 7.5 cm SL and females at 2–3 years and 8 cm. Later, Goryunova (1950) and Maksunov (1953) reported maturity of Balkhash perch from Balkhash Lake and the delta of the Ili River occurred around 4–5 years. Strelnikov (1970) indicated maturity at 4–5 years for the pelagic form and 3–4 years for the coastal form in the Alakol Lakes system.

Trophoplasmic growth (accumulation of vitellus in eggs) of oocytes was not observed, or proceeded slowly during the summer and the beginning of autumn and occurred mostly during winter (Dukravets and Mitrofanov 1989). Usually females outnumber males by 2–3 times. Poorer feeding conditions, predominance of young (2 years old) or older (6 years and older) perch on spawning grounds, or extreme abiotic conditions (e.g., high salinity) resulted in elevated numbers of females and their predominance up to 15–24 times (Mitrofanov 1970). Our recent investigation showed sex ratios of 1:1 for some ponds at the Ili River inflows, but females outnumbered males by 2–3 times in the Alakol, Big Altay, and Sorbulak lakes and Bakanas River and by 5–7 times for the Shynzhaly River, Kurty water reservoir, Yssik Lake and some ponds (Mamilov 2000; Mamilov unpublished data 2000–2011).

Absolute individual fecundity of female Balkhash perch increases with age and body length, and varies from 1280 to 245,360 eggs, which is somewhat lower than European perch fecundity, and appears consistent among water bodies and years (Dukravets and Mitrofanov 1989). Four females of 147 to 177 mm SL had fecundities ranging from 17,000 to 22,000 eggs for the Aleksandrovsky pond and two females of 234 and 251 mm had fecundities of about 65,000 eggs in the Shynzhaly River.

Balkhash perch spawn once a year in early spring, at the time or shortly after the ice breaks-up from March to the end of May, depending on the location and weather conditions. Spawning occurs in fresh and saline waters (up to 9‰), in depths from a few centimetres to 2–3 m, and at any time of day or night.

Duration of spawning is population-specific and depends on the size of the spawning group (density) and habitat characteristics. Berg (1949) mentioned that spawning duration in Balkhash Lake was about one month, beginning in the western (freshwater) side in mid-April at water temperatures of 8–10°C, and half a month later at the east (saline) side. Goryunova (1950) and Maksunov (1953) described spawning in Balkhash Lake beginning earlier at 4–5°C. Dukravets and Mitrofanov (1989) observed spawning commencing March 18, 1966, 3 days after ice breakup, and ending on March 26 in one of the floodplain lakes of the Ili River. In the following year, its ice breakup occurred on March 26 and spawning lasted from March 30th to April 5, at 6–7.5°C.

Females lay eggs in jelly-like spirals or bands that adhere to vegetation or submerged stones, and sometimes directly on the substratum on stormy days. Males release their sperm over several spawning events, staying in the spawning area over a long period and fertilizing eggs from several females. Females usually do not spend a long time at the spawning sites. It appears probable that fractional sperm ripening may allow fertilizing the eggs of several females, which dominate most populations in numbers.

Before significant reductions of their numbers, Balkhash perch went to spawning sites in dense schools, reaching 20 fish per square meter (Dukravets and Mitrofanov 1989), which can still sometimes be observed at some Alakol Lakes spawning sites (our observation). However, at most other spawning locations, only a few specimens appear, which quickly move away from the shore when disturbed, and do not return for 30–60 minutes.

Newly hatched perch concentrate nearshore in the upper water layer, in schools of 2000 to 2500 per square meter, which are about 1 m wide and a few tens of meters long, according to Timirkhanov and Iskakbaev (1999). They observed high larval mortality (99.5–99.8%) during the first two days after hatching in a floodplain lake of the Urdzhar River; perch larvae were consumed by dragonfly larvae, diving beetles, and topmouth gudgeons. One day later, larval density had significantly decreased to 5–10 per square meter.

Young fish remain near the spawning sites until the end of May and then begin to settle in the water column (Dukravets and Mitrofanov 1989). Young-of-the-year Balkhash perch consistently are found in the Alakol, Big Altay and Almaly water reservoir and Tentek, Bakanas, Tekes rivers.

3.5 Growth and Longevity

Growth data for Balkhash perch are presented in Table 8. During their first summer, young perch from the Kapshagay water reservoir ranged from 6–22 mm TL in April–May and 24–67 mm TL in June–July 1971–1973 (Dukravets and Mitrofanov 1989). Pelagic and coastal forms differed in growth rates, in the Balkhash and Alakol lakes (Zhadin 1948; Strelnikov 1974). See Section 1 and Fig. 4 (this chapter) for a discussion of growth patterns in Alakol Lake perch.

Table 8. Body length of Balkhash perch per age class (cm).

Water body	Form, sex	Age, years											Source
		1	**2**	**3**	**4**	**5**	**6**	**7**	**8**	**9**	**10**	**11**	
Lake Balkhash	Pelagic	10.0	13.5	15.5	20.5	23.5	26.5	30.0	33.0	–	–	–	Domrachev 1930
	Pelagic	–	9.5	12.0	14.2	18.7	21.6	–	–	–	–	–	Zhadin 1948
	Coastal	5.5	7.6	9.4	11.8	14.1	15.2	–	–	–	–	–	
	Did not mention	10.0	13.5	15.5	20.5	23.5	26.5	30.0	33.0	–	–	–	Berg 1949a
Lake Alakol 1962	Female	–	12.9	14.9	16.3	19.0	20.2	22.6	25.3	27.0	–	30.0	Tsyba 1965
	Male	–	13.0	13.7	17.1	16.8	17.7	18.5	20.0	30.0	–	–	
Lake Alakol 1970	Pelagic	9.5	–	11.7	–	18.4	–	–	24.1	–	–	–	Strelnikov 1974
	Coastal	–	–	–	–	12.0	–	–	16.2	–	–	–	
Kurlep pond 2005	Did not mention (n = 21)	6.0	8.6	11.5	14.5	17.8	21.5	24.0	–	–	–	–	Dukravets 2005
Panfilovskiy pond. 2004	Did not mention (n = 36)	5.5	9.8	13.9	17.2	20.6	24.2	–	–	–	–	–	Dukravets 2005

Growth in length of Balkhash perch is intensive during the first years of life and slows with sexual maturity, as in many other fish species. In contrast, the rate of body weight increase accelerates after sexual maturation. Dukravets and Mitrofanov (1989) suggested that the growth rate of Balkhash perch depends more on specific habitat factors, such as oxygenation, temperature, flow, size of the water body, population density, etc., and on individual metabolic variations rather than on feeding conditions. Changes in environmental conditions quickly modified growth rate of Balkhash perch in Krakol Lake (Dukravets and Mitrofanov 1989).

The maximal known lifespan for Balkhash perch is 23 years. In 1944, a 21 year-old female was recorded from the lower reach of the Ili River, of 45 cm SL (without the caudal fin) and weighing 2.2 kg (Zhadin 1948). Commercial perch on Balkhash Lake once reached 50 cm TL and 1.5 kg (Berg 1949). The maximal known lifespan for Balkhash perch in the Alakol Lakes system was 16 years (Tsyba 1965), which declined to 9 years in 1990 due to overfishing (Mamilov and Mitrofanov 2002) and now is 10 years. Reduced commercial fishing after 2000 prolonged the maximum lifespan to 15 years for Alakol Lake and to 18 years in Sassykkol Lake (Amirgaliev et al. 2007). Cessation of fisheries in Big Altay Lake increased the maximal lifespan from 8 years in 1980 (Dukravets et al. 1984) to 11 years at present. Currently, the maximum age of Balkhash perch in Alakol Lake is 10 years, 9 years in the Bakanas River, and 8 years in the Sorbulak wastewater reservoir. In many of the smaller water bodies, the fish have a short lifespan (4–6 years), attributed to fluctuating hydrology and poaching (Mamilov 2000; Dukravets 2005; Mamilov unpublished data 2000–2011). Eight years was the maximum age of perch in the Shynzhaly River and in some ponds of the Ili River inflows. A temporal analysis of populations of Balkhash perch reveals a significant overall decrease of longevity due to alien fish introduction and strong human impact.

3.6 Parasites

Many Balkhash perch populations were described to have extensive parasite infections at the end of 1940's, in which many of the helminths were species-specific (Zhadin 1949). In 1980, almost 100% of perch in Big Altay Lake had infested gills and livers, and some had numerous cysts in their muscle (Dukravets et al. 1984). Perch from Sorbulak Lake had large quantities of flukes in their stomachs and livers. After 2000, Balkhash perch had lower infection levels, which might be explained by lower fish density and/or reduction in intermediate hosts for those parasites. Over the last years, we observed helminths only in 4 of 12 small water bodies where Balkhash perch was present (Mamilov 2000; Mamilov unpublished data).

Trematodes were observed in the stomach and intestines of perch from Alakol Lake and from a pond close to the delta of the Ili River. Extensive infestation of Balkhash perch (97.7% of individuals and 4–650 metacercaria per individual) from Sassykkol Lake with grub flukes (*Clinostomum complanatum*) occurred from 2004–2007 (Zhatkanbaeva and Nysambaeva 2011). Despite significantly reduced numbers of Balkhash perch, the rate of infestation in the largest lakes where this species still survives remains the same today as it was 60 years ago.

3.7 Economic Importance and Abundance

Balkhash perch was one of the primary targets of commercial and amateur fishing in many water bodies of the Balkhash Lake basin until the late 1960s, with specimens up to 40 cm standard length and 1.5 kg caught (Dukravets and Mitrofanov 1989). The flesh of Balkhash perch is white, with few bones and fat, but tasty and similar to sander flesh. Local people caught this species by seine and gill nets mainly in winter and spring, and prepared it frozen, or salted and dried. Annual commercial catches of Balkhash perch in Balkhash Lake were 3.5 to 7.4 tonnes in 1930–1933 and 23.9–53.9 tonnes in 1936–1939 (Berg 1949a). During the same period, catches of Balkhash perch in the Alakol lakes system were about 312 to 321 tonnes annually (Nekrashevich 1946), and reached 1000 tonnes in 1967 and 1986 (Mitrofanov 1992; Amirgaliev et al. 2007).

Until the 1960s, Balkhash perch was one of the most abundant commercial fish species in floodplain lakes and plain rivers in the Balkhash watershed. For example in the floodplain of Lake Karakol (middle reach of the Ili River), on some days, 100 to 250 Balkhash perch individuals could be harvested overnight with a single 25 m long and 2 m high gill net. During spawning time, 10 to 15 fish could be caught by scooping with a sweep net (35 cm mouth opening) at depths of 2 to 2.5 m in April 1972 in the Issik River (Dukravets and Mitrofanov 1989). Chusainova et al. (1970) evaluated the numbers of Balkhash perch using a mark-recapture method in Karakol Lake, and estimated it as 32 adult fish having on average 3.3 kg of total biomass per hectare. The number of young of the year fish estimated using data for fecundity, diet, and proportions of different age groups was estimated at >1,000 young fish per hectare. At that time, the number of marketable Balkhash perch was >3 million individuals for lakes on the Ili River and hundreds of millions for the Balkhash and Alakol lakes. Dukravets and Mitrofanov (1989) calculated that up to 5% of the adults were being captured by commercial fishery annually. Nekraschevich (1965b) reported annual capture rates of 1.5–2 million perch out of an estimated 400 million adults in the Alakol lakes system.

Balkhash perch were eliminated from pelagic zones of the Sassykkol and Koashkarkol lakes after sander introduction (Sokolovsky and Galuschak 2000). Significant reductions of sander due to intensive commercial fishing since early 2000's led to the reappearance of Balkhash perch in the pelagic zone (Amirgaliev et al. 2006).

Balkhash perch population size has reached commercial potential in water bodies where it has been introduced, including the Chu and Nura basins and several isolated lakes of Central and Northern Kazakhstan (Skakun et al. 1986; Gayduchenko 1986; Dukravets and Mitrofanov 1989; Dukravets and Mamilov 1992; Goryunova and Dan'ko 2011). However, many of the introduced populations were not sustainable. In recent years, Balkhash perch was not found in the Chu River (Mamilov 2011), or in some lakes of Central Kazakhstan (Goryunova and Dan'ko 2011), where it was abundant about 20 years ago.

Populations of Balkhash perch from the Balkhash Lake and Ili River are now listed in the Red Book of the Republic of Kazakhstan along with several other species, whose numbers are rapidly declining; these populations risk extinction in the near future (Red Book of the Republic of Kazakhstan 2008). Balkhash perch fishing is now prohibited by law in the Ili River watershed and inflows of the Balkhash Lake.

Nowadays, small isolated populations inhabit some left-bank inflows of the Ili River (Dukravets 2005) as well as the rivers Aksu, Ayagoz, Karatal, Tokraun, and Baskan (Timirkhanov and Merkulov 1998; Isbekov et al. 2006). A few recent occurrences in the Balkhash Lake also were mentioned (Isbekov and Timirkhanov 2009).

Now this species needs to adapt to new living conditions in mountain rivers and lakes. The main water bodies where Balkhash perch remain are experiencing unfavourable human impacts. Significant increases of parasitism and changes in food items are linked to those shrinking Balkash perch populations. Favorable living conditions remain in some relatively small water bodies isolated from general hydrological networks and situated in sparsely populated areas, such as Big Altay Lake and the Bakanas and Tokraun rivers. The Alakol Lake population is threatened by commercial fishing, alien fish species, and loss of spawning habitat. High extinction probability faces populations from small water bodies due to various causes, such as drying or other loss of habitat, introduction of alien species and poaching.

Our preliminary investigations show that young Balkhash perch can adapt to cool water living conditions in aquaria, for more than 3 years. However, Balkhash perch aquaculture has not been attempted to date. This is clearly an avenue to explore in the future, as it could allow the restocking of threatened or lost populations and ultimately the survival of the species. Given the similarities in the biology of all three *Perca* species, our knowledge of the reproductive biology (see Chapter 7 in this volume) and population genetics of *P. fluviatilis* and *P. flavescens* (Chapter 2 in this volume) will be invaluable sources of information for any initiative of Balkhash perch aquaculture.

Conservation measures for Balkhash perch include creating special protected areas like national nature reserves and/or protected water bodies in the Tokraun River (Isbekov et al. 2006), part of the delta of the Ili River (Red Book of the Republic of Kazakhstan 2010), Bakanas, Baskan and Tekes rivers, fishery catch limits, and regular evaluations of the population in the Alakol Lakes system. The Balkhash perch is classified in "The IUCN of Threatened Species" (v2013.2) as "data deficient" (http://www.iucnredlist.org/).

The data presented in this chapter highlight similarities in morphology and biology of Balkhash perch with European and yellow perches. A difference is that Balkhash perch appear incompatible with sander, whose introductions have reduced the distribution range of the former. Introductions of exotic species and excessive fishing significantly reduced perch numbers and distribution, and it almost vanished from Balkhash Lake and Ili River (Dukravets and Mitrofanov 1989). Conservation of local native Balkhash perch populations and investigations of prospects for artificial reproduction require more attention from scientists and fishery authorities.

3.8 Acknowledgments

I am grateful to I.V. Mitrofanov, S.S. Galuschak, V.R. Sokolovskiy and G.M. Dukravets for providing part of the material. Special thanks to M.V. Mina and I.V. Mitrofanov for consultation during this investigation and help with data interpretation, and to our editors, Patrice Couture and Gregory Pyle, for their persistence, kind attention and patience during the preparation of this chapter. I also acknowledge the contribution of

Prof. Catherine Alexander from the Department of Anthropology, Durham University (UK) and Dr. Anvar Mamilov. Prof. Carol Stepien extensively edited the manuscript and provided valuable insights into the evolutionary history of the species. A part of the work was supported by the grant #1380/GF4 of MES of the Republic of Kazakhstan.

3.9 References

Amirgaliev, N.A., S.R. Timirkhanov and Sh.A. Alpeysov. 2007. Ichthyofauna and ecology of the Alakol lakes system. Bastau, Almaty 368 p.

Baimbetov, A.A., V.A. Mel'nikov and V.P. Mitrofanov. 1988. *Abramis brama* (Linné)—Bream. pp. 127–159. *In*: E.V. Gvozdev and V.P. Mitrofanov (eds.). Fishes of Kazakhstan, Vol. 3. Nauka, Alma-Ata.

Baimbetov, A.A., L.I. Sharapova, R.Kh. Mamilova, V.P. Mitrofanov and L.O. Pichkily. 1995. Characteristic of hydrobiocenosis of the Sorbulak sewage pond in the early 90-thes. Bull. of Kazakh State University. Series Biology 2: 38–48.

Berg, L.S. 1949. Freshwater fishes of the USSR and adjacent countries.—Moscow, Leningrad: Publishing house of the Academy of sciences of USSR. Part 3: 926–1382.

Berg, L.S. 1949a. Balkhash perch—*Perca schrenki* Kessler. pp. 575–576. *In*: Commercial Fishes of USSR. Pischepromizdat, Moscow.

Berg, L.S. 1958. Changes of relief since dry postboulder-period. *In*: Selected works Publishing house of the Academy of sciences of USSR. Moscow. 2: 76–85.

Brown, T.G., B. Runciman, M.J. Bradford and S. Pollard. 2009. A biological synopsis of yellow perch (*Perca flavescens*). Canadian Manuscript Report of Fisheries and Aquatic Sciences 2883. 28 p.

Chebotareva, J.V., S.P. Savoskul, M. Yu. Pichugin, K.A. Savvaitova and S.V. Maksimov. 1999. Characteristics of abnormalities in external and internal organs of fishes. pp. 142–146. *In*: Diversity of Fishes of Taiymyr. Nauka, Moscow.

Chusainova, N.Z., L.I. Sharapova, R.Kh. Mamilova, T.G. Kriusenko and V.P. Mitrofanov. 1970. On biological productivity of the Karakl Lake. Preliminary papers for UNESCOIBP Symposium on Productivity Problems of Freshwaters. Warszawa 1: 209–215.

Craig, J. 1987. The Biology of Perch and Related Fishes. Timber Press. Portland, OR 333 p.

Dukravets, G.M. 2005. The biological state of some local populations of Balkhash perch *Perca schrenki* Kessler in the basin of the Ili River. pp. 93–109. *In*: Fishery Economical Investigations in the Republic of Kazakhstan: History and Modern State (collected articles). Bastau, Almaty.

Dukravets, G.M. and Yu.A. Biryukov. 1976. Ichthyofauna of the Nura river in Central Kazkahstan. J. Ichthyol. 16(2): 309–314.

Dukravets, G.M., R.Kh. Mamilova, B.Kh. Minsarinova and E.A. Merkulov. 1984. A characteristic of hydrofauna of the Big Altay lake at the lower flow of the Baskan River in the Taldy-Kurgan oblast. 25 p. Deposit manuscript in KazNIINTI 10.05.84б №652-Ка.

Dukravets, G.M. and V.P. Mitrofanov. 1989. Perca schrenki Kessler—Balkhash perch. pp. 157–190. *In*: E.V. Gvozdev and V.P. Mitrofanov (eds.). Fishes of Kazakhstan, Vol. 4. Nauka, Alma-Ata.

Dukravets, G.M. and N.Sh. Mamilov. 1992. Materials on morphometry and biology of perch fishes in the basin of Chu river. J. Ichthyol. 32(6): 49–56.

Froese, R. and D. Pauly (eds.). 2014. FishBase. World Wide Web electronic publication. www.fishbase.org, version 02/2014.

Gayduchenko, L.L. 1986. Short communication about Balkhash perch. pp. 193–194. *In*: E.V. Gvozdev and A.B. Bekenov (eds.). Rare Animals of Kazkahstan. Nauka, Alma-Ata.

Golodov, Yu.V. and V.P. Mitrofanov. 1968. Morphology and biology of Balkhash perch from high-water bed of the Ili river. *In*: Biology and Geography: Collected articles of graduate students and applicants. —KazGU. Alma-Ata 4: 105–114.

Goryunova, A.I. 1950. On biology of Balkhash perch. Proc. Acad. Sci. of KazSSR. Ser. Zoology 84(9): 78–86.

Goryunova, A.I. and E.K. Dan'ko. 2011. The stock of lakes of the Kazakhstan. Part IV. The lakes of Akmolinskay oblast (in the borders 1961–1999). Til. Almaty 108 p.

Isbekov, K.B., S.Zh. Asylbekova and S.R. Timirkhanov. 2006. Prospects of conservation of gene pool of rare and endangered fish species of the Balkhash Lake. Bulletin of Kazakh State University. Series biology 3(29): 226–232.

Isbekov, K.B. and S.R. Timirkhanov. 2009. Rare fish species of the Balkhash Lake. LEM. Almaty 182 p.
Ismukhanov, Kh.K. and V.A. Skakun. 2008. The modern state of biodiversity of the transboundary Ili river and Kapshagay water reservoir, impact of migratory alien fishes on the ecosystems. pp. 273–280. *In*: Ecology and Hydrological fauna of Transboundary Basins of Kazakhstan. Bastau. Almaty.
Karpov, V.E. and S.S. Kaldaev. 2005. Morphobiological characteristics of the bitterling (family Cyprinidae, *Rhodeus* sp.) from the Kapshagay water reservoir and Ili river. pp. 168–173. *In*: Fishery Economical Investigations in the Republic of Kazakhstan: History and Modern State (collected articles). Bastau, Almaty.
Kassin, N.G. 1947. Materials on paleogeography of Kazkakhstan. Alma-Ata 258 p.
Kessler, K.F. 1874. Fishes. *In*: Journey of A.P. Fedchenko to the Turkestan: Bulletin of Society of Amateurs of Natural History, Anthropology and Ethnography 2(3): 1–63.
Korobkin, V.A. 1990. Some results of researches of substantiation works on projects devoted to the Alma-Ata sewage water use for agriculture. Bulletin of Academy of Sciences of the Kazakh SSR 12: 36–48.
Lebedev, V.D. 1959. Neocene freshwater fish fauna in the Zayssan depression and Western Siberia depression. J. Ichthyol. 12: 28–69.
Lebedev, V.D. 1960. Freshwater ichthyofauna of the European part of the USSR in the Quaternary. Moscow. Moscow State University Press 404 p.
Loginovskikh, E.V. and A.S. Strelnikov. 1973. Feeding and food relations of fishes in the Alakol Lakes system. Proc. Cycle of mater and energy in lakes and water reservoirs. Part 1. pp. 160–163.
Malinovskaya, A.S. 1959. Nutritive base of the Alakol Lakes and it use by the fishes. *In*: collected articles on Ichthyology and Hydrobiology. Academy of Sciences of Kazakh SSR, Alma-Ata. 2: 116–144.
Maksunov, V.A. 1953. Seasonal flocks of perch in the Balkhash Lake. J. Ichthyol. 1: 104–108.
Mamilov, N.Sh. 2000. Current state of the Balkhash perch Perca schrenkii (Perciformes, Percidae). Zoologichesky Zhurnal. 79(5): 572–584.
Mamilov, N.Sh. 2011. Modern diversity of alien fish species in the Chu and Talas river basins. Rus. J. Biol. Invas. 2(2–3): 112–119.
Mamilov, N.Sh. and T.P. Kostyuk. 2000. A comparative morphological assessment of states of the two populations of Balkhash perch (*Perca schrenkii*). Bull. of the Kazakh State University. Series of Biology 4: 48–54.
Mamilov, N.Sh. and I.V. Mitrofanov. 2002. The state of population of Balkhash perch (*Perca schrenki* Kessler) in the Alakol lake. Bull. of Kazkah National University, series ecology 2(11): 91–98.
Mamilov, N.Sh., G.K. Balabieva and G.S. Koishybaeva. 2010. Distribution of alien fish species in small water bodies of the Balkhash Basin. Rus. J. Biol. Invas. 1(3): 181–186.
Mamilova, R.Kh. 1982. Dynamics of feeding of the bream in the Kapchagay water reservoir (1976–1979). pp. 124–133. *In*: Investigation of Producers in Water Bodies in Basin of the Ili River. Alma-Ata: Kazakh State University.
Mayr, E. 1969. Principles of Systematic Zoology. McGraw-Hill Book Company, New York.
Mina, M.V. 1974. Some observations on area of distribution of Balkhash perch *Perca schrenki* Kessler and its relationships with European perch *Perca fluviatilis* L. J. of Ichthyology 14(2): 332–334.
Mina, M.V. 1986. Microevolution of Fishes. Nauka, Moscow.
Mitrofanov, V.P. 1970. Gender ratio in fish populations. pp. 96–104. *In*: N.Z. Khusainova and V.P. Mitrofanov (eds.). Biology of Waters of Kazakhstan. Alma-Ata, VGBO-KazGU.
Mitrofanov, V.P. 1986. Formation of the modern fish fauna of Kazakhstan and icthyogeographical demarcation. pp. 20–40. *In*: E.V. Gvozdev and V.P. Mitrofanov (eds.). Fishes of Kazakhstan. V.1. Nauka, Alma-Ata.
Mitrofanov, V.P. 1991. Ecological particularities of fish reproduction in the delta of the Ili river in the modern conditions. p. 48. *In*: collected articles The Topical Problems of Modern Biology. Kazakh State University, Alma-Ata.
Mitrofanov, V.P. 1992. Fishery in Kazakhstan. V. 5: 372–411. *In:* E.V. Gvozdev and V.P. Mitrofanov (eds.). Fishes of Kazakhstan. Gylym, Alma-Ata.
Mitrofanov, V.P. and G.M. Dukravets. 1992. Some theoretical and practical aspects of fish acclimatization in Kazakhstan. pp. 329–371. *In*: E.V. Gvozdev and V.P. Mitrofanov (eds.). Fishes of Kazakhstan. V.5. Gylym. Alma-Ata.
Nekrashevich, N.G. 1946. Fishes of the Alakol Lakes (systematics, biology, catches). Candidate thesis. Tomsk.
Nekrashevich, N.G. 1965a. The materials on ichthyology of the Alakol Lakes. *In*: collected articles: The Alakol Depression and its Lakes: Problems of Geography of Kazkahstan. Nauka. Alma-Ata 12: 236–268.

Nekrashevich, N.G. 1965b. Biological basis and the first results of sander introduction in the Alakol Lakes. *In*: collected articles: The Alakol Depression and its Lakes: Problems of geography of Kazakhstan. Nauka. Alma-Ata 12: 269–279.

Nikolsky, G.V. and N.A. Evtukhov. 1940. Fishes of the plains locus of the Ili River. Bulletin MOIP. Dep. Biology 49(5/6): 57–70.

Nuriev, Kh.N. 1985. Acclimatized fishes in water bodies of the basin of Zeravshan River. FAN, Tashkent 104 p.

Obruchev, V.A. 1949. Boundary Dzungaria. An Geological Outline. Moscow, Leningrad 3: 292.

Pivnev, I.A. 1985. Fishes of Basins of Rivers Talas and Chu. Ilim. Frunze 190 p.

Red Book of the Republic of Kazakhstan—1996, 2008.

Reshetnikov, Yu.S., O.A. Popova, N.A. Kashulin, A.A. Lukin, P.-A. Amundsen and F. Staldvik. 1999. Estimating the favorable state of a fish community using morphologic analysis of fishes. Adv. Curr. Biol. 119(2): 165–177.

Shaiykin, A.V. 1989. Discrimination of fishes intrapopulation groups by body coloration. J. Gen. Biol. 40(4): 491–503.

Sinitsyn, V.M. 1962. Paleogeography of Asia. Publishing House of the Academy of Sciences of USSR. Moscow, Leningrad 268 p.

Skakun, V.A., A.I. Shustov, V.Ya. Gubanova, A.A. Raspopin, N.Z. Aleeva and Sh.M. Aymukanova. 1986. Influence of accidental acclimatization to the fish fauna of the Sary-Oba feeding lake. Proc. Biological grounds of fish industry in the water bodies of Central Asia and Kazakhstan.—Ylym, Ashkhabad 293–295.

Sokolovsky, V.R. and S.S. Galuschak. 2000. The modern state of Balkhash perch Perca schrenki (Percidae) in the lakes of the Alakol system. J. Ichthyol. 40(2): 228–234.

Strelnikov, A.S. 1970. Morphobiological characteristics of commercial fishes of the Alakol Lakes. pp. 113–119. *In*: Biology of Water Bodies of Kazkahstan. Alma-Ata.

Strelnikov, A.S. 1974. Fishes and biological bases of fish industry of the Alakol lakes. Candidat thesis. Tomsk.

Svetovidov, A.N. and Dorofeeva E.A. 1963. Systematic relationships, origin and history of distribution of Europe-Asian and North American perches and sanders (genus *Perca*, *Lucioperca* and *Stizostedion*). J. Ichthyol. 3(4): 625–651.

Timirkhanov, S.R. and E.A. Merkulov. 1998. Ichthyofauna of the Kuraksusky Lakes (the Aksu River, Balkhsh basin). Bull. Kazakh State University. Series biology 5: 67–72.

Timirkhanov, S.R. and A.A. Iskakbaev. 1999. About spawning ecology and natural death rate of Balkhash perch. P. 108. *In*: E.V. Gvozdev, A.B. Bekenov et al. (eds.). Problems of Conservation and Sustainable Use of Animal Biodiversity of Kazakhstan. Tethys, Almaty.

Tsyba, K.P. 1965. On the biology of white perch *Perca schrenkii* Kessler in the Alakol Lakes. pp. 280–287. *In*: collected articles: The Alakol Depression and its Lakes: Problems of geography of Kazkahstan. Nauka. Alma-Ata.

Turkiya, A. 1997. Ichthyofauna of water bodies of Xinjiang and it changes after acclimatization. Candidate thesis. Institute of zoology and animal gene pool, Almaty.

Yakovlev, V.N. 1964. History of forming of complexes of freshwater fish fauna. J. Ichthyol. 4:1(30): 10–22.

Yakovlev, V.N., A.V. Kozhara, Yu.G. Iziumov, A.N. Kasiyanov and N.M. Zalanetskiy. 1988. Phenes of cyprinid fishes and European perch. pp. 53–64. *In*: A.V. Yablokov (ed.). Phenetics of Wild Populations. Nauka, Moscow.

Zakharov, V.M. and G.I. Ruban. 1985. The disturbance of developmental stability as an indicator of anthropogenic influence on animal population in the Baltic Sea Basin. Symp. Ecol. Invest. Baltic Sea Env. (Riga, USSR, 1983). Helsinki: Valkion Painatukeskus pp. 526–536.

Zakharov, V.M., A.S. Baranov, V.I. Borisov, A.V. Valetskiy, N.G. Kryazheva, E.K. Chistyakova and A.T. Chubinichvili. 2000. Environmental health: a method for evaluation. The Center of Ecology Policy of Russia. Moscow 68 p.

Zhadin, B.F. 1948. Balkhash perch. Candidate thesis. Leningrad.

Zhadin, B.F. 1949. On the origin of Balkhash perch (*Perca schrenkii* Kessler). Papers of Academy of sciences of USSR. New series. 66(3): 499–502.

Zhatkanbaeva, D.M. and S.M. Nysambaeva. 2011. On functioning of the Balkhash perch (*Perca schrenki*) population seat of clinostomoz in the Sassykkol Lake. pp. 97–98. *In*: Proc. Zoological Researches of the 20 years of independence of Republic of Kazakhstan: Materials of the International scientific conference devoted to the 20 years of independence of Republic of Kazakhstan. September 22–23, 2011. Almaty.

4

Distribution of Yellow Perch *Perca flavescens* in Lakes, Reservoirs and Rivers of Alberta and British Columbia, in Relation to Tolerance for Climate and other Habitat Factors, and their Dispersal and Invasive Ability

Joseph B. Rasmussen* and Lars Brinkmann

ABSTRACT

The absence of yellow perch from rivers in northwestern Alberta and northern British Columbia is consistent with a climatic boundary of –3°C mean annual air temperature (MAT) and +14°C, mean summer air temperature (MSAT). However, the elevational boundary in river systems east of the Rocky Mountains, while decreasing with latitude as expected, occurred at significantly higher MAT and MSAT values. Thus the elevational limits of the species are not likely determined

Department of Biological Sciences, University of Lethbridge, Lethbridge, Alberta, Canada.
E-mail: lars.brinkmann@uleth.ca.
* Corresponding author

by the temperature regimes, but rather, are limited by the ability of perch to find suitable standing water habitats, with slow moving, navigable and permanent tributaries that can be accessed, and possibly to their ability to navigate the rivers themselves, which become steeper further south and upslope. Yellow perch are absent from prairie portions of the Saskatchewan River system, which likely reflects the paucity of permanent lakes and tributaries on the semi-arid grassland landscape, as well as limits imposed by salinity and winterkill issues. While perch dispersal ability appears to have limited their natural range, perch introductions (both authorized and unauthorized) have in general been highly successful, in both natural lakes and rivers on both sides of the Continental Divide, and we have analyzed data on their present distributions in these areas, with special emphasis on the irrigation reservoirs of southern Alberta.

Keywords: temperature, elevation, range, habitat, irrigation, fisheries, introductions, glacial history

4.1 Introduction: Objectives and Hypotheses

Although yellow perch *Perca flavescens* Mitchill are widely distributed in inland waters across Canada, east of the Rocky Mountains from the boreal forest southward (Scott and Crossman 1973), their ability to colonize northerly and high elevation river systems appears to be limited (McPhail and Lindsey 1970). Like all Canadian fish species their present natural range represents their ability to recolonize a completely glaciated landscape over the last 10–12 millennia by dispersal through postglacial drainage networks (McPhail and Lindsey 1970; Mandrak 1995). The limited northern range is consistent with their preference for warm, weedy littoral habitats of lakes and ponds (Nelson and Paetz 1992; Boisclair and Rasmussen 1996). Although the species has limited capacity to deal with current, yellow perch are sometimes found in rivers, albeit slow moving sections and floodplain habitats. Since river networks are the key to dispersal on the landscape, the limited tolerance of yellow perch for flowing water might be expected to have restricted their range in steep areas such as the eastern slopes of the Rocky Mountains. In this chapter we examine both large and small-scale distribution patterns of yellow perch in Alberta and British Columbia, both native and introduced, by reviewing published and unpublished literature and government databases. We examine these patterns for insights into tolerances for climate, salinity and other habitat factors, and for dispersal and invasive abilities that may impact their future spread and their impacts on other species and fisheries resources.

The distribution of yellow perch across Canada has encompassed most of the Hudson Bay drainage, the St. Lawrence system including the Great Lakes and the Ottawa River, and much of the Atlantic region east of the Appalachians (Scott and Crossman 1973). Thus, although the species is rarely found in the mainstems of fast rivers, it appears to have utilized river systems effectively for dispersal, although the details concerning how they accomplish this, given their limited capacity to deal with current, have not been closely examined. Yellow perch were also among the most abundant and widespread species at early stages of ecosystem recovery from acidification and metal pollution, in the Sudbury region of Ontario (Gunn and Keller

1990; Rasmussen et al. 2008), and in recent years have been observed to be establishing and spreading through interior British Columbia, west of the Continental Divide (Runciman and Leaf 2009; Brown et al. 2009). While these observations suggest that yellow perch disperse effectively through drainage networks, their absence from many apparently suitable habitats, in Alberta well within their overall range (Nelson and Paetz 1992), suggests that their ability to colonize by upstream movements through river systems may, at least in some situations, be fairly limited.

Although the species' native range extends to the Mackenzie drainage, it is neither abundant nor widespread in the north, and the most northerly record for the species is Great Slave Lake (mean annual air temperature (MAT) –3°C, mean summer air temperature (MSAT) +14°C, McPhail and Lindsey 1970; Scott and Crossman 1973), which is likely set by a combination of cold water temperatures, short summers and long winters. The northern boundary of its range in western Canada tracks southeasterly through northern Saskatchewan and Manitoba, and across northwestern Ontario to James Bay, and more or less follows the –3°C contour of MAT. The distribution of yellow perch at its western boundary appears somewhat more complex, in that the elevational limit of the species appears to vary greatly among rivers, and the species appears to be absent, or at least not native to, large portions of Alberta, and the eastern slopes regions of British Columbia. The extent to which the yellow perch has been able to colonize lakes and rivers of the eastern slopes of the Rockies by dispersing up rivers might then also be expected to be a function of climate, and as such, elevational limits might reasonably be expected to conform to the same temperature contours that determine the northern limits. The distribution of yellow perch along the drainages of the prairie regions of Alberta is also of interest, since few natural lakes and ponds are present, and even fewer have permanent outlet streams linking them to the river, posing a significant dispersal challenge. Indeed, many of the habitats suitable for perch in southern Alberta are man-made irrigation reservoirs, and Alberta fish and wildlife stocking programs, aimed at enhancing reservoir fishery potential, resulted in many of these water bodies being stocked with northern pike *Esox lucius* L., walleye *Sander vitreus* (Mitchill), lake whitefish *Coregonus clupeaformis* (Mitchill) and yellow perch (Mitchell and Prepas 1990). Thus authorized introductions appear to have played a significant role in extending the distribution of yellow perch onto the prairie landscape, and such range extensions are likely to continue.

The native range of *P. flavescens* does not extend to the west of the Rocky Mountains (McPhail 2007; Runciman and Leaf 2009). The species was introduced west of the Continental Divide into lakes of the lower Columbia River system over a century ago, and has subsequently expanded its range northward up the Columbia system, becoming widespread in lakes and streams of the Kootenay and Okanagan systems, and a number of other systems. The species appears well suited to the climatic conditions and lake habitats west of the Continental Divide, and the possibility that it has considerable potential for further range expansion, both northward and coastward, in this area is a source of considerable concern (Runciman and Leaf 2009). Many of the native fishes of the Columbia and other coastal drainages have no previous history of contact with yellow perch, and several studies have shown that perch can have significant detrimental impacts on salmonid populations (Walters and Kitchell 2001;

Browne and Rasmussen 2009, 2013), which make up a significant component of the commercial and recreational fisheries of BC.

The objectives of this chapter are to:

1. Review the literature on the postglacial environment and the dispersal of fishes onto the Western Canadian landscape and analyze the historical factors that contributed to the native range of yellow perch in Alberta, British Columbia and northern Canada.
2. Evaluate the potential role of climatic factors and historical processes on the distribution of yellow perch along its northern boundary, and in Alberta and British Columbia. Of particular interest, is the extent to which yellow perch have spread upstream through rivers of the eastern slopes of the Rocky Mountains, and we compare their elevational range in river systems across a broad latitudinal range, and,
3. Test the idea that the elevational limits can be explained by the same climatic factors that appear to set the northern limits. Thus we hypothesize that the limited extent of the yellow perch range in northwestern Alberta and Eastern British Columbia east of the divide would be a function of cold temperatures, and as such, the species would extend to higher elevations in more southerly river systems. Since the northerly range of the species conforms roughly to the –3°C MAT contour and the 14°C MSAT contour, we hypothesize that these contours would also set the elevational limits within eastern slope drainages.
4. Based on these analyses we examine the potential for significant range expansion on both sides of the Continental Divide. We also examine the processes that have contributed to recent yellow perch range expansion on both sides of the Continental Divide, and,
5. Examine the distribution of yellow perch on the prairie landscape in relation to limitations imposed by the semi-arid hydrological regime, and by issues such as salinity and winterkill.
6. Examine the introduced distribution of yellow perch within the irrigation system of southern Alberta in relation to irrigation reservoirs and government stocking programs that have expanded the range of yellow perch on the prairie landscape.

In addition to these objectives, we also briefly discuss the potential for ecological impacts of the expanded yellow perch range in western Canada, and the fisheries significance of yellow perch in Alberta in relation to the potential contribution of introduced populations, in Alberta Irrigation reservoirs.

4.2 Postglacial Dispersal and the Native Range of Yellow Perch in Western Canada

We begin by reviewing the literature on the postglacial environment and the dispersal of fishes onto the Western Canadian landscape and analyze the historical factors that resulted in the native range of yellow perch in Alberta, British Columbia and northern Canada.

Proglacial lakes were a persistent drainage feature, situated mostly along the southern margin of the Laurentide and Cordilleran ice sheets, throughout the Pleistocene era (Pielou 1991). These aquatic environments were cold, often very deep, and carried heavy sediment loads, and were thus presumably unproductive—in all of their attributes, quite different from the warm, weedy ponds that yellow perch prefer. Proglacial lakes formed wherever drainage toward the ice, usually northward, was impounded by the ice; moreover, depression of the landmass under the weight of thousands of meters of ice, contributed greatly to their size and depth. Figure 1 shows the major proglacial lakes that have played a significant role in the zoogeography of western Canadian fishes (Pardee 1910; McPhail and Lindsay 1970; Pielou 1991; Hill 2000).

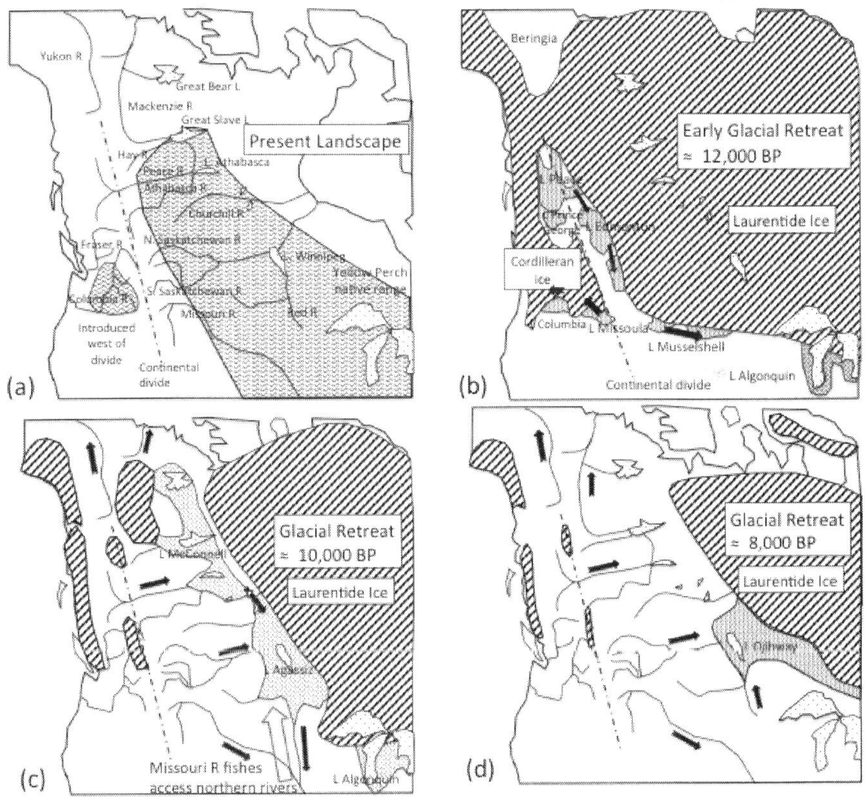

Fig. 1. (a) Native range of yellow perch east of the Continental Divide and introduced range west of the Continental Divide. (b) Proglacial lakes forming during the early glacial retreat phase as Laurentide ice separates from the Cordilleran ice. (c) Lake Agassiz forms and drains into the Mississippi allowing yellow perch and other Mississippian species to access prairie and northern drainages. (d) Lake Ojibway forms and allows fish access to Hudson Bay drainage in Northern Ontario.

East of the Continental Divide rivers that presently flow into the Missouri River formed several lakes in Montana (Hill and Valppu 1997; Hill 2000), and possibly southern Alberta, and these lakes were separated by ice throughout much of the last glaciation from those that formed along the Missouri River in the center of the continent, south of Manitoba. West of the Continental Divide, the two main lakes were Columbia Lake and Missoula Lake, which formed along the Columbia River and were usually separated by a lobe of ice (Pardee 1910; Atwater 1987); these lakes played a major role in the postglacial dispersal of Pacific species. As the retreating Laurentide glacier began to pull away toward the northeast, from the Cordilleran ice ($\approx$12,000 y BP) a number of large glacial lakes formed on the Alberta landscape (Fig. 1b), due to impoundment of the emerging Peace, Athabasca and Saskatchewan Rivers (McPhail and Lindsey 1970; Mitchell and Prepas 1990; Bednarski 1999), and since the drainage along the glacier margin was of necessity southeasterly (despite the uphill gradient of hundred of meters), Lake Peace became so deep that it backed up across the Continental Divide, spilling into Lake Prince George which had formed along Fraser River system (Liverman 2009). The species present in these lakes were those that spread northward from the Upper Missouri proglacial lakes, and presently occupy a restricted range in central British Columbia (McPhail 2007). The most notable of these were lake trout *Salvelinus namaycush* (Walbaum), lake whitefish, burbot *Lota lota* (L.) and white sucker *Catostomus commersoni*, Lacepede; these species all succeeded in crossing the Continental Divide into central British Columbia, and remain a prominent part of the modern fish fauna of the area.

Notably, the yellow perch, walleye, and northern pike did not establish populations west of the Continental Divide (McPhail 2007), which indicates that they were not present in Upper Missouri or Alberta lakes. These species, however, became widespread throughout the interior basin, including the Churchill River, and even the Mackenzie, and thus must have spread through the Lake Agassiz/McConnell systems (Fisher and Smith 1994; Smith 1994), and entered the lakes from the Missouri drainage once Lake Agassiz at its maximum extent had begun to flow southward through that system (Rempel and Smith 1998) (Fig. 1c). This link is supported by genetic studies that show haplotype signatures of yellow perch from Lake Winnipeg match those of fish from the Missouri River system (Sepulveda-Villet and Stepien 2012).

Having colonized Lake Agassiz, yellow perch likely reached Alberta by dispersing upstream through the Saskatchewan River system, and became abundant in the parts of the watershed where suitable lakes and ponds could be accessed. Such water bodies are much more common in the parkland regions of the northern branch of the river, than they are on the prairie landscapes that make up most of the South Saskatchewan system. Climatic conditions have fluctuated considerably during the last 10,000 y (Pielou 1991), and it seems likely that yellow perch may have had a more extensive distribution along the South Saskatchewan River, during wetter postglacial periods, and with many populations becoming extinct as lakes and ponds dried up as the prairies became dryer.

The massive area of Lake Agassiz at its maximum, as well as its drainage link to Lake McConnell (Smith 1994) would explain how the species obtained access to the more northerly Churchill River, where at present it is widespread, and finally to the Mackenzie drainage (Rempel and Smith 1998), the southern portion of which

(Great Slave Lake and Lake Athabasca) represent its northern limit (McPhail and Lindsey 1970).

While several authorities have Great Slave Lake as the northern limit of the range of yellow perch (Ryder 1972; Keleher 1972; Scott and Crossman 1973), only a few individuals have been caught (Scott 1956; McPhail and Lindsey 1970). Indeed, Rawson (1951) who did some of the earliest and most extensive ichthylogical surveys of the lake, which included hundreds of hours of gill netting and extensive beach seining in all of the lake basins, notably, did not find yellow perch. Rawson (1947, 1960a) did list yellow perch as present, though not abundant, in Lake Athabasca, and the Fish and Wildlife Management Information System (FWMIS, http://esrd.alberta.ca/fish-wildlife/fwmis/access-fwmis-data.aspx) database also report the species in a number of lakes near the Athabasca area. Although yellow perch have never become abundant or widespread in the north, other cool water species such as pike and walleye are sufficiently well adapted to low temperature lakes allow this. As the Laurentide ice retreated further to the northeast, Lake Agassiz extended more eastward into across the Laurentian—Hudson Bay divide, as Lake Ojibway (Lajeunesse and St. Onge 2008; Pielou 1991) (Fig. 1d) allowing yellow perch and other central basin species access to Northern Ontario and Québec.

Most warmwater species such as the sunfishes and basses, and many of the least cold-tolerant minnows, likely were not successful in establishing themselves in Lake Agassiz, or in other proglacial lakes, and by the time that climates in the central basin of Canada had warmed to the extent that such species could have survived there, the drainage connections to the refugia had been lost (Fig. 1d). Consequently, these species have never had the opportunity to invade systems like the Saskatchewan River or the Assiniboine or Qu'Appelle systems, where temperature regimes are sufficiently warm to support them at present (Mandrak and Crossman 1992).

4.3 The Present-day Distribution of Yellow Perch in Alberta and British Columbia

Analytical Methodology

By examining the fish and wildlife databases in Alberta and British Columbia, together with published and unpublished literature, and other government website databases, we analyzed the presence and absence of yellow perch in lakes, rivers and reservoirs throughout Alberta and British Columbia.

The main Alberta Resource was the Fisheries and Wildlife Management Information System (FWMIS, http://esrd.alberta.ca/fish-wildlife/fwmis/access-fwmis-data.aspx), which is the Government of Alberta's Fisheries and Wildlife database (Alberta Environment and Sustainable Resource Development) provides a central repository for which government, industry, and the public can store and access extensive and reliable fish and wildlife inventory data. The main British Columbia resource was the Fisheries Inventory Data Queries (FIDQ, http://a100.gov.bc.ca/pub/fidq/main.do), which is the BC government fisheries database (BC Ministry of Environment), which provides access to BC lake, stream and fish data, as well as fish stocking data.

A stratified approach was used to sampling information on water bodies in the databases. In areas where yellow perch were abundant throughout, as indicated by older literature and provincial fishing regulations and guides, only a sample (10–20%) of the lakes and stream data were examined. Near geographical or elevational boundaries of the range, near introduced localities, and within the irrigation district of Alberta, all records were examined. In areas where no yellow perch were reported in provincial fishing regulations and guides, only (10–20%) the data entries were examined.

Latitude, longitude and elevation data were obtained from Google Earth (http://www.google.com/earth/). Google Earth was also used to construct a map of the irrigation reservoirs of southeastern Alberta. Climate data, including mean annual and monthly temperatures for municipalities within the study are obtained from Climatetemps.com (www.climatemps.com).

Although climate data are not provided for the sites listed in the government fisheries database, general models were constructed using linear multiple regression relating temperature variables (mean annual air temperature MAT, mean summer (July/August) temperature MST to latitude LAT, longitude LONG (decimal form) and elevation ELEV (km) based on data obtained for 50 Alberta and British Columbia municipalities from the Climatetemps.com website, that spanned the geographical and elevation range of lakes and rivers in the fish and wildlife databases.

$$\text{MAT °C} = -2.91 - 1.11 * (\text{LAT} - 61) + 0.15 * (\text{LONG} - 115) - 5.57 * (\text{ELEV} - 0.157) \quad \text{Eqn. 1}$$

$$\text{MST °C} = 14.9 - 0.50 * (\text{LAT} - 61) - 0.051 * (\text{LONG} - 115) - 4.10 * (\text{ELEV} - 0.157) \quad \text{Eqn. 2}$$

These regression models accurately predicted MAT and MST from 30 additional municipalities from the region with no significant bias (MAT SE_{est} = 0.48°C; MST SE_{est} 0.52°C). MAT and MST were estimated for each site for which fishery data were examined based on Eqns. 1 and 2. Although data on water temperatures in the littoral zones of the lakes and streams where perch were recorded were not available, air temperatures (MAT and MST) at the site were used to provide a reasonable proxy that would reflect the relative temperature regimes among the water bodies. While such a proxy would of course not suffice for hypolimnetic species, it would be expected to provide a reasonable reflection of the ambient climatic regime experienced by a shallow water species such as yellow perch.

The Northern Limit of Yellow Perch

Aside from the few individuals reported from Great Slave Lake, the most northerly populations of yellow perch are associated with the Slave River, which flows from the Peace Athabasca delta and Lake Athabasca to Great Slave Lake (Fig. 2). Many of the small rivers that enter the east side of the Slave, and the lakes associated with them do support yellow perch populations, together with walleye and pike. These lakes, e.g., Charles Lake, Mercredi Lake, Cornwall Lake, Potts Lake and Andrew Lake, range in elevation from 250–400 m, and our climate estimates (Table 1), place

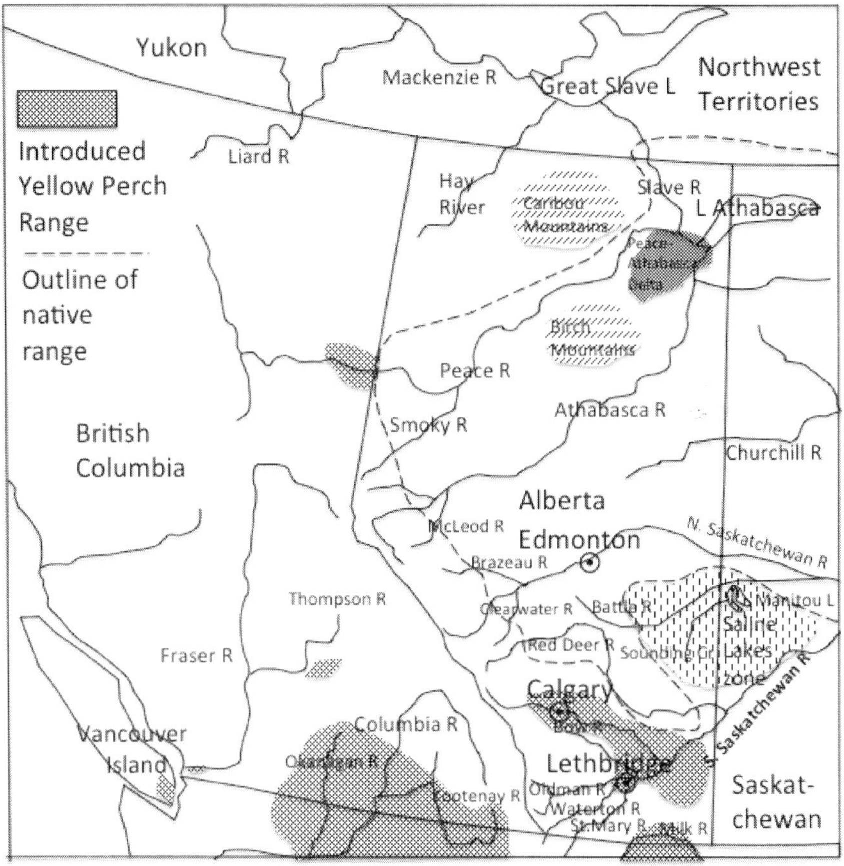

Fig. 2. Distribution of yellow perch in Alberta and British Columbia in relation to the major river systems, based on the results presented in this study.

MAT between –2 and –3.5°C, and the MST between 14.8 and 15.3°C. All of these lakes have similar climate to western end of Lake Athabasca, which has a MAT of –2.5°C and a MST of 15.1°C.

No records of yellow perch were found for any of the lakes or tributaries of the Liard River, which drains northeastern BC and enters the Mackenzie, or the Hay River, which drains northwestern Alberta (e.g., Thurston Lake 615 m and Bitscho Lake 773 m) and enters Great Slave Lake. While the Liard becomes very steep in its mid and upper reaches, there are several lakes in the lower reaches of these river systems that have suitable perch habitat, and all of these contain walleye and pike, but no yellow perch. These lakes are at elevations of 400–773 m, and our estimates of their local climate (Table 1), place the mean annual air temperatures between –3 and –4.5°C, and the mean summer temperatures between 13 and 14°C. These cold temperatures combined with the fact that the rivers drain into the Mackenzie River and Great Slave Lake, where perch are very rare, make their absence from these lakes very plausible.

Table 1. Some water bodies referred to in the text to characterize the range and environmental limits for yellow perch in the River systems of Alberta, British Columbia and the Northwest Territories. Mean Annual Air Temperature (MAT) and Mean Summer Air temperatures for these locations were estimated from Equations 1 and 2.

Water body	River System	Lat (°N)	Long (°W)	Elev (m)	P/A	N/I	MAT (°C)	MST (°C)
			Slave River System					
Charles L	Slave	60.1	110.6	295	P	N	–3.4	15.0
Mercredi L	"	60.0	110.9	298	P	"	–3.2	15.0
Cornwall L	"	59.6	110.5	283	P	"	–2.8	15.4
Potts L	"	59.3	110.5	350	P	"	–2.8	15.2
Andrew L	"	59.9	110.1	386	P	"	–3.5	14.8
Athabasca L	"	58.8	110.7	207	P	"	–1.4	15.5
			Great Slave Lake/Hay River System					
Hay R	Hay	60.1	116.7	157	A	–	–3.4	15.1
Thurston L	"	59.9	118.1	615	A	–	–3.8	13.4
Bitscho L	"	59.8	110.8	556	A	–	–4.4	14.1
Buchan L	Great Slave	60.0	115.0	304	A	–	–2.6	14.8
Pichimi L	"	59.0	114.6	773	A	–	–4.2	13.4
			Peace River System					
Wentzel L	Peace	59.1	114.5	664	A	–	–3.6	13.8
Vermillion R	"	58.5	116.3	289	P	N	–0.7	15.6
Keg R	"	57.7	118.1	560	P	N	–1.1	14.6
Swan L	"	55.5	120.0	726	P	N ?	0.7	15.1
Charlie L	"	56.4	121.1	694	P	I	0.7	14.8
Kelly L	"	55.3	120.0	861	P	"	0.2	14.6
Pouce Coup R	"	55.8	120.1	600	P	"	1.1	15.4
Raspberry L	"	54.5	116.7	792	P	N	1.0	14.4
Iosegun L	"	54.5	116.8	775	P	N	1.1	15.5
Smoke L	"	54.3	116.9	878	P	N	0.7	15.2
Sturgeon L	"	55.0	117.3	678	P	N	1.1	15.7
Musreau L	"	54.5	118.6	880	A	–	0.7	15.0
Victor L	"	53.9	119.1	1190	A	–	–0.2	14.0
Grande Cache L	"	53.9	119.0	1269	A	–	–0.7	13.7

Table 1. contd....

Table 1. contd.

Water body	River System	Lat (°N)	Long (°W)	Elev (m)	P/A	N/I	MAT (°C)	MST (°C)
			Peace-Athabasca Delta					
Claire L		58.5	112.0	210	A	–	–0.9	16.1
Baril L		58.8	110.7	210	A	–	–1.4	16.0
Otter L		58.5	111.5	210	A	–	–1.0	16.1
Richardson L		58.4	111.0	210	A	–	–0.9	16.2
			Upper Athabasca River System					
Blue L	Athabasca	53.5	117.6	1157	P	N	0.2	14.4
Jarvis L	"	53.5	117.8	1151	P	N	0.3	14.5
Gregg L	"	53.5	117.7	1139	P	N	0.3	14.5
Goose L	"	54.3	115.1	725	A	–	1.3	15.9
Edna L	"	53.1	118.0	1041	A	–	1.3	15.9
Talbot L	"	53.1	118.0	1034	A	–	1.3	15.1
Sundance L	McLeod	53.7	117.0	1037	P	N	0.5	14.9
Fickle L	"	53.4	116.8	978	P	N	1.1	15.9
			North Saskatchewan River System					
Wolf L & Cr	Brazeau	53.2	116.1	1051	P	N	0.8	15.1
Jackfish L	N Sask	52.5	115.5	1159	P	N	0.9	15.0
Tay L	Clearwater	52.1	115.1	1214	P	N	1.1	15.0
Battle L	Battle	53.0	114.2	848	P	N	1.9	16.1
Pigeon L	"	53.0	113.1	854	P	N	1.7	16.1
Oliva L	"	53.1	111.5	679	A	–	2.3	16.9
Miquclon L	"	53.2	112.9	766	A		2.0	16.4
Peninsula L	"	52.9	111.5	656	A	–	2.7	17.1
Sounding Cr	"	51.5	110.7	719	A	–	3.8	17.6
			Red Deer River System					
Burnstick L	James	51.9	114.8	1187	P	N	1.3	15.2
Gull L	Red Deer	52.5	114.0	899	P	N	2.2	16.2
Pine L	"	52.1	113.5	900	P	N	2.5	16.4
Glenmore Res.	Bow	51.0	114.1	1078	P	I	2.8	16.2

Table 1. contd....

Table 1. contd.

Water body	River System	Lat (°N)	Long (°W)	Elev (m)	P/A	N/I	MAT (°C)	MST (°C)
			Irrigation System—Bow and Oldman River System					
Eagle L	WID Bow	51.0	115.2	926	P	I	3.8	16.7
Chestermere L	"	51.0	113.8	1027	P	I	3.1	16.4
Newell L	"	50.5	112.0	769	P	I	4.8	17.8
McGregor L	BRID Bow	50.5	112.9	876	P	I	4.4	17.3
Travers Res	"	50.1	113.1	935	P	I	4.5	17.5
Rattlesnake L	"	49.8	111.5	791	P	I	5.4	18.1
Sherburne L	"	49.7	111.7	814	P	I	5.4	18.1
Little Bow R	Oldman	50.2	113.4	950	P	I	4.3	17.1
Knights L	"	49.1	113.5	1288	A	–	3.8	16.3
			Columbia River System					
Swan L	Okanagan	50.3	119.2	390	P	I	8.2	19.1
Pinaus L	"	50.4	119.6	985	P	I	4.8	16.6
Yellow	"	49.3	119.7	750	P	I	7.4	18.1
Trout L	"	49.3	119.7	767	P	I	7.3	18.0
Kamalka L	"	50.2	119.4	392	P	I	8.3	19.1
Lower Arrow Res	"	49.4	118.0	487	P	I	8.5	19.2
Pend d'Oreille R	Columbia	49.0	117.5	494	P	I	8.8	19.4
Kootenay L	Kootenay	49.0	117.5	494	P	I	7.6	18.7
Monroe L	"	49.3	115.9	1071	P	I	5.0	16.0
Jim Smith L	"	49.5	115.8	1060	P	I	4.8	16.9
Wasa L	"	49.7	115.7	773	P	I	6.2	18.0
Tie L	"	49.4	115.2	851	P	I	6.0	17.9
Baynes L	"	49.2	115.2	792	P	I	6.6	18.2
Hiawatha L	"	49.4	115.8	931	P	I	5.7	17.5
Koocanusa Res	"	49.4	115.4	747	P	I	6.6	18.3

Table 1. contd....

Table 1. contd.

Water body	River System	Lat (°N)	Long (°W)	Elev (m)	P/A	N/I	MAT (°C)	MST (°C)
			Fraser/Thompson River System					
Skmana L	Thompson	50.9	119.7	602	P	I	6.4	17.9
Forest L	"	51.1	119.9	597	P	I	6.3	17.8
Skimikin L	"	50.8	119.5	548	P	I	6.8	18.2
Miller L	"	50.6	119.6	877	P	I	5.2	16.9
Shuswap L	"	51.0	119.0	359	P	I	7.6	18.9
Judson L	Fraser	49.0	122.3	48	P	I	12.0	19.6
			Coast Islands					
St. Mary L	Salt Spring	48.9	123.5	62	P	I	12.1	19.4
Loon L	Vancouver	49.2	124.8	365	P	I	10.4	19.4

Yellow perch are also absent from the lakes and rivers of the Caribou Mountains, a small upland plateau in along the northern border of Alberta. The streams that drain them enter Great Slave Lake to the north, the Hay River on the west, the Peace River to the south and the Slave River on the east. Although there are several lakes with suitable depth regimes and littoral habitats in this region, they are above 600 m in elevation and have MAT estimates < –3.5°C and MST <14°C, and thus are likely colder than the lakes at the northern limit of the range.

The Peace River System and its Tributary, the Smoky River

Most of the lakes in the Peace River basin of Alberta have yellow perch populations. The most northerly of these are in the elevation range of 200–350 m and have MAT of –1 to –2°C, and MST of >15°C. The highest lake that drains to the Peace River mainstem is Swan Lake (726 m) on the Alberta-BC border (MAT 0.7°C, MST 15.1°C). This is slightly warmer than the lower-elevation lakes, because of the more southerly latitude (Table 1). The upper Peace River is in British Columbia and has no native yellow perch populations, and those that are present were introduced by government stocking programs. The prevailing climate in the area of Charlie Lake, and several other BC lakes along the Peace River (elevations 694–850 m) that have been stocked (Table 1) have MAT of 0.1–0.5°C, and MST 14.7–15°C. There are likely many more lakes with suitable tributaries, depth regimes and climate in this part of BC that could support yellow perch.

A major tributary that enters the Peace River from the south within Alberta is the Smoky River (Fig. 2), and there are many lakes along the Smoky system that support abundant perch populations. The highest elevation lake in this system that supports perch is Smoke Lake (878 m) but there are several others between 678–792 m. MAT estimates at these lakes are 0.7 to 1.3°C and MST are 15.2–15.7°C, which are all substantially warmer than the water bodies at the northern limit of the range. Three lakes on the upper Smoky system, have apparently suitable small tributaries, and depth regimes, but have no record of ever having held yellow perch; these are Musreau Lake (880 m), Victor Lake (1190 m) and Grande Cache Lake (1269 m). While the temperature parameters estimated for these lakes are lower than those from nearby lakes that have perch (Table 1), they are all substantially warmer than those found at the northern range limit.

The Peace Athabasca Delta Region

The Peace and the Athabasca Rivers converge at the western end of Lake Athabasca (207 m) to form the massive Peace-Athabasca delta (Fig. 2). The hydrology of this delta is very complex and the system includes a network of shallow but very large perched lakes whose levels are above the average water table due to fine clay sediments that seal the basins. The delta lakes (210 m), including Lakes Claire, Baril, Mamawi, Hilda and Otter west of the Athabasca River and Richardson Lake to the east, are linked to the rivers, to each other and to Lake Athabasca through an intricate series of channels, and receive backflow from the rivers intermittently when the rivers are in flood. While the lakes are very productive during the ice-free season, they are so shallow that they almost freeze solid during winter, and their waters are so rich in particulate and dissolved organic matter that oxygen levels in winter below the ice become rapidly depleted causing winter kill (Mitchell and Prepas 1990).

In spite of this, the lakes are heavily used by fishes, especially as nursery habitats, from early spring through fall by goldeye, pike, walleye and whitefish, which exhibit well-timed movements in and out of the lakes through the connecting channels. However, yellow perch although present in Lake Athabasca and in most of the neighbouring lakes east of the delta, do not use the delta lakes; although FWMIS listings for the in the channels near Lake Athabasca do sometimes mention perch. Perch are not known to exhibit pronounced seasonal migrations (Craig 2000b), and so their absence from lakes such as these, with inhospitable winter conditions is not surprising.

The Athabasca River System

Most of the lakes along the Athabasca river system from its upper reaches to the delta, support yellow perch, and the species extends to higher elevations in the Athabasca River than in the Peace River. Blue (1157 m), Jarvis (1151 m) and Gregg (1139 m) Lakes, all discharge into the upper Athabasca mainstem, and on its southern tributary the McLeod River, Sundance (1037 m) and Fickle (978 m) are the highest lakes that contain yellow perch. Although substantially higher in elevation than the corresponding lakes in the Peace system, the MAT and MST estimates are quite similar (Table 1). Although perch extend to quite high elevations along the Athabasca River, they are puzzlingly absent from some of the lakes despite apparently suitable MAT and MST (Table 1). Examples are Edna (1041 m) and Talbot (1034 m) Lakes, which are generally quite shallow with good littoral zones and apparently navigable creeks linking them to the river, should be able to support yellow perch; these lakes do contain pike and lake whitefish.

The Saskatchewan, Red Deer and Battle River/ Sounding Creek systems

The transition zone from the boreal forest to the aspen parkland across central Alberta and Saskatchewan is drained by the Churchill River and the North Saskatchewan River, and these systems have many lakes with excellent populations of yellow perch. The North Saskatchewan River emerges from the Columbia Ice fields and is rather cold, turbid and inhospitable in its upper reaches, with few apparently suitable tributaries and lakes. However, there are a few high elevation lakes containing perch along its Brazeau River tributary to the north (Wolf Lake 1051 m; MAT 0.84°C, MST 15.1°C), its mainstem (Jackfish Lake 1159 m; MAT 0.9°C, MST 15.0°C), and its Clearwater River tributary to the south (1214 m; MAT 1.1°C, MST 15.1°C). These are the highest lakes in the drainage that would appear both suitable and potentially accessible to yellow perch. In the eastern portion of Alberta, the landscape transitions sharply from parkland to prairie as it becomes progressively more arid toward the south. Few water bodies containing yellow perch were found south of the North Saskatchewan River in Eastern Alberta.

The Battle River is a slow moving rather turbid river that emerges in central Alberta at moderate elevation levels and so has many lakes with yellow perch in its upper reaches (e.g., Pigeon Lake 854 m and Battle Lake 848 m); however, its lower reaches drain in a northeasterly direction through a semi-arid prairie landscape in eastern Alberta, where yellow perch have not been recorded either in the mainstem, or in its tributaries or in the few lakes present in the area. These lakes are very shallow, and moderately to strongly saline (conductivity 5000–100,000 $\mu S\ cm^{-1} \approx$ salinity range 2500–50,000 $mg\ L^{-1}$). Oliva (649 m), Peninsula (656 m) and Miquelon Lakes (766 m) are examples of moderately saline lakes in this system that lack yellow perch. Although Miquelon Lakes and some others in the region have had yellow perch introductions, these populations have subsequently disappeared. In the southern portion of this region, Sounding Creek, a moderately saline stream with saline lakes along its course, also lacks perch, being completely fishless over much of its length. The Sounding system is, in fact, an endorheic system that terminates in Manitou Lake, Saskatchewan, a highly saline lake where fish have never been recorded (Rawson and Moore 1944).

The South Saskatchewan River system does not have nearly as many good perch lakes as the North Saskatchewan, which reflects the fact that much of the river's mid-elevation reach flows through semi-arid grasslands that have few lakes, and even fewer permanent tributaries linking them to the river systems. Nonetheless, its most northern branch, the Red Deer River does have several good perch lakes along its course, such as Pine Lake (900 m) and Gull Lake (899 m), and has a few lakes that support perch as high up as 1187 m (Burnstick Lake, 1187 m). However, two main southern branches have a very limited number of lakes and permanent tributaries due to the semi-arid grassland landscape. The elevation range of perch along these rivers is much more limited than in the rivers further north, despite the warmer climate. Thus, the Bow River system has no perch above Glenmore Reservoir (1078 m) in Calgary, and the Oldman River has no perch populations above 850 m (Sherburne Reservoir, 814 m; Little Bow River, 935 m; Fig. 3), and these appear to be derived from introduced populations in the irrigation system.

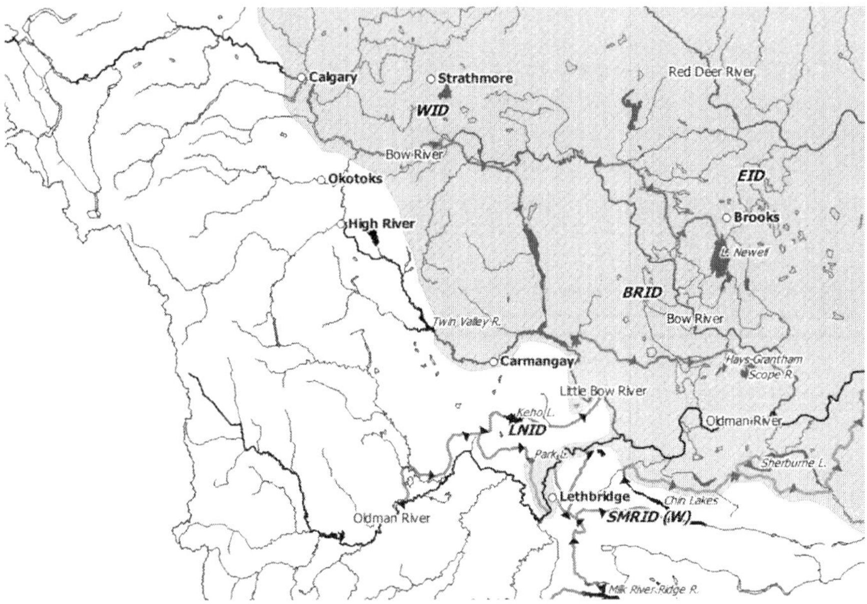

Fig. 3. Map showing the relationships of the irrigation systems of southern Alberta to the Bow and Oldman River systems. Irrigation reservoirs and canals are shown as well as direction of flow and links to the rivers. Abbreviations: WID = Western Irrigation District; EID= Eastern Irrigation District; BRID = Bow River Irrigation District; LNID = Lethbridge Northern Irrigation District; SMRID = St. Mary River Irrigation District.

Irrigation Reservoirs along the Bow and Oldman Rivers

The water bodies that do have perch along the Bow and Oldman Rivers are reservoirs built for the storage and distribution of irrigation water (769–1027 m) and the drainage relationships and distribution of yellow perch within the irrigation systems are shown in Fig. 3. For several decades during the mid 1900's, Alberta Fish and Wildlife raised perch in their hatcheries, and introduced them, along with pike, lake whitefish and walleye, to many of the reservoirs in the irrigation system. Most of these introduced populations continue to be very successful, and perch contributed significantly to the development of the reservoir fisheries. There is nothing to prevent perch from entering the Bow River from the irrigation system; in fact fish rescue operations that return fish to the Bow River from the irrigation canals, when they are being drained in the fall, frequently add yellow perch to the Bow River.

The most northerly, and the oldest reservoirs in the system, known as the Western Irrigation District, including Chestermere Lake and Eagle Lake, store water from the Bow River that enters the system at Calgary. This system dates back to 1910, and although it is known that yellow perch were introduced to these systems some time in the early 1900's, records are poor. Shortly afterward, the Eastern Irrigation system was built, drawing its water via a canal system at the Bassano Dam, downstream of Calgary, and moving it east to a storage reservoir called Lake Newell, to which yellow perch was also introduced. Subsequently perch appear to have spread about 10 km downstream through distribution canals to two reservoirs near Brooks, AB (Stafford and Johnson Lakes, Fig. 3).

The Bow River Irrigation district began in 1920 drawing water from the Bow River at the Carsland wier downstream of Calgary and sending it eastward through a canal to Lake McGregor, a reservoir constructed in a natural outwash channel. From Lake McGregor water flows to two additional storage reservoirs, Travers Reservoir and the Little Bow Reservoir, which connects to the Little Bow River. At present all three reservoirs, plus the smaller reservoirs (Scope, Hays-Grantham, and Badger Lakes) connected by > 30 km of distribution canals, contain perch. In addition, perch can now be found in the lower 60 km of the Little Bow River downstream of Twin Valley reservoir. Alberta Fish and Wildlife stocking records indicate that yellow perch were stocked into Lake McGregor on several occasions between 1938 and 1972, however, there are no records of either Travers, or any of the other reservoirs, or either the Little Bow Reservoir or the river being stocked with perch; however, perch do occur in the downstream reservoirs plus the lower section of the Little Bow River. The Little Bow River empties into the Oldman River at Taber, and there is nothing to prevent perch from the Little Bow from entering the lower Oldman and South Saskatchewan Rivers.

The St. Mary Dam was built in 1950, and supplies water to the St. Mary Irrigation District south and east of Lethbridge. In 1959 and 1960 additional impoundments were built on the Belly and Waterton Rivers, with canals linking them to the St. Mary Reservoir. Thus water impounded from these three southern tributaries of the Oldman River supplies the network of irrigation reservoirs east of Lethbridge. While none of the three rivers supplying water to the St. Mary River Irrigation District (SMRID in Fig. 3) contain yellow perch naturally, and none of the three source reservoirs has been stocked with perch, several of the downstream reservoirs in the system do support abundant perch populations. These include Sherburne, Murray, Grassy, Yellow and Rattlesnake Lakes, and connecting channels. Although detailed records are not available, it is certain that at least Sherburne and possible other downstream reservoirs were stocked with yellow perch by Alberta Fish and Wildlife. It is not possible for fish to swim against the locks and colonize upstream in the system, so the upstream reservoirs including Chin Lakes, the Milk River Ridge Reservoir, Tyrell Lake or St. Mary and the Waterton Reservoirs have remained perch-free. Downstream however, the storage reservoirs connect through channels with intermittent flow to Seven Persons Creek, which enters the Oldman River upstream of Medicine Hat, and this stream has abundant yellow perch, and was probably invaded by fish moving downstream from the storage reservoirs. Ross Creek is another tributary of the Oldman that enters the Oldman at Medicine Hat a few miles downstream of the mouth of Seven Persons Creek, and it too has abundant yellow perch, and was likely invaded by perch from the irrigation system.

West of Lethbridge, the Oldman River Dam was built in 1992 and stores water which enters the Lethbridge Northern Irrigation system through a canal downstream of the dam. This canal moves water to Keho Lake, Park Lake and to other water bodies in Lethbridge, AB, including Nicholas Sherin Pond, Campus Pond and some urban storage ponds. In the last decade yellow perch have appeared in Park Lake and in the Lethbridge ponds downstream. These introductions must have been unauthorized, since there are no records of any authorized perch introductions to these water bodies. It is highly unlikely that these perch arrived from the Oldman River through downstream dispersal, since Keho Lake, the next upstream basin in the system, does not have yellow perch.

Relationship between Elevation Limits and Latitude

In general, the maximum elevation at which perch can be found increases from north to south, as we expected (Fig. 4); however, the pattern breaks in the South Saskatchewan system as the rivers flow into semi-arid grasslands and good lakes with navigable permanent tributaries are scarce, and the perch originated from introductions. Thus the maximum elevational extent of yellow perch populations in these southern rivers is much reduced compared to the North Saskatchewan and the Athabasca systems where suitable lakes are plentiful.

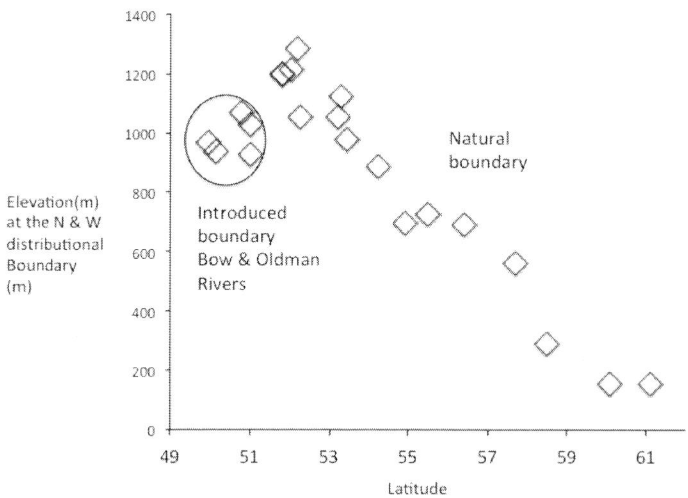

Fig. 4. Elevational limits along river systems of the eastern slopes of the Rocky Mountains and its relationship to latitude.

Although the general pattern with latitude was more or less as expected, with perch reaching higher elevations at more southerly latitudes, the maximum elevations they reached did not appear to be determined by cold temperatures. Thus at their northern limit along the lower Slave River (MAT of –2.5 C to –3°C and MST of 14°C) were the rule, the elevational limit is reached at significantly warmer temperatures in the Athabasca and North Saskatchewan River systems (Fig. 5).

Yellow Perch in British Columbia West of the Continental Divide

Perch were first recorded in BC in the 1950's in Osoyoos Lake and were soon after seen in other nearby lakes along the US border; since then, the distribution has expanded northward through the Okanagan system. The most northerly and highest elevation

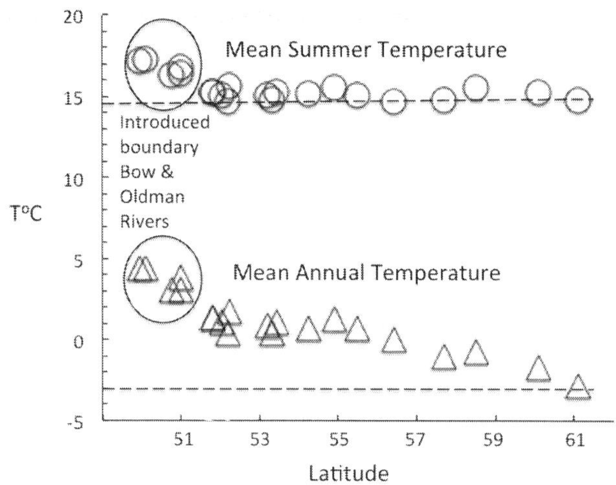

Fig. 5. Mean summer and mean annual temperature estimates at the elevational limits of the eastern slope river systems. Dashed lines represent the MAT and MST estimates for the lakes along the Slave River at the northern limit of the yellow perch distribution.

lakes where perch have been recorded in the Okanagan watershed are Pinaus Lake (985 m), Yellow Lake (750 m), Trout Lake (767 m) all headwater lakes from its watershed. While it is possible that perch could have invaded some of the lower Okanagan lakes such as Kamalka Lake (392 m), or Swan Lake (390 m), it is unlikely that they would have navigated any of the streams linking these higher lakes to the Okanagan, and thus the perch in these lakes all certainly result from unauthorized introductions.

The distribution of perch in the Upper Columbia is centered around four locations: the Columbia mainstem, the Pend d'Oreille River, the Kootenay River (Koocanusa Reservoir 747 m and Kootenay Lake 494 m) and Moyie River. These areas, together with the Arrow lakes (487 m), which were secondarily invaded, make up a contiguous set. There are however, a number of 'satellite' invasions in the area of small lakes and associated streams, such as Tie Lake (851 m), Wasa Lake (773 m), Baynes Lake (792 m) and others that are not geographically contiguous or hydrologically linked, and which clearly must represent unauthorized introductions. The streams linking these higher lakes to the main river systems are not likely to be navigable by perch. The highest elevation lakes where perch have been recorded in the Kootenay watershed are Monroe Lake (1071 m), Jim Smith Lake (1060 m), and Hiawatha Lake (931 m).

Other records of perch in BC west of the Continental Divide are more recent. In 1996 the species was recorded in Skmana Lake (602 m) on the South Thompson River system, and subsequently they have been recorded in several more lakes such as Forest Lake (597 m), Skimikin Lake (548 m), Miller Lake (877 m), Shuswap Lake (359 m) and associated streams in this region (Table 1). In addition to this, yellow perch have been recorded in several water bodies on Vancouver and Salt Spring Islands, and one occurrence has been recorded in the Lower Fraser system in Abbotsford (Table 1). With the exception of a few cases of lakes being invaded through downstream dispersal there is no contiguity to the perch distribution pattern in the South Thompson watershed, and all invasions represent localized unauthorized introductions.

All of the lakes in southern BC are very warm, and are thus well inside the climatic boundaries for perch, and it is unlikely that even the highest lakes in any of these watersheds would be too cold to support yellow perch.

4.4 Factors Limiting the Range of Yellow Perch in Alberta and British Columbia

Yellow perch populations in this study were not found north of –3.5°C MAT and 14.4°C MST, which is consistent with McPhail and Lindsey (1970), and is likely a combination of the effects of long periods of ice cover, delayed spawning, and short summer growing seasons. Thus in the northerly portion of the yellow perch range their absence from the lower Liard and Hay River systems, and from upland region such as the Caribou Mountains appears to be a direct result of cold temperatures. However, the failure of perch to invade lakes of the upper Peace, and Saskatchewan River systems can clearly not be explained by temperature. Therefore, our hypothesis that upper elevational limits along foothill rivers would be set by the same thermal limits that set the northern boundary, was not supported. Many successful introductions have occurred in the upper Peace region around Fort St. John, and Dawson Creek, BC area, and there are many more lakes with suitable temperature regimes on many of these where yellow perch have been introduced and been very successful. This further supports the idea that the upstream extent of the native yellow perch range along the Peace River was not set by climatic limits. In British Columbia west of the Continental Divide, no populations even approach the temperature limits, which indicates that there is a considerable scope for both northerly and upward range expansion of this species.

While the absence of perch from northern river systems was consistent with this climatic boundary, and as expected the elevational limit increased with decreasing latitude, the species range in the foothills of the Rocky Mountains did not reach the predicted climatic limits. Thus rather than being limited by climate, the elevational range is likely limited by the ability of perch to find suitable standing water habitats, with slow moving, navigable and permanent tributaries that can be accessed, and possibly to their ability to navigate the rivers themselves, which become steeper further south and upslope. Thus most of the river systems have lakes at high elevations that would appear to be sufficiently warm, and have enough littoral habitat to support yellow perch; yet, they have never been colonized, most likely due to the limited ability of yellow perch to effectively disperse upstream along steep foothills rivers. The idea that ability to navigate streams may limit their elevational range of yellow perch is not surprising since they are known to have very limited swimming ability and tolerance for current (Thorpe 1977a,b; Craig 2000b).

Yellow perch are also naturally absent from prairie sections of the South Saskatchewan River system, which likely reflects the paucity of permanent lakes on the semi-arid grassland landscape, and the even fewer permanent tributaries linking them to the river systems. Thus the natural range of yellow perch within the South

Saskatchewan River system is limited to the upper reaches of the Red Deer River system where it flows through parkland regions, and suitable lakes with accessible tributaries are present. Similarly, the upper reaches of the Battle River system also contain yellow perch. However, our analysis indicates that the prairie reaches of the Red Deer system, the Bow River, and the Oldman and South Saskatchewan Rivers were not part of the natural range for yellow perch in Alberta, at least during recent times. Thus the prairie portion of the present yellow perch range was added as a result of government-authorized introductions to several of the irrigation reservoirs, followed by downstream dispersal from the points of introduction. While there are few suitable natural lakes for perch along the Oldman system, Knight's Lake (Table 1) a shallow, weedy lake at the headwaters of its Waterton River tributary which contains pike and whitefish, would almost certainly have been colonized by yellow perch, had the species been native to the Oldman River.

The absence of yellow perch from the portion of east-central Alberta where saline lakes and streams predominate, is likely is not a simple a function of salinity since yellow perch are known to tolerate moderate salinity, being frequently found in brackish estuaries (Thorpe 1977a,b; Craig 2000a) although they move to freshwater to spawn (Scott and Crossman 1973). In Saskatchewan, saline lakes Rawson and Moore (1944) recorded yellow perch populations in lakes with average salinity up to 8000 mg L^{-1} ($\approx$16,000 $\mu S\ cm^{-1}$) although they were often absent in lakes considerably less saline than this upper limit. This was attributed to a high likelihood of winterkill due to low under-ice dissolved oxygen values ($<$2 mg L^{-1}) commonly encountered in these lakes (Rawson and Moore 1944). Alberta Fish and Wildlife introduced perch and pike to several of the lakes in this area during the 1930's to 60's (e.g., Miquelon Lake, saslinity 5,500 mg L^{-1}; conductivity 11,000 $\mu S\ cm^{-1}$, under-ice dissolved oxygen $\approx$2 mg Mitchell and Prepas 1990); however, most of these introduced populations, including that of Miquelon Lake have subsequently disappeared. Although yellow perch have reasonable tolerance to hypoxia (Thorpe 1977a,b; Craig 2000a), with a lower winter dissolved oxygen (DO) limit of $\approx$1 mg L^{-1} and summer limit of $\approx$3 mg L^{-1}, they are most successful in lakes that have DO levels that remain above 5 mg L^{-1} (Moore 1942; Magnuson and Karlen 1970). Thus winter oxygen levels commonly found in natural lakes and streams from semi-arid regions are likely to compound salinity issues and severely restrict the distribution of yellow perch. Interestingly throughout this region, there are many lakes and streams where the only fish species recorded are brook sticklebacks *Culea inconstans* (Kirtland) and fathead minnows *Pimephales promelas* (Rafinesque), which are known for their extreme tolerance of the combination of low oxygen and high salinity (Armitage and Olund 1962; Robb and Abrahams 2002; McPhail 2007). Winterkill is also likely an important constraint limiting the distribution of yellow perch in the Peace Athabasca delta. The shallow and productive delta lakes nearly freeze to the bottom during winter, and most residual water becomes anoxic (Mitchell and Prepas 1990). Although migratory fish (e.g., walleye and goldeye) make heavy seasonal use of these lakes and interconnecting channels, perch are not present.

4.5 Range Extension Through Authorized and Unauthorized Introductions

Although the Alberta irrigation systems have significantly expanded the range of perch in prairie regions, some have never been stocked with perch, and there are significant barriers that appear to restrict dispersal throughout the system. The most extensive perch introductions occurred within the Calgary and the Eastern Irrigation District systems. In these systems, the points of introduction were to central storage reservoirs located high within the canal systems. Subsequently, downstream dispersal has led to colonization of most suitable reservoirs within the system, and to the colonization of the lower reaches of the Little Bow River, where no introductions took place. The upper reaches of the Little Bow, including the Twin Valley Reservoir remain free of perch since no introductions have taken place there, and the irrigation water intake to the system from the Highwood River, appears not to contain any perch, although mountain whitefish and longnose dace do gain access at this point. Interestingly, perch are now found in the Little Bow River immediately below Twin Valley Dam, which was built in 2003, several decades after perch introduction to the Little Bow System. The fact that perch were clearly not present in the reach above Twin Valley Reservoir prior to dam-construction suggests that their presence in the reach up to the reservoir may be due to a recent expansion. Although speculative, most likely a change in flow regimes as a result of river regulation by the Twin Valley Dam occurred, which provided conditions suitable for upstream movement of perch from Travers Reservoir.

Significant perch stocking also took place in reservoirs (e.g., Sherburne Lake) from the lower portions of the St. Mary Irrigation district reservoirs. While downstream invasions have also occurred into streams near Medicine Hat, no upstream spread has occurred to Chin Lakes or other neighboring reservoirs, since the control structures, which regulate the flow of water downstream from the storage reservoirs, do not allow upstream fish passage. While all of the reservoirs in this upstream portion of the St. Mary system have suitable temperature regimes, and most of the reservoirs have at least some suitable littoral habitat, any entry by perch would be through the intake from the St. Mary and Waterton Rivers, which lack yellow perch. No authorized yellow perch introductions have occurred in the Lethbridge Northern Irrigation District, but the species is now found in Park Lake, and in all West Lethbridge urban ponds fed by this system. These populations must have resulted from an unauthorized introduction to Park Lake, followed by downstream dispersal, since upstream passage from Lethbridge through the Park Lake control weir would not have been possible. Perch are not found in Keho Lake, or in other water bodies upstream of Park Lake, since they have apparently not been able to access the system via the intake from the Oldman River upstream of Fort McLeod, where no perch are present.

Lakes and associated tributaries have been invaded in five disjunct systems west of the Continental Divide. Four are in mainland BC including: the Columbia mainstem and associated tributaries (Okanagan, Kootenay, Pend'Oreille and Osoyoos Lake), the lower Fraser and the lower Thompson. In addition both Vancouver and Salt Spring islands contain many lakes and streams where yellow perch are now present. For the most part there is little contiguity or pattern to the distribution, although it does appear the large mainstem lakes and rivers associated with the Columbia mainstem and its tributaries (e.g., Osoyoos Lake, Okanagan Lake and River, Kootenay River) provide favourable perch habitat, and have served as avenues for introduction and dispersal from the Western Washington area where the species was first introduced west of the divide. However, apart from this, perch have been recorded in many geographically and hydrologically isolated individual lakes or small groups of lakes and associated inlet and outlet streams, strongly suggesting several unauthorized introductions. These are most likely intentional, although the minnow bait bucket has long been known as a common contributor to fish introductions, even in provinces where fishing with live minnows is illegal.

Potential Ecological Impacts of Yellow Perch Invasions

Yellow perch are generalist feeders whose diet would overlap with most other fish species. As is the case with many other fishes, the larvae and fry of perch feed mainly on Crustacean zooplankton and they are known to significantly reduce the abundance of larger cladocerans and copepods (MacDougall et al. 2001), and, the ecological impact of perch on other fish species appears to be mainly a result of this effect. Experimental removal of perch from a small lake resulted in a significant increase in abundance of large *Daphnia* and copepods, which led to an immediate increase in the size of young-of-the-year brook trout (Browne and Rasmussen 2013).

While there likely are many lakes along the eastern slopes of the rocky mountains in the foothills of Alberta and northern BC that are suitable for yellow perch and could potentially be invaded or inadvertently introduced, it is not likely that significant ecological impacts would accompany such introductions. Most of the lakes that perch could successfully invade have populations of northern pike and/or walleye and/or burbot, and most contain lake whitefish. These species coexist with yellow perch across most of central Canada and the northern United States, and perch have not been known to cause ecological impacts on food webs that contain these species. Small perch serve as excellent forage for pike, walleye and burbot, and lake whitefish appear to have little difficulty competing with yellow perch for zooplankton and benthic invertebrates, possibly due to differences in habitat use, with whitefish more pelagic, and perch more littoral. Whether this idea is actually true or not, it has doubtless been the view of fisheries managers who have not been reluctant to allow perch introduction into waters east of the divide.

The situation is much different when perch gain access to a salmonid lake, since young salmonids seem to be strongly impacted by competition with young perch. In Ontario, a range extension of yellow perch into the Algonquin Park, an upland region north of Toronto, has impacted eastern brook trout *Salvelinus fontinalis* through its competitive impact on the growth and survival of the juvenile trout, despite the fact that the adult trout feed mainly on perch (Browne and Rasmussen 2009, 2013). Moreover, brook trout introductions to lakes containing perch are much less successful than introductions to perch free lakes (Fraser 1978). Walters and Kitchell (2001) discussed many possible examples of key fisheries species having their own juvenile recruitment 'bottlenecked' through competition from their major prey, and salmonid life histories appear to make them particularly prone to this problem, and thus vulnerable to potential invaders, that they have difficulty keeping in check. While introduced perch in most cases, do not appear to be major predators on salmonids (Zimmerman 1999), they can sometimes consume significant numbers of salmon smolts (Bonar et al. 2005). Thus there are valid reasons for concern about potential yellow perch range expansion west of the Continental Divide, which has already resulted in their invasion of lakes that support key salmonid fisheries, and likely could lead to many such invasions, and to invasion of systems serve as nursery environments for anadromous salmon. Lake and reservoir species such as sockeye, and kokanee salmon (*Oncorhynchus nerka*), rainbow trout and steelhead (*O. mykiss*), westslope and coastal cutthroat trout (*O. clarki*) and charrs such as bull trout (*S. confluentus*) and dolly varden (*S. malma*), would likely be the most impacted by yellow perch.

Although salmonids are certainly important components of fish communities and fisheries west of the divide that could potentially be impacted by yellow perch and other invaders 'from the east,' there are also many native fish from other families that have had no previous experiences with perciform fishes, and could potentially be at risk. These include several Columbia River endemic cyprinids that use food and habitat resources similar to yellow perch; e.g., peamouth *Mylocheilus caurinus*, pikeminnow *Ptychocheilusoregonensis*, redside shiner *Richarsonius balteatus*, and many others (McPhail 2007).

4.6 The Importance of Yellow Perch to Fisheries in Alberta Lakes and Reservoirs

Yellow perch are major contributors to both recreational and commercial fisheries, both in Canada and the United States (Craig 2000c). In terms of total monetary value, yellow perch rank second, only to walleye in the Canadian freshwater commercial fishery (DFO 2012), although in Alberta and British Columbia they are not commercially important. The fisheries contribution of perch is, however, often limited by 'stunted' growth, and indeed, perch are well known for their highly variable adult sizes (12–35 cm) and growth rates (Craig 1987; Post and McQueen 1994). As a result, the contribution of yellow perch to fisheries is often limited to their role as forage fish, however, where adult perch reach targetable sizes they can be highly prized.

While perch populations near the northern boundary are rarely abundant or fast growing (Rawson 1960b; McPhail and Lindsey 1970), the largest perch records noted for Canadian waters by Scott and Crossman (1973) were from central Alberta and Saskatchewan. They listed three Alberta lakes, Lac La Biche, Tucker Lake and Baptiste Lake (central Alberta, Athabasca and Churchill River drainages) as having produced fish well over 30 cm, with weights approaching or exceeding 1 kg. Mitchell and Prepas (1990), citing Alberta Fish and Wildlife records, also document many yellow perch populations from this region that support excellent sport fisheries, and while this is by no means common to all lakes in this region, many have populations supporting fisheries where the average perch captured exceeded 300 g. Similarly, FWMIS records indicate large sized perch to be common in several of the introduced populations from irrigation reservoirs in southern Alberta such as Rattlesnake and Sherburne Lakes, and others (Fig. 3). Whether such large sized perch occur in BC is unknown due to insufficient data; the only data available on size and growth of yellow perch in British Columbia is from Chisholm et al. (1989).

Most lakes and reservoirs in parkland regions of central Alberta and Saskatchewan tend to be highly productive compared to those situated in the Appalachian and Canadian Shield regions of central and eastern Canada (Prepas and Trew 1983; Mitchell and Prepas 1990), and plankton productivity is known to be an important contributor to the growth rates of young-of-the year yellow perch (Abbey and MacKay 1991). A major factor determining growth rate of juvenile perch is the nature of the littoral fish community; perch grow much more slowly in species rich communities where fish density is high (Boisclair and Leggett 1989a; Post and McQueen 1994). Lakes and reservoirs in western Canada have species poor littoral fish communities compared to central and eastern Canada, which is related to the post-glacial exclusion of many warm water fish families such as centrarchids, cyprinids and Ictalurids (Scott and Crossman 1973; Nelson and Paetz 1992; Mandrak and Crossman 1992; Mandrak 1995). Rich littoral zone fish communities have invertebrate communities dominated by small size classes and taxa, which also leads to stunted growth (Boisclair and Leggett 1989b; Rasmussen 1993; Boisclair and Rasmussen 1996). Young perch shift progressively from small to large prey as they grow, and the absence of large invertebrates reduces growth efficiency (Sherwood et al. 2002a,b; Kovecses et al. 2005; Rasmussen et al 2008). Western lakes and reservoirs often have littoral prey communities that feature large invertebrates such as *Gammarus lacustris* (Amphipoda) and the Erpobdellid leeches in great abundance, which contribute greatly to the upper end of the community size spectrum (Hanson et al. 1989; Brinkmann and Rasmussen 2010). Thus in addition to their productivity, prairie parkland lakes also tend to have food webs that are conducive to good growth of juvenile yellow perch, allowing them to rapidly reach sizes where piscivory is possible in the adult stage, and this likely is an important contributing factor to the large perch sizes that are commonly reported in this region.

As a result, yellow perch have long been an important contributor to recreational fisheries in central Alberta lakes. While hydrological limitations in the prairie regions of southern Alberta appear to have limited the native range of yellow perch in southern prairie regions of Alberta, the productivity of the southern Alberta reservoirs, especially those that are shallow and have well developed littoral communities, combined with the presence of food webs conducive to good growth has allowed yellow perch to become an important component of recreational fisheries in this area as well. Their fisheries significance will likely increase in the future with the continued spread of yellow perch throughout the irrigation system, despite significant barriers to upstream dispersal.

In northern BC, east of the Rocky Mountains, where fisheries managers introduced yellow perch, their contribution to recreational fisheries is generally appreciated; however, this is not the prevalent view in southern BC. While there is little published information on their potential positive contribution to fisheries in this area, there is considerable concern that they, together with introduced centrarchids, may constitute a major threat to native fish biodiversity, and possibly a major threat to commercial and recreational fisheries.

4.7 Summary

Large and small-scale distribution patterns of yellow perch *Perca flavescens* in Alberta and British Columbia, both native and introduced, were examined by reviewing published and unpublished literature and government databases, providing insights into dispersal history, invasive ability and their tolerances for climate, salinity and other habitat factors. We examined the historical factors leading to the native distribution of yellow perch in the region, and hypothesized that the same temperature limits that appear to limit the northern range, would also determine the upper elevational boundaries along foothills river systems. While the absence of perch from northern river systems was consistent with this climatic boundary, and as expected the elevational limit increased with decreasing latitude, the species range in the foothills of the Rocky Mountains did not reach the predicted climatic limits. Rather, it appeared to be limited by the ability of perch to find suitable standing water habitats, with slow moving, navigable and permanent tributaries that can be accessed, and possibly to their ability to navigate the rivers themselves, which become steeper further south and upslope. Yellow perch are also naturally absent from prairie sections of the South Saskatchewan River system, which likely reflects the paucity of permanent lakes on the semi-arid grassland landscape, and the even fewer permanent tributaries linking them to the river systems. They are also absent from the lower Battle River system, likely due to limits imposed by salinity and winterkill, and they are absent from the Peace-Athabasca delta region, likely due to winterkill issues. While perch dispersal ability appears to have limited their natural range, perch introductions (both authorized and unauthorized) have in general been highly successful, in both natural lakes and rivers on both sides of the Continental Divide, and we have analyzed data on their present distributions in these areas, with special emphasis on the irrigation reservoirs of southern Alberta. The fisheries significance of yellow perch in Alberta is discussed in relation to the potential contribution of introduced populations, in Alberta Irrigation reservoirs. Unauthorized perch introductions in southern British Columbia are considered a major potential threat to salmonid fisheries, and to other native species that have no history of interaction with percid fishes.

4.8 References

Abbey, D.H. and W.C. Mackay. 1991. Predicting the growth of Age-0 yellow perch populations from measures of whole-lake productivity. Freshw. Biol. 26: 519–525.

Armitage, K.B. and L.J. Olund. 1962. Salt tolerance of the brook stickleback. American Midland Naturalist 68: 274–277.

Atwater, B.F. 1987. Status of glacial lake Columbia during the last floods from Glacial Lake Missoula. Quaternary Research 27: 182–201.

Bednarski, J.M. 1999. Quaternary Geology of northeastern Alberta. Geological Survey of Canada Bulletin 535, 29 pp.

Boisclair, D. and J.B. Rasmussen. 1996. Empirical analysis of the influence of environmental variables associated with lake eutrophication on perch growth, consumption and activity rates. Ann. Zool. Fenn. 33: 507–515.

Boisclair, D. and W.C. Leggett. 1989a. Among population variability of fish growth 2. Influence of fish community. Can. J. Fish. Aquat. Sci. 46: 1546–1550.

Boisclair, D. and W.C. Leggett. 1989b. Among population variability of fish growth 3. Influence of prey type. Can. J. Fish. Aquat. Sci. 46: 468–482.

Bonar, S.A., B.D. Bolding, M. Divens and W. Meyer. 2005. Effects of introduced fishes on wild juvenile coho salmon in three shallow Pacific Northwest lakes. Trans. Am. Fish. Soc. 134: 641–652.

Brinkmann, L. and J.B. Rasmussen. 2010. High levels of mercury in biota of a new prairie irrigation reservoir with a simplified food web in southern Alberta, Canada. Hydrobiologia 641: 11–21.

Brown, T.G., B. Runciman, M.J. Bradford and S. Pollard. 2009. A biological synopsis of yellow perch (*Perca flavescens*). Can. Manuscr. Rep. Fish. Aquat. Sci. 2883. + 28 p.

Browne, D.R. and J.B. Rasmussen. 2009. Shifts in the trophic ecology of Brook Trout resulting from interactions with yellow perch: an intraguild predator-prey interaction. Trans. Am. Fish. Soc. 138: 1109–22.

Browne, D.R. and J.B. Rasmussen. 2013. Rapid response of brook trout to removal of its intraguild prey, yellow perch. Env. Biol. Fish. 96: 915–926.

Chisholm, I., M.E. Hensler, B. Hansen and B. Skaar. 1989. Quantification of the Libby Reservoir levels needed to maintain or enhance reservoir fisheries. Summary Report 1981–1987. Bonneville Power Administration, Division of Fish and Wildlife Portland, OR.

Craig, J.F. 1987. The Biology of Perch and Related Fish. Croom Helm Ltd. Kent, UK.

Craig, J.F. 2000a. Ionic and water balance. pp. 108–115. Percid Fishes: Systematics, Ecology and Exploitation. J. Wiley & Sons.

Craig, J.F. 2000b. Swimming, movements and migrations. pp. 116–123. Percid Fishes: Systematics, Ecology and Exploitation. J. Wiley & Sons, Oxford.

Craig, J.F. 2000c. Fisheries and economic importance. pp 168-190. Percid Fishes: Systematics, Ecology and Exploitation. J. Wiley & Sons, Oxford.

Fisher, T.G. and D.G. Smith. 1994. Glacial lake Agassiz—its northwest maximum extent and outlet in Saskatchewan (Emerson Phase). Quaternary Science Reviews 13: 845–858.

Fisheries and Oceans Canada. 2012. Commercial Fisheries: Freshwater. http://www.dfo- mpo.gc.ca/stats/commercial/land-debarq/freshwater-eaudouce/2012-eng.htm.

Fraser, J.M. 1978. The effect of competition with yellow perch on the survival and growth of planted brook trout, splake and rainbow trout in a small Ontario lake. Trans. Am. Fish. Soc. 107: 505–517.

Gunn, J.M. and W. Keller. 1990. Biological recovery of an acid lake after reductions in industrial emissions of sulfphur. Nature 345: 431–433.

Hanson, M.J., E.E. Prepas and W.C. Mackay. 1989. Size distribution of macroinvertebrates in a freshwater lake. Can. J. Fish. Aquat. Sci. 46: 1510–1519.

Hill, C.L. 2000. Pleistocene lakes along the southwest margin of the Laurentide Ice Sheet. Current Research in the Pleistocene 17: 145–147.

Hill, C.L. and S.H. Valppu. 1997. Geomorphic relationships and paleoenvironmental context of glaciers, fluvial deposits and glacial lake Great Falls, Montana. Current Research in the Pleistocene 14: 159–162.

Keleher, J.J. 1972. Great Slave Lake: effects of exploitation on the salmonid community. J. Fish. Res. Board Can. 29: 741–753.

Kovecses, J., G.D. Sherwood and J.B. Rasmussen. 2005. Impacts of altered benthic invertebrate communities on the feeding ecology of yellow perch (*Perca flavescens*) in metal-contaminated lakes. Can. J. Fish. Aquat. Sci. 62: 153–162.

Lajeunesse, P. and G. St. Onge. 2008. The subglacial origin of the lake Agassiz-Ojibway final outburst flood. Nature Geoscience 1L 184–188.

Liverman, D.G. 2009. Sedimentology and history of a Late Wisconsinan glacial lake, Grande Prairie, Alberta, Canada Issue. Boreas 20: 241–257.

MacDougall, T.M., H.P. Benoit, R. Dermot, O.E. Johannsson, T.B. Johnson, E.S. Millard and M. Munawar. 2001. Lake Erie 1998: Assessment of abundance, biomass and production of the lower trophic levels, diets of juvenile yellow perch, and trends in the fishery. Can. Tech. Rep. Fish. Aquat. Sci. 2376: xvii + 190.

Magnuson, J.J. and D.J. Karlen. 1970. Visual observations of fish beneath the ice in a winterkill lake. J. Fish. Res. Board Can. 27: 1059–1068.

Mandrak, N.E. 1995. Biogeographical patterns of fish species richness in Ontario Lakes in relation to historical and environmental factors. Can. J. Fish. Aquat. Sci. 52: 1462–1474.

Mandrak, N.E. and E.J. Crossman. 1992. Postglacial dispersal of freshwater fishes into Ontario. Can. J. Zool. 70: 2247–2259.

McPhail, J.D. 2007. The Freshwater Fishes of British Columbia. The University of Alberta Press, Edmonton.

McPhail, J.D. and C.C. Lindsey. 1970. Freshwater Fished of Northwestern Canada and Alaska. Fish. Res. Board Can. Bull. 173, Ottawa.

Mitchell, P. and E.E. Prepas. 1990. Atlas of Alberta Lakes. University of Alberta Press, Edmonton. http://alberta-lakes.sunsite.ualberta.ca/.

Moore, W.G. 1942. Field studies on the oxygen requirements of certain freshwater fishes. Ecology 23: 319–329.

Nelson, J.S. and M.J. Paetz. 1992. The Fishes of Alberta. University of Alberta Press, Edmonton.

Pielou, E.C. 1991. After the Ice Age: The Return of Life to Glaciated North America, University of Chicago Press, Chicago.

Post, J.R. and D.J. McQueen. 1994. Variability in first year growth of yellow perch (*Perca flavescens*); predictions from a simple model, observations and an experiment. Can. J. Fish. Aquat. Sci. 51: 2501–2512.

Prepas, E.E. and D.O. Trew. 1983. Evaluation of the phosphorus chlorophyll relationship for lakes off the Precambrian shield in western Canada. Can. J. Fish. Aquat. Sci. 40: 27–35.

Rasmussen, J.B., J.M. Gunn, G.D. Sherwood, A. Iles, A. Gagnon, P.G.C. Campbell and A. Hontela. 2008. Direct and indirect (foodweb mediated) effects of metal exposure on the growth of yellow perch (*Perca flavescens*): implications for ecological risk assessment. Hum. Ecol. Risk Assess. 14: 317–350.

Rasmussen, J.B. 1993. Patterns in the size structure of littoral zone macroinvertebrates. Can. J. Fish. Aquat. Sci. 50: 2192–2207.

Rawson, D.S. 1947. Lake Athabaska. Fish. Res. Board Can. Bull. 72: 69–85.

Rawson, D.S. 1951. Studies of the fishes of Great Slave Lake. J. Fish. Res. Board Can. 8: 207–240.

Rawson, D.S. 1960a. A limnological comparison of twelve large lakes in northern Saskatchewan. Limnol. Oceanogr. 5: 195–211.

Rawson, D.S. 1960b. Five Lakes on the Churchill River near Stanley, Saskatchewan. Fisheries Branch, Department of Natural Resources, Province of Saskatchewan. Fisheries Report No. 5.

Rawson, D.S. and J.E. Moore. 1944. The saline lakes of Saskatchewan. Can. J. Res. 22D (6): 141–201.

Rempel, L.L. and D.G. Smith. 1998. Postglacial fish dispersal from the Mississippi refuge to the Mackenzie River Basin. Can. J. Fish. Aquat. Sci. 55: 893–899.

Robb, T. and M.V. Abrahams. 2002. The influence of hypoxia on risk of predation and habitat choice by the fathead minnow, *Pimephales promelas*. Behav. Ecol. Sociobiol. 52: 25–30.

Runciman, J.B. and B.R. Leaf. 2009. A review of yellow perch (*Perca flavescens*), smallmouth bass (*Micropterus dolomieu*), largemouth bass (*Micropterus salmoides*), pumpkinseed (*Lepomis gibbosus*), Walleye (*Sander vitreus*) and northern pike (*Esox lucius*) distributions in British Columbia. Can. Man. Rep. Fish. Aquat. Sci. 2882: xvi + 123.

Ryder, R.A. 1972. The limnology and fishes of oligotrohic glacial lakes in North America (about 1800 A.D.). J. Fish. Res. Board Can. 29: 617–628.

Scott, D.C. 1956. Record of perch from Great Slave Lake. Can. Field Nat. 70: 90.

Scott, W.B. and E.J. Crossman. 1973. Freshwater fishes of Canada. Fish. Res. Board Can. Bull. 184, Ottawa.

Sepulveda-Villet, O.J. and C.A. Stepien. 2012. Waterscape genetics of the yellow perch (*Perca flavescens*): patterns across large connected ecosystems and isolated relict populations. Molecular Ecology 21: 5795–5826.

Sherwood, G.D., J. Kovecses, A. Hontela and J.B. Rasmussen. 2002. Simplified food webs lead to energetic bottlenecks in polluted lakes. Can. J. Fish. Aquat. Sci. 59: 1–5.

Sherwood, G.D., I. Pazzia, A. Moeser, A. Hontela and J.B. Rasmussen. 2002. Shifting gears: enzymatic evidence for the energetic advantage of switching diet in wild-living fish. Can. J. Fish. Aquat. Sci. 59: 229–241.

Smith, D.G. 1994. Glacial lake McConnell: paleogeography, age, duration, and associated river deltas, Mackenzie river basin, western Canada. Quaternary Science Reviews 13: 9–10.

Thorpe, J.E. 1977a. Morphology, physiology, behavior, and ecology of *Perca fluviatilis* L. and *P. flavescens* Mitchill. J. Fish. Res. Bd. Can. 34: 1504–1514.

Thorpe, J.E. 1977b. Synopsis of biological data on the perch, *Perca fluviatilis* Linnaeus, 1785 and *Perca flavescens* Mitchill, 1814. FAO Fisheries Synopses no. 113, vii + 138 pp.

Walters, C. and J.F. Kitchell. 2001. Cultivation/depensation effects on juvenile survival and recruitment: implications for the theory of fishing. Can. J. Fish. Aquat. Sci. 58: 39–50.

Zimmerman, M.P. 1999. Food habits of smallmouth bass, walleye and northern pike minnow in the Lower Columbia River Basin during outmigration of juvenile anadromous salmonids. Trans. Am. Fish. Soc.128: 1036–1054.

5

Yellow Perch (*Perca flavescens*) in the St. Lawrence River (Québec, Canada)

Population Dynamics and Management in a River with Contrasting Pressures

Yves Mailhot,[1] Pierre Dumont,[2] Yves Paradis,[3,a] Philippe Brodeur,[4] Nathalie Vachon,[5] Marc Mingelbier,[3,b] Frédéric Lecomte[3,c] and Pierre Magnan[6,*]

ABSTRACT

The St. Lawrence River flows for more than 350 km in the province of Québec (Canada). Its freshwater portion exhibits high habitat heterogeneity, alternating between large shallow fluvial lakes, archipelagos, and long narrow corridors. Yellow perch was or is still a dominant species of the fish community depending on the sector. Tagging, life history traits, and genetic studies indicate that there are several independent populations in this system. The results of large-scale sampling surveys between 1988 and 2012 have allowed the comparison of yellow perch population dynamics in four sections of the St. Lawrence River; it was found that populations have remained abundant and balanced in the upstream sections

Authors' affiliations given at the end of the chapter.

of the river (lakes Saint-François and Saint-Louis) while they have collapsed downstream (Lake Saint-Pierre and the Bécancour–Batiscan area). In less than 15 years, yellow perch became a marginal species in the fish communities of the downstream sectors, and both the commercial and sport fisheries were closed. The effects of fish exploitation and habitat alterations as well as the fisheries management measures implemented during the past three decades are discussed. The synthesis presented in this chapter provides knowledge and tools that should be helpful when elaborating future management decisions.

Keywords: Yellow perch, St. Lawrence River, Fluvial lakes, Population dynamics, Limiting factors, Habitat alteration, Sport fishery, Commercial fishery, Fisheries Management

5.1 Introduction

"Suzanne takes you down to her place near the river...
Then she gets you on her wavelength
And she lets the river answer"

Suzanne by Leonard Cohen[1]

Leonard Cohen, our famous poet, singer, and song writer, was born and lived near the St. Lawrence River in Montréal. It is easy for artists like him—or for Suzanne from his well-known song—to be inspired. They get close to the river, simply ask, and let the St. Lawrence River answer.... For scientists, answers must be drawn from data, from facts. Understanding what might have happened under the water is not an easy task, especially when observed changes are complex and have evolved over decades.

This chapter benefits from the experience acquired by three generations of biologists who have worked to document and understand yellow perch biology and population dynamics, and to preserve its status along the Québec portion of the St. Lawrence River.

Yellow perch is usually thought of as an abundant, dominant, and resilient species of the riverine fish communities. From the 1940s to the present, the yellow perch status was good in the upstream part of the St. Lawrence River (Lake Saint-François and Lake Saint-Louis), while major population failures were observed in the 1990s in comparable habitats in the downstream part (Lake Saint-Pierre and Bécancour–Batiscan area) (Fig. 1). In less than 15 years, yellow perch became a minor and marginal species, and both commercial and sport fisheries were closed. In this chapter, we present yellow perch life history dynamics in different areas of the St. Lawrence River. We use available scientific data and examine the different pressures exerted on the perch populations and their habitats to reveal the causes behind the strong differences that we see between the up- and downstream sectors. Finally, we offer future perspectives for these yellow perch populations.

[1] "Excerpted from *Stranger Music: Selected Poems and Songs* by Leonard Cohen. Copyright © 1993 Leonard Cohen. Reprinted by permission of McClelland & Stewart, a division of Penguin Random House Canada Limited, a Penguin Random House Company".

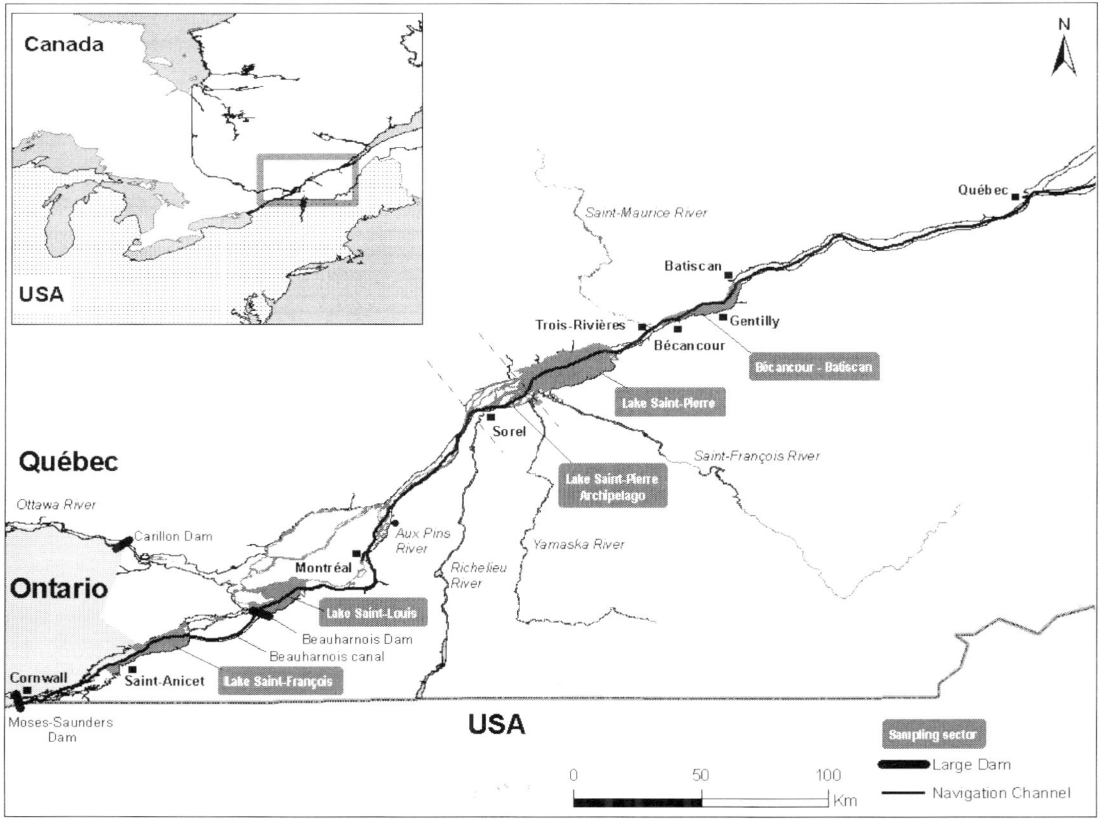

Fig. 1. Québec portion of the St. Lawrence River and locations of the various sectors sampled by the fish monitoring network (*Réseau de Suivi Ichtyologique* [RSI]).

5.2 The St. Lawrence River: Natural Characteristics and Human Pressures

5.2.1 A Complex Landscape Leading to Diverse Fish Habitats

The St. Lawrence River is the outflow of the Great Lakes to the Atlantic Ocean (Fig. 1). It ranks among the largest river systems of the world, with a mean annual outflow of 12,600 m^3/s at Québec City and a length of 1,140 km; its watershed covers 720,000 km^2 (Frenette et al. 1989). Construction of the St. Lawrence Seaway, an important waterway within the river, has had a major effect on the economy of the whole St. Lawrence—Great Lakes area: an average of 4,000 ships traverse the seaway every year. While the sheer size of the system may seem sufficient to buffer the ecological impact of the 30 million inhabitants of its watershed, numerous studies have shown that anthropogenic alterations have strongly affected the ecology and the physical characteristics of the river (Morin and Leclerc 1998; Vincent and Dodson 1999; Morin et al. 2000).

Because of the topography and complex hydrological network, the river can be described as a mosaic of fluvial lakes separated by islands/archipelagos and narrow sections (La Violette 2004). The elevation difference between Lake Ontario (~75 m above sea level) and Montréal Harbour (~8 m above sea level) originally resulted in several sections of rapids alternating with still waters (fluvial lakes). Most of the rapids were eliminated or greatly modified by large dams built during the 1950s. Habitat heterogeneity is modified by the input of several large tributaries (e.g., the Ottawa River has a mean annual outflow of 1,800 m^3/s) that progressively increases water level variations in a downstream direction (Morin and Bouchard 2000). Moreover, these tributaries generate contrasting water masses in the river given their various origins and physicochemical characteristics: they flow side by side, with little lateral mixing over hundreds of kilometers (e.g., green and transparent water from Lake Ontario, brown and turbid water from several agricultural watersheds; Frenette et al. 1989). There is also a marked tidal influence along more than 200 km in the freshwater estuary.

The biodiversity of the St. Lawrence River is the result of the large geographic extent of the river, its position between the Great Lakes and the Atlantic Ocean, its connections with southern river systems such as the Mississippi River via Lake Michigan and the Hudson River via Lake Champlain, and the introduction, over more than two centuries, of up to 160 non-native aquatic species in the Great Lakes—St. Lawrence system (de Lafontaine and Costan 2002). Variations in local fish communities reflect the overall habitat heterogeneity along the river.

5.2.2 Description of the Main River Sections

Lake Saint-François

Lake Saint-François (272 km^2) is located downstream from Lake Ontario, at the intersection of the provinces of Ontario and Québec (Canada) and the state of New York (USA) (Fig. 1). There are two medium-sized urban centres located at each end of the lake along with several small villages in between, intensive agriculture practices, and few tributaries (~2% of total discharge). The main part of its water is the clear and green-coloured water from the Great Lakes (water transparency reaches up to 7 m). Its maximum depth is 26 m, average depth 5.7 m, and average water flow 7,500 m^3/s. With the construction of hydroelectric dams upstream and downstream of the lake in the 1950s, the water level increased by 0.4 m and is strongly regulated (annual variations <0.15 m compared to ~0.50 m before 1952), the littoral zone has shrunk, and shores have been artificialized (Morin and Leclerc 1998). The construction of the St. Lawrence Seaway (opened in 1959) at the downstream extremity of the lake led to the diversion of most of the discharge to the 25 km long man-made Beauharnois Canal. The pristine running water section, which historically included four cascading steps between Lake Saint-François and Lake Saint-Louis, used by many migratory fish species, was replaced by a single vertical drop of 25 m at the Beauharnois Dam (Fig. 2a). Except for two ladders exclusively used by American eel (*Anguilla rostrata*), there is no device allowing upstream movement of fish between the two lakes. Even

Fig. 2. Representative images of the St. Lawrence River: (a) Beauharnois hydro power dam separating Lake Saint-François (upstream) and Lake Saint-Louis (downstream); (b) commercial fishermen targeting yellow perch using hoop-nets in Lake Saint-Pierre; (c) sport fishermen near the navigation channel of the St. Lawrence River in Lake Saint-Pierre; (d) yellow perch ice sport-fishing in Lake Saint-Pierre; (e) confluence of the Saint-François and Yamaska river outlets draining farmland sediments and nutrients to the south shore of Lake Saint-Pierre; (f) flooded farmlands surrounding Lake Saint-Pierre during the spring freshet.

though the current fish community in Lake Saint-François is relatively diverse (58 species), some migratory species have disappeared (for example, the American shad, *Alosa sapidissima*) and the biomass of several others has decreased (for example, the lake sturgeon, *Acipenser fulvescens*, and the sauger, *Sander canadensis*) over the last few decades (La Violette et al. 2003; Vachon et al. 2013), likely because of habitat fragmentation, water level regulation, and increasing water transparency since the zebra mussel (*Dreissena polymorpha*) and quagga mussel (*Dreissena bugensis*) were introduced into the Great Lakes (Vachon 2002; Bechara et al. 2003).

Lake Saint-Louis

Lake Saint-Louis (208 km^2) is a natural hydrological crossroad between the St. Lawrence River corridor and its main tributary, the Ottawa River. The high discharge of green waters from the Great Lakes mainly flows into the lake through the Beauharnois Dam (Figs. 1 and 2a) and has moderate seasonal variation. On the north shore, the Ottawa River carries brown, nutrient-rich water (Morin and Bouchard 2000). The Ottawa River strongly contributes to water level increases in Lake Saint-Louis (~1.4 m annual variation; Mingelbier and Morin 2003) and in the downstream portion of the St. Lawrence River during spring (up to 50% of the water discharge). The maximum depth of the lake is 27 m with an average of 3.4 m when the channel is excluded, and average water flow is 8,400 m^3/s. While the Lake Saint-Louis shores are almost completely artificialized, the lake hosts the most diverse fish community of the St. Lawrence River (80 species; La Violette et al. 2003). Because this lake is close to Montréal, the various impacts associated with human activities are magnified, with increased sport fishing, boating, industrial and domestic pollution, wetland loss, shore alteration, and so forth. The watershed along the south shore is mostly agricultural and industrial while the watershed to the north is now a typical North American suburb (densities of >500 inhabitants per km^2). Nevertheless, the degradation is not as severe as the transformation of the drainage basin might suggest: the lake still supports high fish diversity, several key fish spawning grounds, and large populations of waterfowl, which has led to the creation of bird sanctuaries.

Lake Saint-Pierre and its Archipelago

Lake Saint-Pierre (318 km^2) and its archipelago (63 km^2 excluding the islands) constitute the next sector of the river. Lake Saint-Pierre is the last and largest fluvial lake of the St. Lawrence River (Fig. 1). It appears to be a huge marshland due to its shallowness (mean depth 2.7 m, maximum depth 13.7 m; Carignan and Lorrain 2000), the large macrophyte beds, and its temporary floodplain during the spring freshet, which can increase the marshy surface by 140 km^2. The archipelago is located at the upstream end of the lake and is composed of 103 islands and numerous channels (total distance of 155 km). A large, deep channel (width >230 m, depth >11.3 m), which was progressively dredged between 1851 and 1999 for navigation, now divides the north and south parts of the lake and is maintained through periodic dredging. Several important tributaries increase water input (average discharge at the mouth of Lake Saint-Pierre is 10,500 m^3/s) and water level variations (annual variation is ~2.5 m;

Mingelbier and Morin 2003). From an aerial view, one can distinguish the navigation channel, which drains almost all the green water from the Great Lakes, flowing side-by-side with sharply contrasting water originating from the different tributaries (Frenette et al. 2006). The overall physico-chemical gradients likely contribute to the diversified fish community of about 60 species in the archipelago and 47 in Lake Saint-Pierre (La Violette et al. 2003). The vast floodplain once comprised thousands of hectares of spawning habitat for several fish species, including yellow perch and northern pike (*Esox lucius*). Because of the uniqueness of this area, Lake Saint-Pierre together with the archipelago was granted the status of wetland of international importance by the Ramsar Convention in 1998 and UNESCO Biosphere Reserve in 2000. Despite indications of its ecological importance, a highway was built in the floodplain along the northern shore in the late 1960s. Intensive agriculture is still maintained in the floodplain and watershed to grow corn and soy; this strongly impacts the ecology of the lake and leads to the loss of valuable natural wetlands.

The Bécancour–Batiscan Sector

The Bécancour–Batiscan Sector (90 km^2) is located in the upstream portion of the freshwater estuary (Fig. 1). It is a narrow sector with a tidal movement of about 0.8 m at Bécancour (compared to ~6 m at Québec City) and an ebb/flow current inversion starting nearby, at Batiscan. The north and south shores habitats are quite different. The northern banks of the river were largely artificialized during the 1950s, when the navigation channel was built, while large vegetated shoals occur along the southern shore, especially in the Gentilly area, which is the most productive of the sector. A nuclear power plant (Gentilly-2) had a significant impact on local water temperatures between its construction in 1983 and its closing in 2013. The yellow perch habitats in Bécancour–Batiscan are restricted to a narrow strip along the northern shore. The southern shore appears to be the most productive and still supports a diverse fish community (40 species; La Violette et al. 2003).

5.2.3 Main Anthropogenic Pressures on the St. Lawrence River

The degree of anthropogenic pressures on the various areas of the St. Lawrence River varied tremendously from the 1940s to the 2000s. Table 1 summarizes these pressures and their impacts on the fish community, and more specifically, on yellow perch stocks. Since the beginning of the industrial era, human activities have strongly affected native fish and their habitats. These activities include dredging the navigation channel, dam construction in the river and in many of its tributaries, water level regulation (Mingelbier et al. 2008a), development resulting in the loss of large wetland areas, intensive agricultural practices, the introduction of exotic fish species (de Lafontaine and Costan 2002), the increasing double-crested cormorant (*Phalacrocorax auritus*) population in Lake Saint-Pierre and its archipelago due to a man-made habitat modification, and the release of untreated wastewater with pharmaceuticals, which has led to reduced reproductive capacities due to oestrogenic molecules and anxiolytics (Aravindakshan et al. 2004; Lajeunesse et al. 2011). In addition to human pressures, fish and habitats in the river are exposed to the effects of climate warming (Hudon

Table 1. Major anthropogenic pressures affecting yellow perch populations in each sector of the St. Lawrence River.

St. Lawrence River Sector		Anthropogenic pressure					
		Agriculture[A]	Shore artificialization[B]	Water level regulation[C]	Urbanization[D]	Commercial Fishery[E]	Sport Fishery[F]
Lake Saint-François		1	89%	5	1	0	4
Lake Saint-Louis	North	1	86%	3	3	0	5
Lake Saint-Louis	South	1	45%	3	4	0	4
Archipelago		3	34%	1	3	1	3
Lake Saint-Pierre	North	3	1%	1	1	3	2
Lake Saint-Pierre	South	5	1%	1	1	5	3
Bécancour–Batiscan		5	46%	Tidal influence	2	3	1

A- Total phosphorus load from agricultural sources combined with the surface areas cultivated in the floodplain on a scale of 1 (lowest) to 5 (highest)
B- Artificial shore (%)
C- On a scale of 1 (lowest) to 5 (highest) (adapted from Mingelbier and Morin 2003)
D- Joint index of demographic and industrial pressures on a scale of 1 (lowest) to 5 (highest)
E- Commercial fishing pressure in the late 1990s – early 2000s on a scale of 0 (absent), 1 (lowest) to 5 (highest)
F- Sport fishing pressure in the late 1990s – early 2000s on a scale of 1 (lowest) to 5 (highest)

(*See APPENDIX for a detailed definition of these parameters*) et al. 2010), a spring flood that may occur three weeks earlier than normal (Boyer et al. 2010), an anticipated reduction in water discharge by 20–40% in the near future (Mortsch and Quinn 1996), and more frequent climate extremes. For example, extreme water temperatures resulted in massive fish mortality in 2001 (Ouellet et al. 2010).

5.3 Historical Importance of Yellow Perch, Chronology of the Major Studies, and Modifications of Fishery Management Practices

In ancient documents on fishes of the St. Lawrence River, yellow perch is cited much less frequently than American eel, lake sturgeon, or Atlantic salmon (*Salmo salar*). This is likely related to the higher nutritional value of these latter species for European newcomers. Because of their small bones, yellow perch has rarely been detected in archaeological excavations done in the St. Lawrence lowlands (M. Courtemanche, Ostéothèque de l'Université de Montréal, personal communication). However, recent archaeological digs exploring a 500-year-old St. Lawrence Iroquoian settlement on Lake Saint-François near Saint-Anicet (Fig. 1) revealed that yellow perch was an important food item (Courtemanche 2012). Montpetit (1897) noted that yellow perch was considered as a cheap and tasty food during its spawning period. Scott and Crossman (1973) wrote that yellow perch has long been a species of prime importance for commercial and sport fisheries in Canada. Because it was an abundant species in the St. Lawrence River, with the highest numbers found in fluvial lakes, lucrative commercial fisheries developed overtime. As sport fishermen became increasingly interested in this species, strong conflicts arose between sport and commercial fishermen during the second half of the last century (Cuerrier 1962). This situation was especially tense in Lake Saint-Pierre and its archipelago.

The first scientific studies focusing on St. Lawrence River fish communities, and more specifically on yellow perch, were published by the Institut de Biologie de l'Université de Montréal in the early 1940s. These studies described the fish community in lakes Saint-Louis and Saint-Pierre and provided baseline information for fisheries management. The first publications described the growth pattern of yellow perch in Lake Saint-Louis and characterized the commercial fishery in Lake Saint-Pierre (Fry et al. 1941; Cuerrier et al. 1946). At this time, even though yellow perch was already one of the main species targeted for commercial fishery, there were no restrictions or control. After World War II, the capacity of the commercial fishery greatly increased with the mechanization of the industry (use of outboard motors, availability of new synthetic materials for fishing gears, faster and more efficient shipping) (Lacasse 1987). With the increasing demand for fish and the higher prices paid to fishermen, commercial landings gradually increased and—according to the sport fishermen—impacted stocks and reduced the fishing potential.

The first action towards more efficient fishery management practices occurred 20 years later. In 1963, the provincial government of Québec created the *Ministère du Tourisme, de la Chasse et de la Pêche* (MTCP) to manage fish and wildlife. Since then, biologists have defined specific regulations for different St. Lawrence River sectors (for example, fishing seasons, gears, number of licenses, quotas). In the western part of the St. Lawrence River (lakes Saint-François and Saint-Louis), hundreds of minor commercial licenses, mostly for personal use, were eliminated because they were in competition with the expanding sport fishing activities (Pluritec 1982). Such reductions were not applied in Lake Saint-Pierre, where the proportion of professional commercial fishermen was much higher than elsewhere in the St. Lawrence River (Bourbeau et al. 1992). The number of commercial fishermen was also high in the downstream Bécancour–Batiscan sector (Leclerc 1956).

These changes in fishery management practices led to some uncertainty with respect to the future of the profession and the loss of income, resulting in mistrust among the remaining commercial fishermen in Lake Saint-Pierre and its archipelago. In addition, there was already open conflict between commercial fishermen (Fig. 2b) and many local sport-fishermen's clubs (Figs. 2c and 2d) (Cuerrier 1962). At the same time, Magnin (1966) published the first exhaustive description of the freshwater commercial fishery in the province of Québec. This assessment showed that commercial exploitation of yellow perch in the St. Lawrence River was an important activity: its landings ranked fifth in terms of biomass among other commercial fish species caught in the St. Lawrence River. The importance of the commercial exploitation in Lake Saint-Pierre was revealed, with 92% of all yellow perch landings coming from this lake alone.

During the 1960s and the 1970s, several studies on yellow perch population dynamics were conducted to improve the species' management in the St. Lawrence River. The first major study focused on Île Perrot cove in the north sector of Lake Saint-Louis (Fortin 1970; Fortin and Magnin 1972a,b,c). Large-scale surveys were also performed in lakes Saint-François, Saint-Louis, and Saint-Pierre (Massé and Mongeau 1974; Mongeau and Massé 1976; Mongeau 1979). Due to growing complaints from anglers, the spring commercial landings of yellow perch in Lake Saint-Pierre were examined in 1978, 1979, and 1980 to characterize the biological parameters of the south shore stock and to estimate the level of exploitation (Leclerc 1987).

In 1983, new legislation resulted in profound modifications to fishery management practices. First, the *Ministère du Loisir, de la Chasse et de la Pêche* (MLCP, formerly MTCP) became responsible for producing annual commercial and sport fisheries management plans while being exclusively responsible for the sportfishing industry. Second, the *Ministère de l'Agriculture, des Pêcheries et de l'Alimentation* (MAPAQ) became responsible for the commercial industry and for managing commercial licences according to MLCP's management plan. Since that time, the main objective of the management plan has been to optimize the socio-economic benefits resulting from exploitation of the resource. Management guidelines state that choices related to resource allocation are prioritized according to criteria defined by the provincial government: (1) protection of the reproductive stock, (2) First Nations harvesting, (3) sport fishery, and (4) commercial fishery. These changes exacerbated conflicts between users, mostly in Lake Saint-Pierre, which led to the development of strong sport and commercial political lobbies to protect divergent interests. Since then, exploitation of the species by First Nations in the St. Lawrence River has been marginal.

These new fishery management practices were first applied in the Lake Saint-Pierre area in 1987. The two departments (MLCP and MAPAQ) produced a new scientific advisory report on the status of the yellow perch, confirming the high exploitation and fragility of the Lake Saint-Pierre north and south shores stocks. The majority of fish caught by commercial fishermen on the spawning grounds belonged to a single year-class, and the total mortality rate (80%) was very high (Mailhot et al. 1987). However, because recruitment variability could not be considered in the mortality estimates, it is possible that the mortality rate was biased. To circumvent this limitation, a five-year monitoring program was implemented to estimate total and fishing mortality rates, model the population dynamics, and ultimately produce management scenarios (Guénette et al. 1994). At the same time, a first study was performed in the Bécancour–Batiscan sector to characterize the commercial fishery (G.D.G. Environnement Ltée 1986). Following the complaints of sport fishermen, who suspected that there had been a significant decrease in the perch population in Lake Saint-Louis, Dumont (1996) initiated a multi-mesh sampling protocol in 1988, 1989, and 1990 that was similar to the one used by the Ontario Ministry of Natural Resources (OMNR) in the upper St. Lawrence River. The objective of this study was to evaluate the influence of various factors (habitat, fishing, and chemical contaminants) on yellow perch population dynamics.

In the 1990s, a standardized survey program was implemented in the Québec portion of the St. Lawrence River. This program, known as the *Réseau de suivi ichtyologique* (RSI), allows unbiased temporal and spatial comparisons of fish communities and population dynamics of different species (La Violette et al. 2003). The standardized RSI database covering the years from 1995 through 2012 is used in this chapter to assess yellow perch population dynamics (see Section 5.5).

In the early 2000s, following a major decline of the Lake Saint-Pierre yellow perch, an independent expert was mandated by the Québec government to produce a scientific advisory report to review the available data and make recommendations for the sustainability of the perch stocks (Magnan 2002). Because yellow perch was still declining, two additional scientific advisory reports were produced (Magnan et al. 2004, 2008). Even though strong actions were implemented following the publication of these two reports, the population continued to decline until it collapsed completely. A five-year moratorium was imposed on both commercial and sport fishing in Lake Saint-Pierre in 2012 and in the downstream Bécancour–Batiscan sector in 2013, as recommended by Magnan et al. (2014).

5.4 Yellow Perch Populations in the Québec Portion of the St. Lawrence River

5.4.1 Distribution and Habitat

Yellow perch occur all along the Québec portion of the St. Lawrence River, from the Ontario–Québec border in Lake Saint-François to the brackish waters about 25 km downstream of Québec City (Fig. 1). While the species uses a wide range of habitats in the flood plain and the main channel (Bertrand et al. 2011), it is mainly associated with shallow (<3 m) vegetated areas of the littoral zone with little current (<40 cm/s) (Leclerc 1984). In this area, except in Lake Saint-Pierre where the species collapsed in the last decade, it is among the most abundant members of the fish community (Massé and Mongeau 1974; Mongeau and Massé 1976; Mongeau 1979; La Violette et al.

2003). Yellow perch is scarcer in the lower reaches of the Upper Estuary (downstream of Gentilly) (La Violette et al. 2003), where pronounced semidiurnal tidal amplitudes, high water velocity, and sparse vegetation offer unsuitable habitat for this species.

5.4.2 Stock Composition and Movements

A system-wide analysis of population genetics at the landscape scale (Leclerc et al. 2008) suggested that yellow perch are genetically structured into at least four distinct populations over a 400 km stretch: (1) the Lake Saint-François population, (2) the northern Lake Saint-Louis population, (3) a population from southern Lake Saint-Louis and the fluvial corridor downstream of Montréal, and (4) a population including all individuals downstream from Lake Saint-Pierre to the freshwater limit. The large dams between Lake Saint-François and Lake Saint-Louis and two water masses—the brown waters from the Ottawa River system on the north shore of Lake Saint-Louis and the green waters from the Great Lakes on the south shore, which are separated by the St. Lawrence Seaway—present man-made and natural boundaries to the first three populations. In Lake Saint-François, a comparison of population dynamics and life history traits between the north and south shores suggests the presence of only one stock, at least on the Québec side of the lake (Vachon and Dumont 2007). In Lake Saint-Louis, similar comparisons (Dumont 1996; Vachon and Dumont 2007) and past intensive tagging (Fortin and Magnin 1972a; Dumont 1996) provided evidence for sedentary and independent demographic populations between the north and south shores. More than 30,000 yellow perch were tagged from Lake Saint-Louis and the great majority of the 2,449 recaptures were reported within 10 km of their tagging sites. Movements between the north and the south shores were also limited: only 2% and 3.7% of yellow perch were recaptured on the other shore over a three-year period (Dumont 1996).

The situation in Lake Saint-Pierre is quite different. Based on a mark–recapture study of 20,000 tagged yellow perch, Leclerc (1987) reported extensive movements between the north and south shores of the lake, involving more than 9% of the recaptures. De Lafontaine et al. (2002) also provided evidence for seasonal movements of yellow perch between Lake Saint-Pierre and Québec City. These movements—downstream during the spring and upstream during the fall—were not necessarily associated with reproductive behaviour, but possibly driven by seasonal variations in habitat accessibility in the upper St. Lawrence Estuary. These observations support the genetic homogeneity and the absence of a significant population structure among all samples collected between Lake Saint-Pierre and Québec City (Leclerc et al. 2008). However, different studies revealed significant differences in life history traits and population dynamics parameters among groups from the archipelago, the north and the south shores of Lake Saint-Pierre, and the Bécancour–Batiscan sector (Guénette et al. 1994; Dumont 1996; G.V.L. Environnement Inc. 2001; Magnan et al. 2004, 2008; Vachon and Dumont 2007). For monitoring and management purposes, these four groups have been considered as distinct stocks. This finer stock definition compared to previous genetic evidence (Leclerc et al. 2008) was recently supported by two studies: (1) a comparison of the $\delta^{13}C$ variability in yellow perch with that of primary consumers revealed that the adult yellow perch feeding range was restricted to about

2 km along the shore–channel axis (Bertrand et al. 2011), and (2) using more recent molecular techniques, Leung et al. (2011) observed a finer genetic structure involving multiple sympatric populations of yellow perch coexisting in Lake Saint-Pierre.

5.4.3 Ecological Importance

As observed in the Great Lakes (Thorpe 1977), yellow perch is a key species in the St. Lawrence River food chain. It is prey for many larger species (Reyjol et al. 2010), and it feeds on a wide variety of aquatic organisms, including Cladocera, amphipods, isopods, chironomids, Ephemeroptera, Trichoptera, mollusks, and fishes in Lake Saint-Louis (Fortin and Magnin 1972b) and Lake Saint-Pierre (Harnois et al. 1992). In Lake Saint-Pierre, a high degree of food overlap was reported among fish species feeding on benthic prey when this resource was more abundant; in contrast, yellow perch diet did not overlap with that of other fish species when benthic prey abundance was low (Harnois et al. 1992).

5.4.4 Growth

The yellow perch growth season ranges from May to September (Fortin and Magnin 1972c). As generally observed, females grow faster than males and achieve a larger size (Fortin and Magnin 1972c; Guénette et al. 1994; Dumont 1996). For example, during the 1980s, back-calculated total lengths of males at age 2 and 4 were respectively 130 and 171 mm on the south shore of Lake Saint-Pierre and 119 and 156 mm on the north shore; at the same ages, females averaged 140 and 196 mm on the south shore and 130 and 184 mm on the north shore (Guénette et al. 1994). In Lake Saint-Louis, males at age 2 and 4 averaged respectively 118 and 172 mm on the south shore and 121 and 171 mm on the north shore during the same period while females of the same age were respectively 126 and 186 on the south shore and 128 and 195 mm on the north shore (Dumont 1996). In the 1980s and 1990s, the faster growth rate observed in Lake Saint-Pierre was correlated with a higher exploitation rate by commercial and recreational fisheries (Guénette et al. 1994). More recently, Glémet and Rodríguez (2007) showed that short-term growth of yellow perch appeared to be influenced by a combination of non-linear effects of temperature and water level in Lake Saint-Pierre. According to their observations, both high temperatures and low water levels have negative effects on yellow perch growth.

Overall, yellow perch growth is slower in the Québec part of the St. Lawrence River than in the Great Lakes. In lakes Erie (Knight et al. 1991) and Michigan (Brazo et al. 1975), average total length exceeds 200 mm for males and 230 mm for females at age 4. However, the condition factor of the St. Lawrence River populations is generally high (Guénette et al. 1994; Dumont 1996) and among the highest in North America (Willis and Guy 1991).

5.4.5 Reproduction

In Lake Saint-Pierre, spawning extends over the last two weeks of April in the flood plain, where water warms rapidly (MLCP 1984), but it is later in Lake Saint-Louis, where it occurs in shallow protected bays during the first two weeks of May (Fortin and Magnin 1972a; Dumont 1996). In the 1980s, fecundity was highest in Lake Saint-Pierre. The number of eggs for a 200 mm female was evaluated at 12,350 compared to 9,060 in Lake Saint-François, 8,840 in the Aux Pins River (a small tributary on the south shore of the St. Lawrence River, south of Montréal), and respectively 7,360 and 9,650 for the south and north shores of Lake Saint-Louis (Dumont 1996). At the same size, fecundity exceeded 15,000 eggs in lakes Ontario (Clady 1976), Erie (Hartman et al. 1980), and Michigan (Brazo et al. 1975).

5.4.6 Year-Class Strength

In the 1980s, yellow perch year-class strength (YCS) was evaluated from the age composition of annual experimental fishing on the north and south shores of Lake Saint-Louis (Dumont 1996), Lake Saint-Pierre (Guénette et al. 1994), and in the Aux Pins River sector (Dumont 1996). During this period, YCS varied by a factor of 10 in Lake Saint-Louis, four in Lake Saint-Pierre, and seven in the Aux Pins River sector. In the Aux Pins River, the 1980 year-class remained dominant in the population for a period of 10 years, and no significant correlation was found between YCS and available climatic or hydrological variables. On both shores of Lake Saint-Louis, YCS showed a regular alternating pattern between strong and weak year-classes. Two multivariate models were developed to explain YCS variability: (1) on the north shore ($R^2 = 0.99$; $p < 0.001$), YCS was negatively correlated with the strength of the previous year-class and the maximal wind strength in May and positively correlated with temperatures in May and June, while (2) on the south shore ($R^2 = 0.68$; $p < 0.05$), the negative effect of the previous year-class and the positive effect of temperature in May and June were the only factors selected by the model (Dumont 1996). In this sector, spawning sites on the south shore are well protected from the dominant winds by the Îles-de-la-Paix archipelago. In Lake Saint-Pierre, YCS was positively correlated with thermal conditions in June and July ($r = 0.74$ and $r = 0.81$; $p < 0.05$) (Guénette et al. 1994). Comparison of trends in these YCS variations with those observed in three upstream reaches of the St. Lawrence system—lakes Saint-François (Hendrick 1991) and St. Lawrence (Gordon 1991) as well as the central corridor (Hendrick 1993)—showed that they were only partially synchronized (Dumont 1996). These comparisons suggest that yellow perch recruitment in the St. Lawrence River is dependent on the temperature regime during its early life, as already reported for other percids (Koonce et al. 1977). However, the influence of temperature is modulated by other (intrinsic) factors, such as the interactions between two successive year-classes.

5.4.7 Total Annual Mortality

High YCS variability may prevent accurate estimations of the annual mortality rate (A) using the age composition of experimental or commercial catch (Ricker 1975). However, this bias may be at least partly compensated by the wide range of ages observed in lakes Saint-François and Saint-Louis (Table 2). In Lake Saint-François, A was estimated at 56% (ages 4 to 10) in the Ontarian sector of the lake in the 1980s (Cholmondeley 1989) while it was 66% and 73% (age 3 to 9) during late 1990s and

Table 2. Summary of life history characteristics of St. Lawrence River yellow perch populations from 1988 to 2012. ND: not determined.

Sector	Year	Sample size		Population				Age 1+		Females	
		No. of stations	No. of yellow perch	CPUE	BPUE	Mean age	Max age	CPUE	Mean total length	BPUE	Age at maturity
				(no./station/night)	(g/station/night)	(years)	(years)	(no./station/night)	(mm)	(g/station/night)	(years)
Lake Saint-François	1996	70	1,558	22.3	1,247	3.0	11	5.1	108	730	3.3
	2004	66	2,502	37.9	1,989	3.0	9	8.5	107	539	2.5
	2009	61	2,753	45.1	3,099	3.1	13	8.4	110	2,373	3.4
Lake Saint-Louis											
North shore	1988	33	1,088	33.0	3,548	3.6	15	8.5	143	2,650	ND
	1989	33	936	28.4	2,839	3.5	15	0.5	138	1,875	3.0
	1990	31	532	17.2	1,856	3.5	14	0.7	136	1,386	ND
	1997	49	2,569	52.4	3,640	2.8	9	1.9	120	1,786	ND
	2005	53	1,147	21.6	1,639	2.4	9	3.0	146	951	2.4
	2009	18	502	27.9	1,994	2.5	7	2.6	134	1,260	2.5
	2011	53	3,158	59.6	2,668	2.0	12	34.4	112	1,297	3.2
South shore	1988	18	117	6.5	666	4.4	11	1.1	118	487	ND
	1989	18	195	10.8	973	3.9	13	0.2	139	544	3.5
	1990	18	185	10.3	956	3.8	13	0.3	114	722	ND
	1997	29	871	30.0	1,793	2.6	10	1.0	115	897	ND

Table 2. contd....

Table 2. contd.

Sector	Year	Sample size		Population				Age 1+		Females	
		No. of stations	No. of yellow perch	CPUE (no./station/night)	BPUE (g/station/night)	Mean age (years)	Max age (years)	CPUE (no./station/night)	Mean total length (mm)	BPUE (g/station/night)	Age at maturity (years)
	2005	19	489	25.7	2,023	2.5	11	1.5	141	1,303	2.6
	2009	8	244	30.5	2,034	2.7	8	1.6	118	1,295	1.9
	2011	19	1,174	61.8	2,380	2.0	10	39.9	106	1,552	3.6
Archipelago	1995	40	285	7.1	358	2.7	10	1.1	109	220	2.5
	2003	64	381	6.0	295	2.4	7	0.5	115	197	1.8
	2010	54	420	7.8	606	3.2	8	1.0	115	435	2.7
Lake Saint-Pierre											
North shore	2002	52	774	14.9	810	1.8	7	6.9	130	640	1.6
	2007	53	551	10.4	446	2.3	12	1.5	106	342	1.9
	2009	19	214	11.3	672	2.8	7	0.5	96	496	2.3
	2011	19	261	13.7	1,119	3.1	8	3.2	115	847	2.3
South shore	2002	60	1,280	21.3	878	1.6	8	12.6	121	559	1.4
	2007	58	720	12.4	524	2.1	6	3.1	109	374	1.8
	2009	22	191	8.7	782	3.2	9	0.1	103	613	ND
	2011	22	31	1.4	158	3.7	9	0.3	140	118	ND
Bécancour–Batiscan	2001	60	231	3.9	302	2.0	6	1.1	138	235	ND
	2008	63	105	1.7	110	2.2	5	0.6	116	87	2.3
	2012	64	135	2.1	159	2.4	7	0.2	133	126	1.9

Data are presented for the north and south shores separately when major differences occurred within a lake. Age 0 yellow perch were excluded from all analyses.

early 2000s, respectively, in the Québec sector (Vachon and Dumont 2007; N. Vachon, unpublished data). In Lake Saint-Louis, A was estimated at 46% (ages 6 to 15) in the 1960s on the north shore (Fortin and Magnin 1972a) and 33% and 37% for the north and the south shores, respectively (ages 3 to 13), during the late 1980s (Dumont 1996), 71% and 44% (ages 4 to 10) in 1990s (Vachon and Dumont 2007), and 60% and 61% (ages 3 to 9) during mid-2000s (N. Vachon, unpublished data).

In Lake Saint-Pierre, all samples collected since the 1980s have revealed a low maximum age, with few individuals older than six years. The range over which a catch curve may be used to estimate A rarely exceeds four age groups, which is low. Annual estimates of A varied from 56% to 86% (mean 78%) in the 1986 to 1991 commercial catch samples (Guénette et al. 1994) and from 67% to 80% (mean 74%) in the 1997 to 2000 samples (Mailhot 2001). Values ranging from 62% to 80% were also found in 2002 (Magnan et al. 2004), indicating that total mortality in Lake Saint-Pierre never decreased since the mid-1980s and also remained higher compared to values estimated in upstream populations.

5.5 Yellow Perch Life History Dynamics in the St. Lawrence River

Since 1995, the yearly standardized sampling survey (RSI) has been conducted in late summer/early fall by Québec's governmental authorities. Each year, one or two enlargements of the St. Lawrence River are sampled, i.e., Lake Saint-François, Lake Saint-Louis (upstream sectors), Lake Saint-Pierre archipelago, Lake Saint-Pierre, or Bécancour–Batiscan (downstream sectors). In each sector, sampling sites are systematically distributed in all habitats except for the navigation channel. At each site, two experimental multi-mesh monofilament gillnets (60 m long by 1.8 m deep; eight panels of 25, 38, 51, 64, 76, 102, 127, and 152 mm stretched mesh) are anchored on the bottom, parallel to the shore, for approximately 24 h. A detailed description of the sampling protocol can be found in La Violette et al. (2003).

The analyses presented here are based on the RSI information as well as some previous standardized surveys targeting the same species/areas. In addition to the RSI data ranging from 1995 to 2012, we included fish collected from 1988 to 1990 in Lake Saint-Louis (Dumont 1996). Thus the situation we describe is based on over 25,000 yellow perch collected in the St. Lawrence River from 1988 to 2012 (Table 2).

Life history traits were analyzed for each individual sector and separated by time period to allow a graphical overview of the temporal trends and spatial contrasts. The following time periods were chosen according to key historical events in yellow perch fisheries management: (1) 1988–1990, corresponding to the pre-RSI period, when data are available only for Lake Saint-Louis; (2) 1995–2002, corresponding to the beginning of the RSI and to the first modifications to regulations of the yellow perch fisheries; and (3) 2003–2012, corresponding to a period of important changes in fisheries regulations in the downstream sectors (see Table 3). As detailed in Section 5.4, there are major differences in yellow perch life history characteristics between the south and the north shores of Lake Saint-Louis and Lake Saint-Pierre, so north and south shore data were analyzed separately for these lakes (Table 2).

Table 3. Major management decisions implemented to the yellow perch commercial and sport fisheries between 1997 and 2013.

Period	Management decision (Minor details omitted for clarity)	Lake Saint-François	Lake Saint-Louis	Lake Saint-Pierre		Bécancour–Batiscan	
		Sport	Sport	Comm.	Sport	Comm.	Sport
1997	- Minimum size limit 165 mm			X	X	X	
	- Daily bag limit applied (50 perch)	X	X		X		X
	- Selling sport catch prohibited	X	X		X		X
	- Yellow perch as baitfish prohibited	X	X	X	X	X	X
1999	- Fishing delayed to 18 April			X		X	
	- Fishing delayed to 5 May to protect spawning	X	X		X		X
2000	- Fishing ahead to 10 April			X		X	
	- Minimum size limit increased to 190 mm			X		X	
2001	- Fishing ahead to 10 April				X		X
2002	- *Buyback of 6 commercial licences (42 to 36)*			X			
2005–2006	- Min. size limit increased to 190 mm				X		
	- *Buyback of 18 commercial licences (36 to 18)*			X			
	- Total catch restricted to 53.5t			40t	13.5t		
	- Daily bag limit reduced to 10 perch				X		
	- Sport fishing season length reduced				X		
2008	- Fishing delayed to 9 May to protect spawning			X	X		
	- *Buyback of 12 commercial licenses (18 to 6)*			X			
	- Total catch restricted to 12.3t and allocation in favour of sport fishery			4.3t	8t		
2010	- Fishing delayed to 9 May to protect spawning					X	X
	- Daily bag limit reduced to 10 perch						X
	- Minimum size limit increased to 190 mm						X
2012	- Fisheries closed for 5 years			X	X		
2013	- Fisheries closed for 5 years					X	X

Before 1997, commercial fishing was permitted from 1 April to 30 November and sport fishing was year-round.
X = applied to the lake.

Fish abundance was expressed in capture per unit effort (CPUE: no./station/night) and biomass per unit effort (BPUE: g/station/night). CPUE and mean length of age 1+ yellow perch were used as proxies of recruitment while BPUE and mean length of females at age 3+ were used as proxies for abundance and growth potential of the spawning stock. Median length and age at maturity of female yellow perch were calculated within each sector and sampling year using the arcsine root method (Chen and Paloheimo 1994). Median length and age at maturity of yellow perch sampled from 1988 to 1990 in Lake Saint-Louis were obtained from Dumont (1996). Stage III yellow perch captured during fall were considered sexually mature (Craig 2000); these fish would have been able to reproduce during the following spring. Age was reassigned to some yellow perch using age–length keys when subsampling was performed in the laboratory. Specific age–length keys were developed for each year, sector, and sampling station when enough data were available. Age 0 yellow perch, which were sporadically captured in the 25 mm stretched mesh, were excluded from all analyses due to inadequate abundance estimates. Temporal comparisons for each parameter and each sector were made using nonparametric Kruskal-Wallis tests based on the chi-square statistical distribution. All statistics were computed using R (R Development Core Team 2007). For each sector, statistical comparisons were performed using the first and last sampling years (see Table 2 for details of years sampled for each sector). Statistical comparisons of mean age and mean total length were performed on a random subset of 100 yellow perch/year when sample size was too large to estimate a meaningful probability value (e.g., more than 1000 yellow perch/year).

5.5.1 Abundance and Biomass

Surveys conducted during the past two decades have shown that yellow perch abundance was generally higher in the upstream portion of the St. Lawrence River, and this pattern was stable over this period (Fig. 3a). Since 1995, the mean annual CPUE has varied between 21.6 and 61.8 yellow perch/station in lakes Saint-François and Saint-Louis (north and south shores), while the average catch rarely exceeded 15 yellow perch/station in the downstream sectors (Table 2).

Mean yellow perch CPUE increased progressively and doubled between 1996 and 2009 in Lake Saint-François (Chi-squared = 11.19, $P < 0.001$). On the north shore of Lake Saint-Louis, the yellow perch abundance observed during the 1988–2011 period did not show a clear temporal trend while a progressive increase was noted for the same period on the south shore (Table 2). On both shores of Lake Saint-Louis, the most recent survey (in 2011) indicated the highest yellow perch abundance of the last 25 years. Compared to 1988, the 2011 yellow perch abundance in Lake Saint-Louis was ten times higher on the south shore (Chi-squared = 8.64, $P = 0.003$) while the difference was not significant (Chi-squared = 1.43, $P = 0.232$) on the north shore despite the recent increase (by a factor of two).

Mean CPUE remained relatively low and stable in the Lake Saint-Pierre archipelago between 1995 and 2010 (Chi-squared = 0.31, $P = 0.580$) and on the north shore of Lake Saint-Pierre between 2002 and 2011 (Chi-squared = 0.59, $P = 0.443$). Yellow perch abundance on the south shore of Lake Saint-Pierre dropped by a factor of 15 between 2002 and 2011 (Chi-squared = 13.80, $P < 0.001$). In the

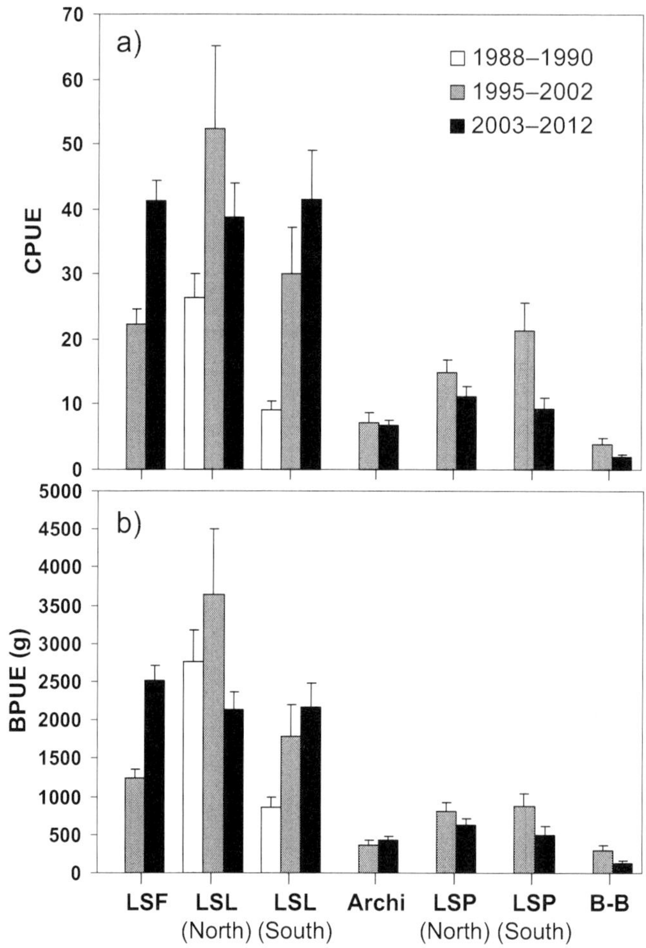

Fig. 3. Longitudinal gradient of (a) mean capture per unit effort and (b) mean biomass per unit effort (±SE) of yellow perch (males, females, and unsexed specimens combined) along the St. Lawrence River during three time periods. LSF: Lake Saint-François; LSL: Lake Saint-Louis; Archi: archipelago of LSP; LSP: Lake Saint-Pierre; B-B: Bécancour–Batiscan. Note that data for 1988–1990 were available only for LSL.

Bécancour–Batiscan sector, no temporal trend was observed between 2001 and 2012 (Chi-squared = 2.74, $P = 0.098$), but overall the CPUE remained the lowest in the study area (Table 2; Fig. 3a).

In general, the mean annual BPUE followed similar patterns (Fig. 3b). However, the differences observed in the BPUE in 1995–2002 and 2003–2012 in Lake Saint-Pierre were less pronounced than those observed in CPUE, which was probably caused by a greater proportion of older and larger yellow perch in recent years.

5.5.2 Age Structure

The widest age ranges of yellow perch were observed in lakes Saint-François and Saint-Louis (both shores), with the maximum ages being 13 and 15 years, respectively (Table 2). In the downstream sectors of the St. Lawrence River, very few specimens older than age 6+ have been sampled since 2002, and specimens older than 10+ were extremely rare (Fig. 4).

In Lake Saint-François, the mean age remained stable at around 3 years old during the 1996–2009 period (Chi-squared = 0.19, $P = 0.662$). In Lake Saint-Louis, the mean age decreased progressively from 3.6 and 4.4 years on north and south shore respectively in 1988 to 2 years old on both shores in 2011 (north: Chi-squared = 12.95, $P < 0.001$; south: Chi-squared = 28.35, P < 0.001). However, the decline in the mean age of yellow perch in Lake Saint-Louis in 2011 is likely biased because of the high abundance of age 1+ specimens on both shores.

Significant increases in mean age were observed in all downstream areas: (1) in the Lake Saint-Pierre archipelago between 1995 and 2010 (Chi-squared = 10.35, $P = 0.001$), (2) on both shores of Lake Saint-Pierre between 2002 and 2011 (north: Chi-squared = 115.12, P < 0.001; south: Chi-squared = 42.16, P < 0.001), and (3) in the Bécancour–Batiscan sector between 2001 and 2012 (Chi-squared = 19.06, $P < 0.001$; Table 2). In Lake Saint-Pierre, the increase in mean age was associated with a strong reduction in age 1+ to 3+ yellow perch abundance on both shores and to an increase in age 4+ and older specimens, especially on the north shore (Fig. 4), where the abundance of age 4+, 5+, and 6+ perch increased by factors of 4, 9, and 20, respectively, between 2002 and 2011. These observations occurred at the same time as the gradual reduction in fishing effort (Table 3). Similar trends in abundance were not observed on the south shore, where the increase was only for the age 6+ group (which increased by a factor of eight), while the abundance of age 4+ and 5+ yellow perch remained very low. In the Bécancour–Batiscan sector, the increase in mean age was mainly associated with the sharp reduction in abundance of age 1+ fish: the average CPUE decreased by almost a factor of six from 2001–2012 (Table 2; Fig. 4).

5.5.3 Recruitment and Growth at Age 1+

Yellow perch recruitment, expressed as mean CPUE of age 1+ fish, showed opposite temporal trends in the upstream and downstream sectors of the St. Lawrence River (Fig. 5a). Since 2003, high recruitment years (2004, 2009 and 2011) have been observed in lakes Saint-François and Saint-Louis, with values ranging from 8.4 to 39.9 age 1+ fish/station (Table 2). In the downstream sectors, data indicated a severe recruitment failure during the same period, with values most often ≤ 1 age 1+ yellow perch/station (Table 2). On the north and south shores of Lake Saint-Pierre, the abundance of age 1+ fish was significantly lower in 2011 than in 2002, which was the last year of high recruitment in the downstream sectors (north: Chi-squared = 8.14, P = 0.004; south: Chi-squared = 17.87, P < 0.0001; Table 2). The mean CPUE of age 1+ fish was also significantly lower in 2012 than in 2001 in the Bécancour–Batiscan sector (Chi-squared = 5.13, $P = 0.024$; Table 2). In 2011, when samplings were synchronized

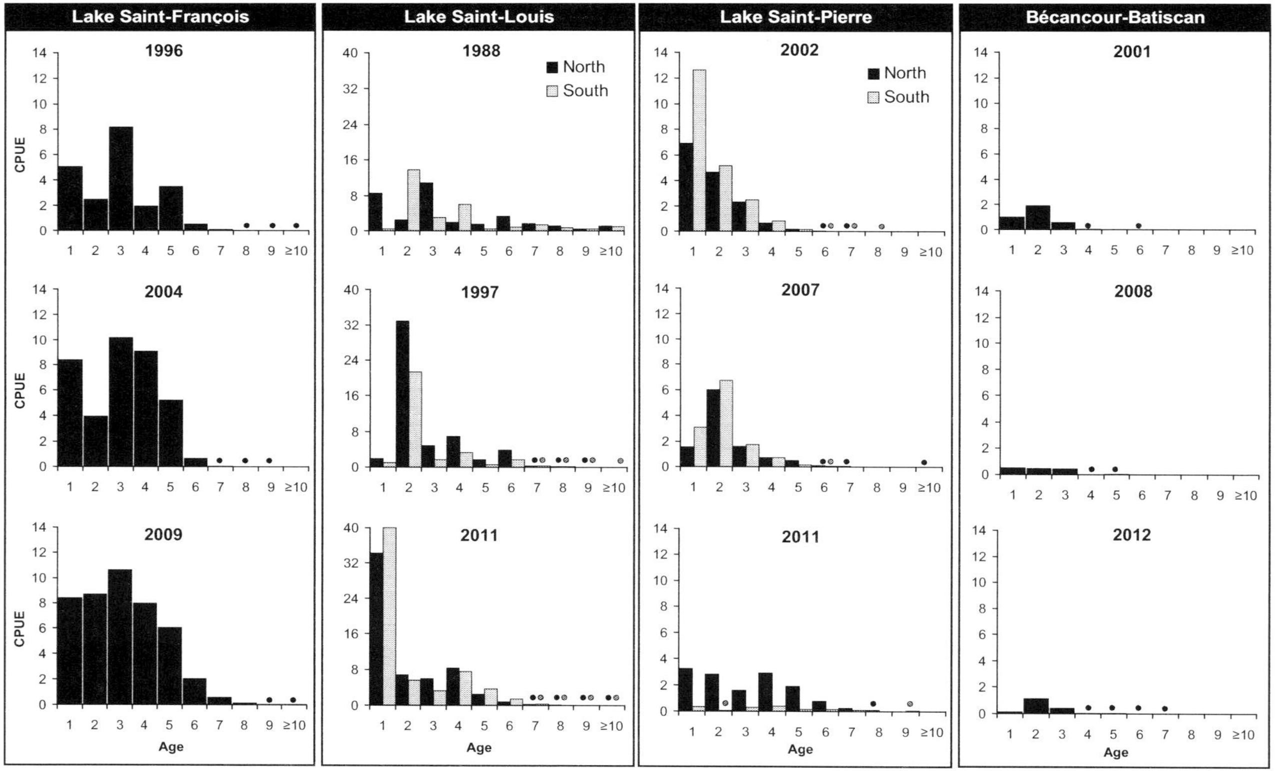

Fig. 4. Age distribution of yellow perch (males, females, and unsexed specimens combined) at various time periods for each sector of the St. Lawrence River. Capture per unit effort for Lake Saint-Louis are on a different scale. Dots represent age groups with low CPUE unseen due to scale magnification. Note that black and grey dots are used to represent low CPUE by shore for Lake Saint-Louis and Lake Saint-Pierre.

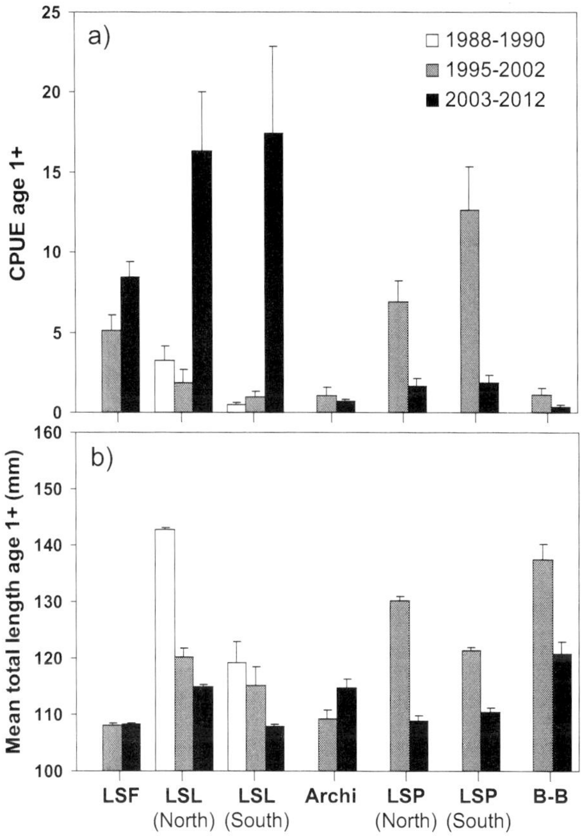

Fig. 5. Longitudinal gradient of (a) mean capture per unit effort and (b) mean total length (±SE) of age 1+ yellow perch (males, females and unsexed specimens combined) along the St. Lawrence River during three time periods. LSF: Lake Saint-François; LSL: Lake Saint-Louis; Archi: archipelago of LSP; LSP: Lake Saint-Pierre; B-B: Bécancour–Batiscan. Note that data for 1988–1990 were available only for LSL.

in the upstream and downstream sectors of the St. Lawrence River, recruitment was lower in Lake Saint-Pierre compared to Lake Saint-Louis by a factor ranging between 11 and 133, depending on the shore.

Growth at age 1+ showed a general decreasing trend over time in most sectors of the St. Lawrence River, except in Lake Saint-François and the Lake Saint-Pierre archipelago (Fig. 5b). While the highest mean lengths at age 1+ were observed in Lake Saint-Pierre and Bécancour–Batiscan during 1995–2002, growth of age 1+ yellow perch in Lake Saint-Pierre was among the lowest of all sectors during 2003–2012 (Fig. 5b), which in turn coincided with low recruitment in the same period. In Lake Saint-Louis, a growth increase was observed in 2005, but mean length at age 1+ was lower in 2011 compared to 1988 on both the north and south shores (north: Chi-squared = 139.22, $P < 0.001$; south: Chi-squared = 4.39, $P = 0.036$; Table 2).

5.5.4 Life History Characteristics and Female Abundance

As seen with other life history characteristics, female yellow perch showed differences between upstream and downstream sectors of the St. Lawrence River. In lakes Saint-François and Saint-Louis, yellow perch populations were characterized by high BPUE of females reaching sexual maturity later than those from the downstream sectors (Figs. 6a and 6b). Globally, yellow perch median age and length at sexual maturity

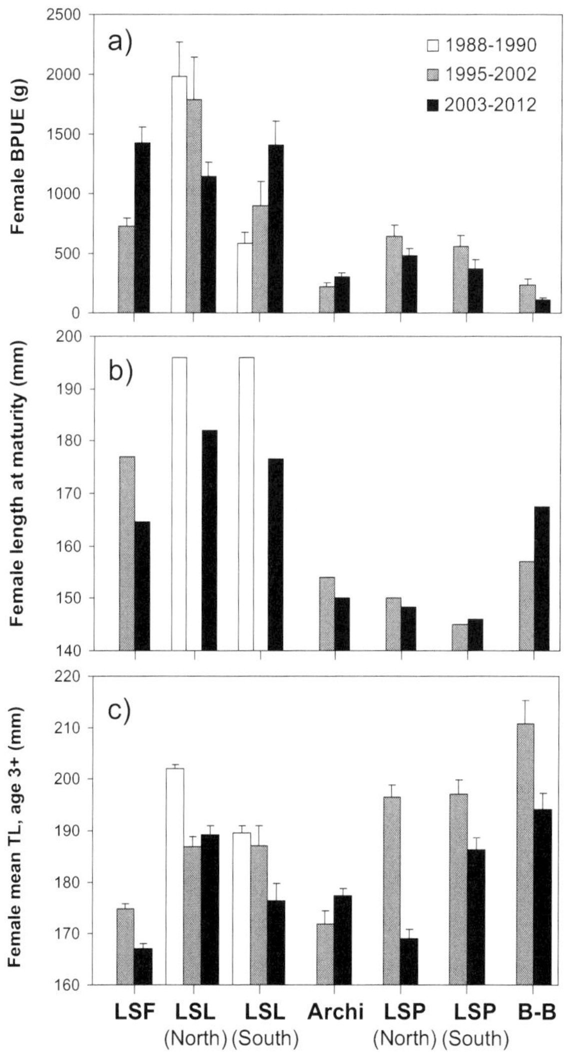

Fig. 6. Longitudinal gradient of (a) mean female yellow perch biomass per unit effort (±SE), (b) mean female yellow perch length at maturity and (c) mean total length (±SE) of age 3+ female yellow perch along the St. Lawrence River during three time periods. Length at maturity was not available for Lake Saint-Louis during the 1995–2002. LSF: Lake Saint-François; LSL: Lake Saint-Louis; Archi: archipelago of LSP; LSP: Lake Saint-Pierre; B-B: Bécancour–Batiscan. Note that data for 1988–1990 were available only for LSL.

are higher in lakes Saint-François and Saint-Louis compared to Lake Saint-Pierre and Bécancour–Batiscan (Table 2; Fig. 6b).

While the biomass of female yellow perch remained stable or increased over the years in most of the upstream sectors, a decline was observed in Lake Saint-Pierre and Bécancour–Batiscan over the last decade (Fig. 6a). Female yellow perch BPUE tripled in Lake Saint-François between 1996 and 2009 (Chi-squared = 28.28, $P < 0.001$) while it doubled in the Lake Saint-Pierre archipelago between 1995 and 2010 (Chi-squared = 4.63, $P = 0.031$; Table 2). No significant trend was observed in female BPUE on the north shore of Lake Saint-Louis between 1988 and 2011 (Chi-squared = 1.07, $P = 0.300$). During the last decades, female BPUE has declined in downstream sectors. This is especially true on the south shore of Lake Saint-Pierre, where female biomass showed a fivefold decline between 2002 and 2011 (Chi-squared = 10.37, $P = 0.001$). Female BPUE in the Bécancour–Batiscan sector was the lowest of all the areas studied, and no significant temporal trend was observed between 2001 and 2012 (Chi-squared = 2.07, $P = 0.150$).

5.5.5 Female Growth Potential

The growth potential of female yellow perch, expressed by mean total length at age 3+, declined over the years in all sectors of the St. Lawrence River except the Lake Saint-Pierre archipelago (Figs. 6c and 7). The decline was more pronounced in the downstream sectors. During the 1995–2002, females in Lake Saint-Pierre and Bécancour–Batiscan had the highest growth of the St. Lawrence River. However, between 2002 and 2011 in Lake Saint-Pierre and between 2001 and 2012 in Bécancour–Batiscan, length at age 3+ showed a sharp decline (Lake Saint-Pierre north: Chi-squared = 5.97, P = 0.015; Lake Saint-Pierre south: sample size too low

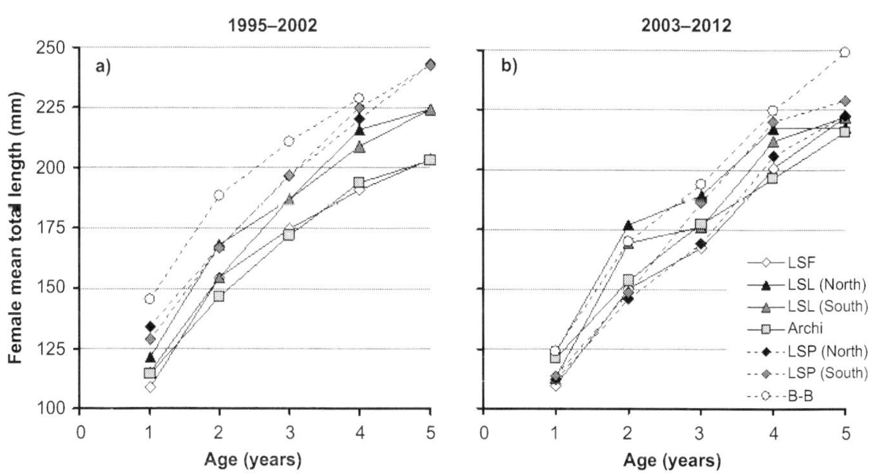

Fig. 7. Mean length at age for female yellow perch age 1+ to 5+ for the different sectors of the St. Lawrence River. Data are presented separately for the (a) 1995–2002 and (b) 2003–2012 time periods. LSF: Lake Saint-François; LSL: Lake Saint-Louis; Archi: archipelago of LSP; LSP: Lake Saint-Pierre; B-B: Bécancour–Batiscan.

for a statistical comparison; Bécancour-Batiscan: Chi-squared = 15.34, $P < 0.001$). This lower growth potential of female yellow perch is most likely due to the decrease in growth observed at younger life stages (Fig. 7). In Lake Saint-Louis, the growth of age 3+ females observed in 2011 was lower than in 1988 (north: Chi-squared = 28.11, $P < 0.001$; south: Chi-squared = 25.98, $P < 0.001$; Table 2), indicating that even though growth potential is still high in this lake, female growth was much higher in the 1980s.

5.6 Yellow Perch Stock Status Since 1995

5.6.1 Yellow Perch in Lake Saint-François: A Dominant Species Favoured by Changes in the Fish Community and the Reduction of Sport Fishing Pressure

Lake Saint-François has been profoundly impacted by anthropogenic modifications. Water level stabilization, the construction of large dams at each end, and man-made modifications of river banks and the littoral zone are the main pressures imposed on the ecosystem. These cumulative alterations have had a negative effect on the aquatic community, resulting in the lowest species diversity observed along the St. Lawrence (Vachon 2002; Vachon et al. 2013). The index of biotic integrity (IBI), which was developed by the provincial government to assess the health of fish communities, is lower here than in other areas of the St. Lawrence River and has decreased since the late 1990s (La Violette et al. 2003; Mingelbier et al. 2008b). The introductions of zebra mussel, quagga mussel and round goby (*Neogobius melanostomus*) have altered the habitat characteristics, fish community, and trophic dynamics (Vachon 2002; Bechara et al. 2003; Vachon et al. 2013).

Yellow perch had long dominated the Lake Saint-François fish community (Mongeau 1979). Its continued importance was confirmed by recent results from the RSI (Vachon et al. 2013) and the Fish Community Index Netting Surveys of the Ontario Ministry of Natural Resources (2013), which were conducted in the Québec and Ontario sectors of the lake. Compared to other sections of the St. Lawrence River, yellow perch growth is generally slower in Lake Saint-François, and this trend appears to be relatively stable through time (Table 2; Figs. 4 and 5). It was also found that females reach sexual maturity at a later age (between age 2.5 to 3.4 years) when compared to the downstream sectors (Table 2).

Until the middle of the last century, yellow perch was commercially exploited in the Québec portion of Lake Saint-François, although landings were low (Magnin 1966; Pluritec 1982). With the increasing interest in recreational fishing, restrictions were gradually imposed on the commercial fishery until it was completely closed in the mid-1960s (Magnin 1966; Pluritec 1982), except for the commercial exploitation that continued in the Ontario sector of the lake. Even though recent annual commercial quotas (12.8 t in 2012) and landings (12.2 t in 2012) are low, this is much higher than values reported from the 1990s and early 2000s (2.7 to 3.6 t; Ontario Ministry of Natural Resources 2013).

Since the decline of yellow perch in the Great Lakes during the 1990s, there has been an increasing demand for the species along the St. Lawrence River and a strong increase in the recreational catch and in its trade in regional and American markets due to the high price paid for it. The relatively high annual mortality rates (calculated for fish between 3 and 8 years of age) estimated from the RSI surveys in 1996 (63.2%) (Vachon and Dumont 2007) and in 2004 (72.8%) (N. Vachon, unpublished data) are likely associated with the high level of exploitation from recreational fishing during this period.

To prevent overfishing, sport fishing rules were strengthened between 1997 and 1999 (Table 3). A daily bag limit of 50 fish was imposed in the Québec sector of the lake. Moreover, yellow perch fishing was closed from the Ontario–Québec border (Lake Saint-François) to the Lake Saint-Pierre archipelago during the spawning season to prevent excessive catch during this period of high vulnerability to fishing (April 1 through the second week of May). Finally, it was prohibited for sport fishermen to sell their yellow perch catch in Québec. This regulation was also implemented to improve the earnings of commercial fishermen in the lower reach of the St. Lawrence River. To further protect the yellow perch stock, similar restrictions on recreational fishing were implemented in the late 1990s in the Ontario part of Lake Saint-François (Ontario Ministry of Natural Resources 2004). More restrictive measures for anglers over the last 15 years and their rigorous enforcement have probably contributed to the recent increase of yellow perch abundance in the Québec part of the lake. Similar observations were made in the Ontario part as well: after a period of continuous decline between 1984 and 2002, the abundance of yellow perch has gradually returned to levels observed in the beginning of the 1980s (Ontario Ministry of Natural Resources 2013).

The major changes in the fish community that occurred over the past two decades may also be an important factor favouring yellow perch in Lake Saint-François. Since the introduction of the zebra and the quagga mussels in the upper St. Lawrence system in the late 1980s, the abundance of the two major yellow perch predators has declined. Sauger disappeared from the lake while walleye (*Sander vitreus*) is still present but at very low abundances compared to past surveys. These changes were related to a marked increase in water transparency, resulting in habitat degradation for these species (Vachon 2002; Bechara et al. 2003; Vachon et al. 2013). Yellow perch is an important prey for walleye (Nelson and Walburg 1977; Forney 1980), and its recruitment is inversely correlated with walleye density in Oneida Lake (New York, USA) (Rutherford et al. 1999). The link between the increase in yellow perch and the decline of walleye abundance in Lake Saint-François is plausible but likely not the only biotic factor involved. Recent surveys have shown that northern pike is also dramatically declining in this lake (Ontario Ministry of Natural Resources 2013; Vachon et al. 2013) while the abundance of two other predators, the smallmouth bass (*Micropterus dolomieu*) and the largemouth bass (*Micropterus salmoides*), is high and increasing. Furthermore, the demographic explosion of another invasive species, the round goby, which was first detected in 2000 in Lake St. François and is now an abundant species in the littoral zone, has led to major changes in the benthic fish community.

5.6.2 Yellow Perch in Lake Saint-Louis: a Dominant Species Positively Impacted by the Reduction of Chemical Contaminants and Sport Fishing Pressure

Delimited by Montréal to the north and by a highly industrialized zone to the south, the land surrounding Lake Saint-Louis is the most urbanized and densely populated of the St. Lawrence River. The yellow perch commercial fishery here has been closed since the 1960s to support the increasing popularity of recreational fishing during both the open-water and ice fishing periods (Tremblay and Dumont 1990). The annual exploitation rate of the north shore population was estimated to be 3.4% at the end of the 1960s (Fortin and Magnin 1972a). Twenty years later, the annual exploitation rate was estimated at 9–11% on the north shore and 6–7% on the south shore, almost equally split between the open-water and ice fishing periods (Dumont 1996). In the 1990s—as was observed for the Lake Saint-François stock—the collapse of many yellow perch fisheries in the Great Lakes led to a major increase in the recreational catch and to its trade in regional and American markets.

The management measures implemented from 1997 to 1999 (Table 3; see Section 5.6.1) likely contributed to the increased yellow perch densities observed during the 2000s (Table 2 and Fig. 6). However, fishing was probably not the major threat to yellow perch stocks in Lake Saint-Louis over the last 50 years. From the late 1930s to the early 1990s, the south shore had been severely polluted by heavy metals and various organic compounds, and it experienced increasing rates of sedimentation (Carignan et al. 1994; Dumont 1996). At least until the 2000s, the south shore persistently exhibited a higher level of chemical contamination than the north shore: mercury showed the most pronounced gradient, but significant differences were also observed for other potentially toxic inorganic and organic substances in the sediments as well as in fish muscle and liver (Hontela et al. 1995; Dumont 1996; Laliberté 2003; Marcogliese et al. 2005).

A study conducted in Lake Saint-Louis between 1988 and 1990 revealed the almost complete ecological independence between the south and north shore yellow perch stocks. The two stocks were characterized by strong and statistically significant differences in their life history traits and population dynamics (Dumont 1996). The south shore population exhibited a relative abundance two to four times smaller, lower body condition, delayed age at sexual maturity in both females (Table 2) and males, and lower relative fecundity than the north shore population. A slower growth rate during the first year of life was also observed, which was opposite to the trend observed in the 1940s (Préfontaine 1941; Grimaldi 1967), when the first two years of growth were found to be slower on the north than the south shore. Total annual mortality rates (age 3 to 13) were comparable (south shore: 37.4%, north shore: 33.6%), but natural mortality rates appeared higher on the south (30–33%) than on the north (22–24%) shore. During the 1980s, a smaller proportion of the year-class strength variability (south shore: 68%; north shore: 99%) was explained by natural factors. The integration of these differences in density, age structure, growth, sexual maturity, and fecundity showed that the relative reproductive potential of the south shore population was about 4.5 times lower than that of the north shore population (Dumont 1996).

The pattern of variation in the south shore's population dynamics was atypical. In similar conditions, one or several of the following traits usually compensate for lower density: faster growth rate, higher fecundity, and/or earlier sexual maturity (Munkittrick and Dixon 1989), but this was not observed in the south shore population. Several factors may explain this absence of density-dependent compensation, and they were compared by Dumont (1996). Thermal differences were excluded: perch habitat on both shores had similar thermal characteristics. The effect of differences in sport fishing effort on yellow perch density was also considered, but the exploitation rate from recreational fishery, while relatively low on both shores, being higher on the north shore (9–11%) than on the south shore (5–7%), this hypothesis was rejected. Overall, the observed differences suggested an adaptation of the south shore population to either lower energetic supplies or to metabolic disturbances that could have resulted from a lack of food or from direct sublethal effects of chemical contamination (Munkittrick and Dixon 1989).

The hypothesis that densities of zooplankton, macro-benthic invertebrates, and phytofauna were higher in the vegetated areas of the north shore was tested for two periods (April–July and August–November 1982). No significant statistical difference was detected between the two shores for zooplankton densities, which are the dominant food for early stages of yellow perch. However, contrary to the hypothesis, average macro-invertebrate densities and biomasses were generally higher on the south shore, and statistically significant differences were observed for various taxonomic groups considered to be major dietary components for young-of-the-year (Dumont 1996). The assumption that lower density, growth rate, relative fecundity, and body condition were related to lower food availability thus appears unlikely. A comparison of the fish communities on both shores, based on three years of multi-mesh gillnetting in the fall from 1988 to 1990, showed comparable densities and biomasses of potential competitors and predators. Introduced species such as alewife (*Alosa pseudoharengus*) and white perch (*Morone americana*), which are considered problematic to yellow perch in the Great Lakes (Wells 1977; Parrish and Margraf 1990), were found to be very rare in Lake Saint-Louis (Dumont 1996).

Biochemical and physiological indicators of exposure to contaminants were also studied. During fall, the south shore population was characterized by lower muscle lipid reserves and lower gonadosomatic (GSI) and hepatosomatic indices (Dumont 1996). Hontela et al. (1995) reported similar results for GSI during the spring. Significant differences were observed for the mechanical resistance of vertebrae, an index of quality of bone development (Hamilton et al. 1981). Hepatic concentrations of retinoids (dehydroretinol and retinylpalmitate) were correlated with perch length on the north but not on the south shore, suggesting that the storage capacity of vitamin A was impaired on the south shore (Dumont 1996). Based on standardized handling stress tests, neither immature nor mature yellow perch from the south shore exhibited the expected physiological stress response, which was observed in the north shore population (Hontela et al. 1995): fish sampled on the south shore showed lower levels of blood cortisol, glucose, and thyroxine (T4) as well as higher levels of liver glycogen than those from the north shore. Blood cortisol concentrations in samples from the south shore population were also abnormally low. Other indices of exposure to chemical contaminants, including asymmetrical fin development, phagocytic activity of the pronephros blood cells (Dumont 1996), and hepatic activity of metallothionein (Doyon et al. 1993) and of two mixed-function oxygenases (AHH and EROD) (Doyon et al. 1993; Ménard et al. 1993) showed no significant differences. In 2002, Marcogliese et al. (2005) observed that yellow perch from the southern shore exhibited higher levels of oxidative stress than those from the reference site on the northern shore. Oxidative stress was also exacerbated on yellow perch affected by a higher prevalence of parasites in the southern shore, but not in the northern shore.

The study of Dumont (1996) suggested that the lower energetic balance of the south shore's population in the 1980s could likely be attributed to chemical contamination of the habitat. In the last two decades, many measures have been implemented to reduce contaminant loading of the St. Lawrence River. Since the 1980s, there has been a downward trend, by 50 to 90%, in the concentrations of several contaminants (PCBs, PAHs, heavy metals) in sediments and fish along the entire length of the St. Lawrence River, including both the south and north shores of Lake Saint-Louis. Current concentrations are now generally below the levels producing toxic effects in benthic organisms but there has been little or no decrease in mercury concentrations over the past 20 years on the south shore of Lake Saint-Louis (Laliberté 2003; State of the St. Lawrence Monitoring Committee 2008, Ministère du Développement durable, de l'Environnement, de la Faune et des Parcs, unpublished data). The fact that south shore population densities (Table 2) were at their highest level and comparable to north shore densities during the last decade ($P > 0.05$ in the 2005, 2009, and 2011 RSI surveys; Kruskal-Wallis tests by sampling year) suggests a positive response of the yellow perch population to the decrease inorganic contamination on the south shore.

5.6.3 Yellow Perch in Lake Saint-Pierre: A Severely Impacted Population

Lake Saint-Pierre yellow perch biomass was high from the 1970s until the mid-1990s, thus this ubiquitous species played a central role in structuring the fish community (Massé and Mongeau 1974; Guénette et al. 1994; Fournier et al. 1996). Its abundance led to the establishment of an important commercial and later a sport fishery. These fisheries had been sustainable since the 1870s (see Magnan 2002), but the population experienced dramatic collapse beginning in the mid-1990s and recruitment failure since the mid-2000s. Yellow perch abundance has decreased by 70 to 86% between 1972 and 2011, depending on the age group (Fig. 8). Yellow perch population dynamics in Lake Saint-Pierre reflect those of a highly stressed population, now characterized by rapid sexual maturation, low female biomass, and low growth of young stages.

Until 1987, yellow perch landings had been totalling about 300 t per year and mostly came from the commercial fishery (commercial: 225 t, sport: 75 t; Fig. 9). The first management plan for both sport and commercial fisheries was put in place in 1987 by MLCP and MAPAQ. At that time, commercial and sport fishermen were in open conflict to exploit this resource. Both groups maintained strong political lobbies in order to influence governmental decisions, which resulted in maintenance of the status quo.

Between 1987 and 1991, the five-year monitoring program (see Section 5.3) enabled estimations of the total mortality and fishing mortality rates, and scenarios were presented to manage this stock (Guénette et al. 1994). This study indicated that the total annual mortality remained high (77%), as reported in the first estimation by

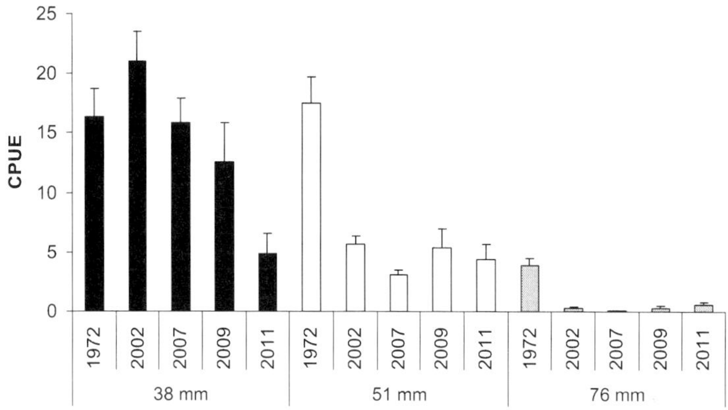

Fig. 8. Comparison of historical and recent abundances (±SE) of yellow perch in Lake Saint-Pierre. Historical data were obtained from Massé and Mongeau (1974) and recent data from the RSI monitoring network. Abundances are expressed for three gill net panels (38, 51, 76 mm stretched mesh) to allow standardization of sampling efforts. Catch per unit effort are expressed as the number of yellow perch/30.5 m of gill net/night.

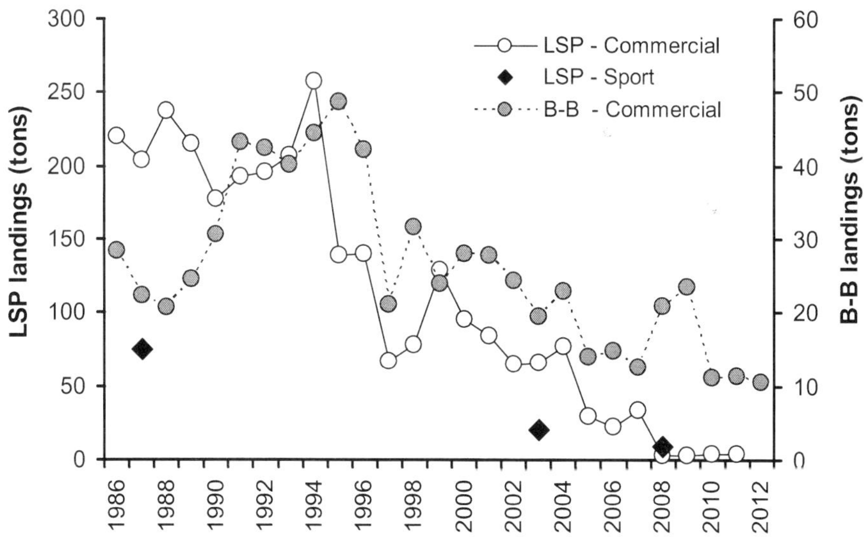

Fig. 9. Yellow perch commercial and recreational landings in Lake Saint-Pierre (LSP) and in the Bécancour–Batiscan sector (B-B). Commercial landings are based on annual declared catches of commercial fishermen and recreational landings are based on creel surveys. Fishing effort and fisheries regulations changed between 1997 and 2010, and both fisheries were closed for a five-year period in 2012 for Lake Saint-Pierre and in 2013 for the Bécancour–Batiscan sector.

Leclerc (1987), that the age structure of the spawning yellow perch was quite narrow (mostly only two abundant age classes), and that the main contribution to total mortality was fishing (63%). Guénette et al. (1994) concluded that the population would not be able to support the commercial catch if there were several years of poor recruitment and recommended that the total exploitation (from sport and commercial fisheries) should not increase. The population was at risk, considering potential threats such as introductions of exotic species, habitat modification, unfavourable climate conditions, and overexploitation. These factors could impede the emergence of strong year-classes, as observed in Lake Ontario (Hoyle 1991), or could reduce the growth rate, as observed in Lake Erie (Hartman et al. 1980).

Following the scientific advice of Guénette et al. (1994), a management plan was proposed to reduce the exploitation rate of the Lake Saint-Pierre yellow perch stock. Biologists from MLCP estimated that the most efficient measure was to protect adults that were exploited during the spawning period. However, to comply with requests from commercial fishermen, who were not in favour of any restriction, the proposed measure was to delay the opening of both the commercial and sport fisheries until the middle of the spawning period (Mailhot 1997). Despite the recommendation of government biologists, fishing was maintained during the spawning period because of political considerations. Instead, a sport fishing bag limit was set to 50 perch, sport fisherman could not sell their catch, and a minimum size limit of 165 mm was set for commercial and sport fishing (Table 3). However, this latter measure had little effect since this size corresponds to the normal minimum length suitable to produce a small fillet.

Between 1995 and 1999, commercial fishing experienced a severe decline and sport fishing deteriorated. For example, the total declared commercial landings dropped from 213 t between 1986 and 1994, to 140 t in 1996 and 1997, and to 74 t in 1998, which represents 33% of what it had been between 1986 and 1994 (Fig. 9). Mailhot and Dumont (2003) concluded that the high exploitation rates combined with years of poor recruitment from 1988–1998 initiated the onset of the stock collapse. Between 1987 and 2005, all management decisions favoured the commercial fishery because it was difficult for politicians to make decisions having a negative impact on the income of the fishermen. For example, fishing was maintained for both industries before and during the spawning period for more than 10 years after the beginning of the failure of the perch population. In Lake Saint-Pierre and the Bécancour–Batiscan sector, yellow perch are more vulnerable to hoop-nets than they would be in the upper sectors because they aggregate in the warmer flood plain waters for spawning, where they could be more easily caught in large numbers (commercial catch during the spawning period represented about the two-thirds of the annual declared catch).

The abundance of the yellow perch population continued to decrease until 2007. This dramatic decline led to the creation of several scientific and management committees and working groups, and to the publication of many reports. Consultative working groups that included stakeholders and managers and were led by external scientists and management experts contributed to the process (Magnan 2002; Magnan et al. 2004, 2008; Thibault 1999, 2004, 2008). Management decisions that were needed to reduce the catch (changes in size limit, duration of the fishing season, number of hoop nets allowed, buyback of commercial licences, sport fishing bag limit) were often delayed or only progressively applied (Table 3) because the priority was to avoid impacting the income of commercial fishermen. Despite a major reduction in fishing effort, mainly between 2005 and 2010, yellow perch abundance continued to decrease, likely as a consequence of continuous low recruitment levels at age 1+, habitat modification, the introduction of exotic species, and the increase in cormorant abundance. The yellow perch stock has been considered as collapsed since the beginning of the 2010s. Commercial and sport fisheries were finally closed for a five-year period in 2012. The reduction in yellow perch abundance was similar in the Bécancour–Batiscan sector, downstream of Lake Saint-Pierre, and a five-year moratorium on both sport and commercial fisheries was also imposed here one year later.

In 2011, it became clear that a severe recruitment failure had begun in Lake Saint-Pierre starting from 2007 (Fig. 10; Magnan et al. 2014). The decreasing carrying capacity of the Lake Saint-Pierre ecosystem in the middle of the 2000s was identified as a major factor explaining this recruitment failure and ultimately the stocks collapse. Along the southern sector, tributaries draining farmlands brought in high concentrations of nutrients (Fig. 2e), resulting in the excessive growth of submerged aquatic vegetation at the mouths of the Richelieu, Yamaska, and Saint-François rivers (Hudon et al. 2012). The slow flow of the tributary water through the extensive vegetative mats increases water transparency, reduces dissolved inorganic nitrogen concentrations, and leads to a downstream shift in dominance from submerged macrophytes to benthic mats of filamentous cyanobacteria (*Lyngbya wollei*) (Hudon and Carignan 2008; Vis et al. 2008). This new equilibrium created a ~20 km long nitrogen-deficient zone that exerted negative effects on fish habitat quality, food quantity, and availability of traditionally

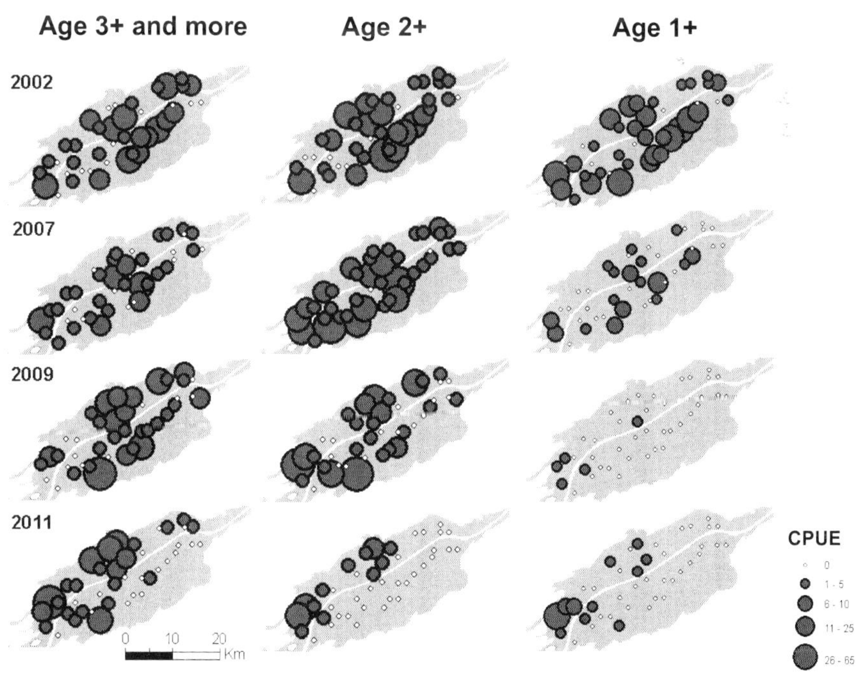

Fig. 10. Spatial distribution and abundance of age 1+, age 2+, and ≥ age 3+ yellow perch in Lake Saint-Pierre for 2002, 2007, 2009, and 2011 RSI surveys (see Section 5.5 for sampling methods). Catch per unit effort are expressed as the number of yellow perch/station/night. The white line across the lake represents the navigation channel.

important spawning and nursery habitats. In this large area, nutrient reduction led to a decrease in the biomass of submerged aquatic vegetation (four-fold) and the biomass of invertebrate prey (nine-fold) as well as a 26% reduction in young-of-the-year yellow perch growth and a five-fold drop in overwinter survival (Hudon et al. 2012). The reduced young-of-the-year yellow perch growth rate would thus translate into a reduction in the amount of energy stored to withstand winter starvation (Huss et al. 2008). Based on back-calculations of yellow perch length at age 1, mean fish length decreased from 87 mm in the 1980s to 81 mm in the 1990s and to 78 mm in the 2000s (Fig. 11). Since 2002, back-calculated length at age 1 was under 80 mm in nine out of 11 years and reached a historical minimum of 72 mm in 2011.

In the mid-2000s, two new exotic fish species colonized Lake Saint-Pierre. Round goby was first identified in the Québec portion of the St. Lawrence River in 1998 and is now present throughout the river. It was first observed in Lake Saint-Pierre in 2006 (Brodeur et al. 2011). The introduction of round goby to North America caused major changes in the structure of native fish communities (Vanderploeg et al. 2002). The second new exotic, tench (*Tinca tinca*), was introduced into the upper Richelieu River in the late 1990s and recently expanded its range to the whole St. Lawrence River between Montréal and Québec City in 2011 (Masson et al. 2013). Tench feeds on invertebrates in shallow vegetated water, where it presumably outcompetes yellow perch for food and habitat; this may impede the restoration of yellow perch stocks to previous densities if tench abundance increases from its current low levels. Another exotic species, the common carp (*Cyprinus carpio*), which was introduced in North America from Europe near the end of the nineteenth century, was unknown in the Lake Saint-Pierre area in 1924 (McCrimmon 1972) but was already abundant in the archipelago in the 1960s (De Koninck 1970). This species is still abundant in Lake Saint-Pierre and its archipelago, but it is not thought to have specifically impacted the yellow perch population.

A strong yellow perch predator, the double-crested cormorant, recently appeared in Lake Saint-Pierre. Colonies have rapidly expanded since the end of the 1990s, from

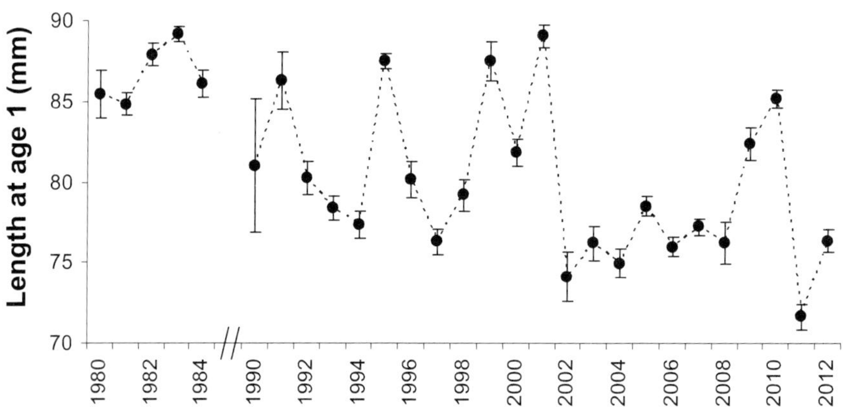

Fig. 11. Mean back-calculated length at age 1 (±SE) of yellow perch in Lake Saint-Pierre based on opercula for year-classes between 1980 and 2012.

66 nesting pairs in 1998 to 575 in 2001 and 948 in 2004. In addition, a few thousand migrating cormorants arrive in the lake in late summer. Predation on the perch stock was not responsible for initiating the collapse, but it represents another source of mortality likely reducing recruitment and hindering the effectiveness of management plans (Magnan et al. 2008). Since 2008, cormorant eggs have been systematically oiled to reduce recruitment, and the population has since stabilized at about 625 nesting pairs.

Another factor likely impeding the rehabilitation of perch stocks is the net loss of the best spawning areas due to changes in land use in the floodplain (Fig. 2f). By the early 1990s, the extensive agriculture practiced in the floodplain increasingly included intensive annual crops (i.e., corn and soy) (Richard et al. 2011). Agriculture development has resulted in a net loss of 5000 ha of key fish spawning and nursery habitats, rendering yellow perch more vulnerable to recruitment failure. This loss can represent more than 50% of the potential spawning habitat during years of high spring floods (Lecomte et al. 2012). Historically, the fact that Lake Saint-Pierre yellow perch spawned earlier in the large floodplain was considered an advantage compared to the upstream populations, which spawn in deeper water along the shores. Such an advantage was associated with the exploitation of a warmer and more productive habitat during larval development and a longer growing season for young-of-the-year. This advantage was lost after the development of intensive agriculture practices in the floodplain and the resulting transformation of the most productive spawning habitats into plowed fields. In addition, connectivity between the lake and its floodplain has been reduced because of inadequate culverts installed on roads and agriculture lands.

Two mechanisms are suggested to explain the absence of recovery following population collapses. The first, called the "Allee effect," refers to inverse density dependence at low densities (Courchamp et al. 1999). Factors generating such an inverse density dependence include (1) genetic inbreeding and a loss of heterozygosity, leading to decreased fitness, (2) demographic stochasticity (including sex-ratio fluctuations), as illustrated by the failure of many biological control programs that released control insects in too small numbers, and (3) reduction of mate encounters during the reproductive periods when density is too low (Courchamp et al. 1999). The second mechanism, known as the "cultivation/depensation effects" on juvenile survival and recruitment (*sensu* Walters and Kitchell 2001), refers to interspecific interactions among fish species involved in a fishery: dominant species that are the basis of many fisheries may be successful due partly to "cultivation effects," where adults crop down other species that are potential competitors/predators of their own juveniles. Such effects imply a reverse impact when adult abundance is severely reduced by fishing, i.e., increases in these other species may then cause lagged, apparently depensatory decreases in juvenile survival. Cultivation/depensation effects can then delay or prevent stock rebuilding and may explain low recruitment success following severe declines in some major marine stocks, such as Atlantic cod (*Gadus morhua*) off Newfoundland (Walters and Kitchell 2001).

It is unlikely that an Allee effect would prevent recovery of yellow perch stocks in Lake Saint-Pierre and the Bécancour–Batiscan sector. Sampling done since the 2000s has proven that adults are still present in the system and contribute to larvae production in May–June. In contrast, the cultivation/depensation effects could be involved but in a slightly different way than that described by Walters and Kitchell

(2001). A potential scenario might involve a slow growth and a very low survival of yellow perch larvae due to habitat degradation (shift from submerged macrophytes to benthic mats of filamentous cyanobacteria) as a first step, which is purely abiotic. Afterward, the cultivation/depensation effects would depress the growth and survival of the "larval stage survivors" (after June of the first year), through competition for food (round goby) and an overall reduction of the amount of energy stored to withstand winter starvation. The end result would be an aging population and very low yellow perch recruitment (Figs. 4 and 10).

5.6.4 Yellow Perch in the Bécancour–Batiscan Sector: Another Severely Impacted Population

Yellow perch was a dominant species of the Bécancour–Batiscan fish community in 1996 (Fournier et al. 1997). Since then, its abundance has continuously decreased to the point that it became a marginal species about 15 years later. This situation is similar to that of Lake Saint-Pierre and probably occurred because the management decisions required to reduce fishing pressure were not implemented in time. As is the case in Lake Saint-Pierre, the yellow perch population dynamics in the Bécancour–Batiscan sector reflect those of a highly stressed population with rapid sexual maturation, low female biomass, and low recruitment over the past decade.

As presented in Sections 5.4 and 5.6.3, the Bécancour–Batiscan yellow perch stock is related to that of Lake Saint-Pierre in many biological and socio-economic aspects. However, the fishery potential of this stock is lower because the surface areas supporting it are smaller and located along a small fringe of the south shore between Bécancour and Gentilly. The north shore of the river was largely artificialized in the 1950s, during construction of the seaway. Approximately 30 commercial fishermen had fishing licenses between 1995 and 2012. When both fisheries were permitted, the yellow perch landings from commercial fishing were several orders of magnitude higher than those of the sport fishery.

There was a clear synchronization in the decline in yellow perch catches in the Bécancour–Batiscan sector and Lake Saint-Pierre (Fig. 9), suggesting close ecological and demographic links between the two stocks. Unlike the situation in Lake Saint-Pierre, the double crested cormorant does not currently represent a threat to perch in the Bécancour–Batiscan sector. Shore artificialization, the development of the floodplain for industrial and agricultural purposes, and the proliferation of benthic cyanobacteria in nursery habitats could contribute to the lack of yellow perch recovery. The demographic explosion of round goby in the late 2000s may have impacted the yellow perch population.

The status of the Bécancour–Batiscan yellow perch stock is not as well documented as that of Lake Saint-Pierre. The survey of this population started only in 2001, although some restrictions were also applied to this sector in 1997 (Table 3) to facilitate implementation of the Lake Saint-Pierre management plan, since many commercial fishermen owned fishing licences for both sectors. Complete closure of the sport and commercial fisheries was imposed in this sector in 2013.

5.7 Conclusions

The scientific data gathered since the 1970s have revealed that yellow perch was an abundant, dominant, and resilient species in all sectors of the fluvial St. Lawrence River until the mid-1990s. Despite the central role played by yellow perch in the St. Lawrence fish community, its population dynamics varied greatly among sectors. Globally, the two upper sectors of the river, lakes Saint-François and Saint-Louis, still host abundant stocks. Their population dynamics (abundance, age structure, recruitment, female biomass) are more representative of balanced stocks than those of the downstream part of the river (Lake Saint-Pierre and its archipelago, Bécancour–Batiscan). The populations of the downstream sectors were in a stressed and fragile state from the late 1970s to the middle of the 1990s. Between 1995 and the late 2000s, a dramatic collapse of these stocks occurred in both Lake Saint-Pierre and Bécancour–Batiscan, where commercial and sport fisheries were finally closed in 2012 and 2013, respectively, in an ultimate effort to preserve the potential to rebuild the stock to a sustainable level. In the downstream sectors of the St. Lawrence River, the status of yellow perch went from a fully exploited dominant species to a marginal contributor to the fish community in about 15 years. Since this major change was well documented and reasons for the stock collapse were understood, the lessons learned from this case have to be taken into account before implementing future management decisions on yellow perch.

While detrimental anthropogenic pressures such as shore artificialization, urbanization, and sport fishing are quite high in the two upper sectors, the local perch populations are surprisingly the healthiest of the St. Lawrence. Past management decisions implemented in the early 1960s shut down the commercial fishery and left sport fishing as the only harvest. The 1997–1999 management measures (complete protection of the spawning period, 50 perch bag limit, and a ban on selling the catch) contributed to minimizing the impact of sport fishing. In Lake Saint-François, this picture may be misleading: recent changes in the fish communities, such as decreasing abundance of sauger and walleye associated with habitat degradation and the demographic explosion of round goby since the beginning of 2000, could eventually impact yellow perch populations.

Before the 2000s, the abundance, growth, condition, and reproductive potential of yellow perch were lower on the south than the north shore of Lake Saint-Louis. These differences were associated with the higher chemical contamination on the south shore. In the last decade, however, both the north and south shore population densities reached their highest levels. Since yellow perch densities along the south shore exhibited values comparable to those observed on the north shore, we suggest that the south shore population improved following decreasing organic contamination coupled with the reduction of the sport fishing pressure. Considering this recent improvement and the fact that levels of mercury are still high in sediments and in the food chain, we must pay close attention to the south shore fish communities in the forthcoming years.

In contrast, the yellow perch populations of Lake Saint-Pierre and Bécancour–Batiscan collapsed over the last decade. This decline occurred despite the implementation of major reductions in fishing effort from the 2000s. Yellow perch populations of the Great Lakes recovered from collapses after fishing pressure was reduced (Lake Ontario: Lake Ontario Management Unit 2007; Lake Erie: Belore et al. 2007; Lake Michigan: Francis et al. 2007; Makauskas and Clapp 2007). In the downstream sectors of the St. Lawrence River, socio-economic factors and habitat degradation have likely prevented the recovery so far. In these sectors, the most prominent pressures were the two fishing industries. The two downstream yellow perch populations were the only ones that were still exploited after the 1960s by both the commercial and sport fisheries. The total yellow perch mortality rate (generally > 75% annually) was always higher than that of upstream populations. Models predicted that the observed fishing yields could be sustainable only if recruitment could be maintained at high levels (Guénette et al. 1994). However, since fishing mortality was the primary source of mortality (~60%), the fishing industry likely played a major role in imperilling the local yellow perch populations by keeping them in such precarious balance. At this time, the possibly most effective management decision—reducing the exploitation rate—was continuously postponed or ineffectively applied because of the fishing lobby. Similarly, restricting fishing to earlier during the spawning period and markedly decreasing the global fishing pressure (commercial and sport) would have certainly increased the capacity of yellow perch populations to withstand the increasing habitat degradation of the shore and floodplain.

Starting in the 2000s, the increasing environmental and anthropogenic pressures resulting mainly from agriculture development further burdened the fragile equilibrium between recruitment and fishing. At this point, these additional stresses restricted recruitment and stocks collapsed. Malfunctioning culverts installed under roads and fields have also affected the connectivity between the lake (= growing habitat) and its floodplain (= spawning habitat). In addition to the major loss of spawning areas in the floodplain, the abundance of submerged macrophytes declined and benthic cyanobacteria proliferated in nursery areas. These major changes in such essential habitats are believed to be a consequence of excessive nutrients and sediment loads originating from tributaries draining the agricultural zone of the St. Lawrence lowlands. Further pressure on young stages of yellow perch came from the predation threat by a new predator, the double-crested cormorant. Finally, two new exotic fish species, round goby and perhaps tench, may have an additional pressure on yellow perch dynamics.

Lessons learned from the recent stock collapse within Lake Saint-Pierre (1995–present) remind fisheries biologists and managers that it is unwise to maintain a strong exploitation rate on a fragile fish stock. The Lake Saint-Pierre stock is now in such a poor state that a sustainable fishery will not be possible until major actions and regulations can be implemented to restore a healthy yellow perch stock. Unfortunately, the nature of the pressures preventing a recovery cannot be dealt with only through the usual fishery management tools that have been used in the past, such as modulating fishing pressure. For this reason, three major measures are in the process of being implemented: (1) improving water quality through the modification of agricultural practices in the watershed (e.g., reducing nutrient and sediment loading, lowering turbidity), (2) restoring essential habitat in the floodplain by changing land use (e.g., removing the corn and soy fields in the floodplain), and (3) enhancing the connectivity between the floodplain and the river (e.g., replacement, maintenance, and building of new culverts). The implementation of such wide-reaching measures will require the cooperative efforts of many different departments of the Québec Government as well as non-governmental organizations. Since yellow perch is considered by many as a ubiquitous species that has all the characteristics of a potentially highly invasive species, it is hoped that the measures implemented will allow the famous Lake Saint-Pierre fishery to come back sooner rather than later.

5.8 Future Perspectives

The availably of large-scale historic surveys and studies allowed us to further analyze the population dynamics of the yellow perch in a new light. Past scientific surveys were planned to aid management decisions concerning the St. Lawrence River yellow perch fisheries by revealing key processes. In the same way, it is essential that we continue working to better our *understanding* of the *mechanisms* affecting recruitment and the effects of the habitat alterations that negatively impacted the Lake Saint-Pierre stock. The progressive disappearance of submerged macrophytes in Lake Saint-Pierre and the perturbations observed in its floodplain were not only associated with the decreasing abundance of yellow perch but of many species exploiting these key habitats during their life cycle. Control of the double-crested cormorant has also proven to be successful in improving some yellow perch fisheries (Van Guilder and Seefelt 2012), but this will need to be evaluated in Lake Saint-Pierre, where control has been implemented only recently. Such information is certainly of interest from a scientific point of view, but more importantly it informs stakeholders and politicians as to which factors are driving these fish communities and how efforts could be maximized to reach goals.

After having reduced fish mortality (using regulation of fishing and finally a moratorium), the next step is to take actions for habitat restoration. The situation in Lake Saint-Pierre, and presumably in Bécancour-Batiscan sector, will improve only when yellow perch are provided with a healthy environment in which to spawn, develop, and grow. In lakes Saint-François and Saint-Louis, the reduction of fishing mortality (allowing only sport fishing in the 1960s and implementing conservative management measures at the end of the 1990s) likely helped yellow perch maintain its dominance despite profound changes in the fish community in the former and following a major reduction of chemical contaminants in the latter. These examples are encouraging and indicate that yellow perch may be able to once again dominate the fish community when fishing pressure is controlled and habitat has improved. The joint efforts devoted to the reestablishment of yellow perch in Lake Saint-Pierre represent a new attempt to rehabilitate fish habitat despite past conflicting use. While fishermen were directly targeted to restore the population of yellow perch in Lake Saint-Pierre, habitat restoration will require the involvement of a broad range of stakeholders. Mobilization will inevitably be costly. In return, it will result in significant socio-economic benefits at the local as well as regional level. Habitat rehabilitation will require stakeholders with varied expertise, and consistent effort must be applied in an effective and sustainable manner if the habitats of Lake Saint-Pierre and their fish populations, including perch, are to be restored.

From a scientific point of view, the St. Lawrence River is an open-air laboratory revealing the different responses of its perch populations according to the pressures prevailing in each section. This is a unique situation that we must continue to survey in order to improve our understanding of the yellow perch. The St. Lawrence River is like an arena in which important societal choices will be made that promote natural and human needs.

5.9 Acknowledgements

We first want to highlight the work of J.-P. Cuerrier, E. Magnin, R. Fortin, and J.-R. Mongeau, which still continues to inspire our collective effort to develop a comprehensive approach to the conservation of the St. Lawrence River yellow perch populations. We also wish to acknowledge the contribution of all biologists, wildlife technicians, game wardens, and sport and commercial fishermen for their involvement over the last 25 years in this long-term objective. L. Devine provided helpful comments and suggestions during text revisions.

5.10 Appendix: Anthropogenic Pressures Definition

The major anthropogenic pressures affecting yellow perch populations in each sector of the St. Lawrence River (summarized in Table 1) were defined and estimated as follows: (A) The agriculture pressure was determined using a combined measure of phosphorus loading from agriculture (P [mg/l] of tributaries × outflows) and the surface areas cultivated in the floodplain (× hectares cultivated in the floodplain/total "wet" surfaces during the high spring flood of 10 years recurrence). Elevated phosphorus loading has been identified as a primary cause for algal and cyanobacteria blooms and increased growth of underwater macrophytes, both of which have deleterious impacts on yellow perch recruitment and habitat quality (Hudon et al. 2012). In addition, areas cultivated for corn and soy in the floodplain directly reduce the availability of the best spawning habitats, which are located on the floodplain during years of high water levels. (B) Shore artificialization pressure was calculated using the percentage of shoreline altered. Shoreline artificialization affects water velocities (hence advection), increases local erosion and turbidity, and in most areas results in a net loss of spawning habitats/nursery grounds (e.g., through loss of macrophytes). (C) Water-level regulation was calculated using the coefficient of variation of water levels (standard deviation/mean). Alterations in natural water-level variations modify vegetation succession, the removal of silty sediment deposits, and, most importantly, the availability of areas used as spawning sites and nursery grounds that naturally occur in the floodplains. (D) Urbanization pressure was calculated using the density of inhabitants per km^2 in the watershed immediately surrounding the area of interest (i.e., not considering the contribution of a tributary polluted by a distant city). The impacts are multiple and may include pollutants directly entering the river (e.g., sewage systems, emerging pollutants like pharmaceuticals, industrial discharge) or indirectly through conflicting uses (e.g., construction in the watershed, sedimentation, increase of recreational uses, loss of wetlands). (E) Even though other fishing gears were used, commercial fishing pressure during the late 1990s and the early 2000s was calculated using the fishing effort by hoop-nets because it is the only one having the capacity to significantly impact the littoral fish community. Hoop-net selectivity is low. They were numerous and spread over large areas in the portions of the St. Lawrence River where commercial fishing was permitted. The St. Lawrence River was divided into 5 km segments, and the mean number of hoop-nets observed in each of the three fishing seasons (spring, summer, fall) was used to determine the commercial fishing pressure as absent (0), low (1) (mean number of hoop-nets [MNHN] between 1 and 20), average (3) (MNHN 21–100), or high (5) (MNHN 101–350). (F) Sport fishing pressure during the late 1990s and the early 2000s was estimated for each sector by three specialists. Each specialist (two senior fishery managers and one experienced fishing guide in the study area) independently provided his evaluation on a scale of 1 (lowest) to 5 (highest), taking into account open-water and ice fishing. Their individual estimates were based on their own observations, surveys of recreational fishing and outfitting services, and statistics from sport fishery creel censuses or yellow perch catch when available. For most sectors, the estimations were quite similar and were averaged. Differences were discussed and estimates were decided on according to data from available studies.

5.11 References

Aravindakshan, J., V. Paquet, M. Gregory, J. Dufresne, M. Fournier, D.J. Marcogliese and D.G. Cyr. 2004. Consequences of xenoestrogen exposure on male reproductive function in spot tail shiners (*Notropis hudsonius*). Toxicol. Sci. 78: 156–165.

Bechara, J., J. Morin and P. Boudreau. 2003. Évolution récente de l'habitat du doré jaune, de la perchaude, du grand brochet et de l'achigan à petite bouche au lac Saint-François, fleuve Saint-Laurent. Rept. 640, INRS-Eau, Terre & Environnement, for the Comité ZIP du Haut Saint-Laurent.

Belore, M., A. Cook, D. Einhouse, T. Hartman, K. Kayle, R. Kenyon, C. Knight, T. MacDougall and M. Thomas. 2007. Report of the Lake Erie Yellow Perch Task Group. Presented to the Great Lakes Fishery Commission, Lake Erie Committee, Standing Technical Committee.

Bertrand, M., G. Cabana, D.J. Marcogliese and P. Magnan. 2011. Estimating the feeding range of a mobile consumer in a river-flood plain system using $\delta^{13}C$ gradients and parasites. J. Anim. Ecol. 80: 1313–1323.

Bourbeau, D., Y. Mailhot and J.C. Bourgeois. 1992. Historique de la gestion et de l'effort de pêche commerciale au lac Saint-Pierre de 1961 à 1989. Ministère du Loisir, de la Chasse et de la Pêche, Direction régionale Mauricie, Bois-Francs, Service de l'aménagement et de l'exploitation de la faune. Trois-Rivières-Ouest, Québec.

Boyer, C., D. Chaumont, I. Chartier and A.G. Roy. 2010. Impact of climate change on the hydrology of St. Lawrence tributaries. J. Hydrol. 384: 65–83.

Brazo, D.C., P.I. Tack and C.R. Liston. 1975. Age, growth, and fecundity of yellow perch, *Perca flavescens*, in Lake Michigan near Ludington, Michigan. Trans. Am. Fish. Soc. 104: 726–730.

Brodeur, P., Y. Reyjol, M. Mingelbier, T. Rivière and P. Dumont. 2011. Prédation du gobie à taches noires par les poissons du Saint-Laurent: contrôle potentiel d'une espèce exotique? Nat. Can. 135: 4–11.

Carignan, R., S. Lorrain and K. Lum. 1994. A 50-year record of pollution by nutrients, trace metals, and organic chemicals in the St. Lawrence River. Can. J. Fish. Aquat. Sci. 51: 1088–1100.

Carignan, R. and S. Lorrain. 2000. Sediment dynamics in the fluvial lakes of the St. Lawrence River: accumulation rates and characterization of the mixed sediment layer. Can. J. Fish. Aquat. Sci. 57 (Suppl. 1): 63–77.

Chen, Y. and J.E. Paloheimo. 1994. Estimation of fish length and age at 50% maturity using a logistic type model. Aquat. Sci. 56: 206–219.

Cholmondeley, R. 1989. The 1987 Thousand Islands warm water assessment. Ontario Ministry of Natural Resources, St. Lawrence River Fisheries Management Unit Report 1989–01, Brockville, Ontario.

Clady, M.D. 1976. Influence of temperature and wind on the survival of early stages of yellow perch, *Perca flavescens*. J. Fish. Res. Board Can. 33: 1887–1893.

Courchamp, F., T. Clutton-Brock and B. Grenfell. 1999. Inverse density dependence and the Allee effect. Trends Ecol. Evol. 14: 405–410.

Courtemanche, M. 2012. Site Droulers-Tsiionhiakwatha (BbgFn-1): identification des restes squelettiques récoltés dans une fosse de la maison longue #2, dite la structure #57. Université de Montréal, Ostéothèque de Montréal, Inc., Rept. 292 prepared for C. Chapdelaine, Québec.

Craig, J.F. 2000. Percid Fishes: Systematics, Ecology and Exploitation. Blackwell Science, Oxford.

Cuerrier, J.-P. 1962. Aperçu général sur l'inventaire biologique des poissons et des pêcheries de la région du lac Saint-Pierre. Nat. Can. 89: 193–213.

Cuerrier, J.-P., F.E.J. Fry and G. Préfontaine. 1946. Liste préliminaire des poissons de la région de Montréal et du lac Saint-Pierre. Nat. Can. 73: 17–32.

De Koninck, R. 1970. Les cent-îles du lac Saint-Pierre. Les presses de l'université Laval, Québec.

de Lafontaine, Y. and G. Costan. 2002. Introduction and transfer of alien aquatic species in the Great Lakes drainage basin. St. Lawrence River. pp. 73–91. *In*: R. Claudi, P. Nantel and E. Muckle-Jeffs (eds.). Aliens Invaders in Canada's Waters, Wetlands, and Forests. Canadian Forest Service, Natural Resources Canada, Ottawa.

de Lafontaine, Y., F. Marchand, D. Labonté and M. Lagacé. 2002. The hydrological regime and fish distribution and abundance in the St. Lawrence River: are experimental trap data a valid indicator? Report submitted to the International Joint Commission Study Board. Environnement Canada, Centre Saint-Laurent, Montréal, Québec.

Doyon, N., C. Langlois, L. Lapierre, Y. de Lafontaine and G. Walsh. 1993. Contaminations et indicateurs d'effets sous-létaux chez les poissons du fleuve Saint-Laurent. Environnement Canada, Centre Saint-Laurent, Montréal, Québec.

Dumont, P. 1996. Comparaison de la dynamique des populations de perchaudes (*Perca flavescens*) soumises à des niveaux différents de stress anthropique. Ph.D. Thesis, Université du Québec à Montréal, Québec.

Forney, J.L. 1980. Evolution of a management strategy for the walleye in Oneida Lake, New York. N.Y. Fish Game J. 27: 105–141.

Fortin, R. 1970. Dynamique de la population de *Perca flavescens* (Mitchill) de la Grande Anse de l'Île Perrot, au lac Saint-Louis. Ph.D. Thesis, Université de Montréal, Québec.

Fortin, R. and É. Magnin. 1972a. Dynamique d'un groupement de perchaudes, *Perca flavescens* (Mitchill), dans la Grande Anse de l'Île Perrot, au lac Saint-Louis. Nat. Can. 99: 367–380.

Fortin, R. and É. Magnin. 1972b. Quelques aspects qualitatifs et quantitatifs de la nourriture des perchaudes, *Perca flavescens* (Mitchill), de la Grande Anse de l'Ile Perrot, au lac Saint-Louis. Ann. Hydrobiol. 3: 79–91.

Fortin, R. and É. Magnin. 1972c. Croissance en longueur et en poids des perchaudes *Perca flavescens* de la Grande Anse de l'Ile Perrot au lac Saint-Louis. J. Fish. Res. Board Can. 29: 517–523.

Fournier, D., F. Cotton, Y. Mailhot, D. Bourbeau, J. Leclerc and P. Dumont. 1996. Rapport d'opération du réseau de suivi ichtyologique du fleuve Saint-Laurent: Échantillonnage des communautés ichtyologiques des habitats lentiques du lac Saint-Pierre et de son archipel en 1995. Ministère de l'Environnement et de la Faune, Direction de la faune et des habitats, Direction régionale de la Mauricie – Bois-Francs, Direction régionale de la Montérégie, Québec, Québec.

Fournier, D., Y. Mailhot and D. Bourbeau. 1997. Rapport d'opération du réseau de suivi ichtyologique du fleuve Saint-Laurent : Échantillonnage des communautés ichtyologiques du tronçon Gentilly-Batiscan en 1996. Ministère de l'Environnement et de la Faune, Direction de la faune et des habitats, Direction régionale Mauricie – Bois-Francs, Québec, Québec.

Francis, R.C., M.A. Hixon, M.E. Clarke, S.A. Murawski and S. Ralston. 2007. Ten commandments for ecosystem-based fisheries scientists. Fisheries 32 (5): 217–233.

Frenette, M., C. Barbeau and J.-L. Verrette. 1989. Aspects quantitatifs, dynamiques et qualitatifs des sédiments du Saint-Laurent. Hydrotechinc. Experts-conseils, pour Environnement Canada et gouvernement du Québec, Projet de mise en valeur du Saint-Laurent, Québec.

Frenette, J.-J., M.T. Arts, J. Morin, D. Gratton and C. Martin. 2006. Hydrodynamic control of the underwater light climate in fluvial Lac Saint-Pierre. Limnol. Oceanogr. 51: 2632–2645.

Fry, F.E.S, J.-P. Cuerrier and G. Préfontaine. 1941. Première croissance de la perchaude (*Perca flavescens*) Mitchill, dans le lac St-Louis, en 1941. pp. 181–187. *In*: Travaux du Ministère de la chasse et de la pêche de Québec, Rapport de la Station Biologique de Montréal et de la Station Biologique du Parc des Laurentides pour l'année 1941, Fascicule 2, Appendice IX.

G.D.G. Environnement Ltée. 1986. Situation de la pêche commerciale dans le tronçon fluvial Trois-Rivières/Québec (août 1983 à juillet 1984). Révisé en février 1989. Présenté au Ministère de l'Agriculture, des Pêcheries et de l'Alimentation du Québec, Québec.

Glémet, H. and M.A. Rodríguez. 2007. Short term growth (RNA/DNA ratio) of yellow perch (*Perca flavescens*) in relation to environmental influences and spatio-temporal variation in a shallow fluvial lake. Can. J. Fish. Aquat. Sci. 64: 1646–1655.

Gordon, W.H. 1991. Lake St. Lawrence warmwater fish stock assessment - 1990. St. Lawrence River Sub-Committee, Ontario Ministry of Natural Resources and New State Department of Environmental Conservation, 1991. Ann. Rep. 12.1–12.8.

Grimaldi, J. 1967. Comparative growth of the yellow perch, *Perca flavescens*, in lakes and rivers in Quebec. M.Sc. thesis, McGill University, Montréal, Québec.

Guénette, S., Y. Mailhot, I. McQuinn, P. Lamoureux and R. Fortin. 1994. Paramètres biologiques, exploitation commerciale et modélisation de la population de perchaudes (*Perca flavescens*) du lac Saint-Pierre. Québec, Ministère de l'Environnement et de la Faune et Université du Québec à Montréal, Montréal, Québec.

G.V.L. Environnement Inc. 2001. Évaluation des impacts des mesures de gestion de la pêche commerciale à la perchaude au lac Saint-Pierre, saison 2000. Prepared for the Association des Pêcheurs Commerciaux du lac Saint-Pierre.Nicolet, Québec.

Hamilton, S.J., P.M. Mehrle, F.L. Mayer and J.R. Jones. 1981. A method to evaluate mechanical properties of bone in fish. Trans. Am. Fish. Soc. 110: 708–717.

Harnois, E., R. Couture and P. Magnan. 1992. Variation saisonnière dans la répartition des ressources alimentaires entre cinq espèces de poissons en fonction de la disponibilité de proies. Can. J. Zool. 70: 796–803.

Hartman, K.J., S.J. Nepszy and R.L. Scholl. 1980. Minimum size limits for yellow perch (*Perca flavescens*) in western Lake Erie. Great Lakes Fish. Comm. Tech. Rep. 22.

Hendrick, A. 1991. 1990 Lake St. Francis index netting survey. St. Lawrence River fisheries management unit, Ontario Ministry of Natural Resources, Brockville, Ontario.

Hendrick, A. 1993. 1992 Middle corridor warm water fisheries assessment. St. Lawrence River Sub-Committee, Ontario Ministry of Natural Resources and New State Department of Environmental Conservation, 1992. Ann. Rep. 6.1–6.8.

Hontela, A., P. Dumont, D. Duclos and R. Fortin. 1995. Endocrine and metabolic dysfunction in yellow perch (*Perca flavescens*) exposed to PAHs, PCBs and heavy metals in the St. Lawrence River. Environ. Toxicol. Chem. 14: 725–731.

Hoyle, J.A. 1991. Status of selected fish populations of eastern Lake Ontario. *In*: Lake Ontario Fisheries Unit 1991 Annual Report. Prepared for the Lake Ontario Committee meeting, March 25–26 1992, Great Lakes Fish. Comm.

Hudon, C. and R. Carignan. 2008. Cumulative impacts of hydrology and human activities on water quality in the St. Lawrence River (Lake Saint-Pierre, Quebec, Canada). Can. J. Fish. Aquat. Sci. 65: 1165–1180.

Hudon, C., A. Armellin, P. Gagnon and A. Patoine. 2010. Variations in water temperatures and levels in the St. Lawrence River (Québec, Canada) and potential implications for three common fish species. Hydrobiologia 647: 145–161.

Hudon, C., A. Cattaneo, A.-M. Tourville Poirier, P. Brodeur, P. Dumont, Y. Mailhot, Y.-P. Amyot, S.-P. Despatie and Y. de Lafontaine. 2012. Oligotrophication from wetland epuration alters the riverine trophic network and carrying capacity for fish. Aquat. Sci. 74: 495–511.

Huss, M., P. Byström, A. Strand, L.-O. Eriksson and L. Persson. 2008. Influence of growth history on the accumulation of energy reserves and winter mortality on young fish. Can. J. Fish. Aquat. Sci. 65: 2149–2156.

Knight, R.L., D.H. Davies and M.W. Turner. 1991. Status of Lake Erie Fishes. Ohio Department of Natural Resources, Lake Erie Fisheries Research, Project F-35-R29.

Koonce, J.F., T.B. Bagenal, R.F. Carline, K.E.F. Hokanson and M. Nagiec. 1977. Factors influencing year-class strength of Percids: a summary and a model of temperature effects. J. Fish. Res. Board Can. 34: 1900–1909.

Lacasse, J. 1987. La mécanisation de la pêche commerciale à Notre-Dame-de-Pierreville: évolution du geste traditionnel. M. A. thesis, Université Laval, Québec.

Lajeunesse, A., C. Gagnon, F. Gagné, S. Louis, P. Čejka and S. Sauvé. 2011. Distribution of antidepressants and their metabolites in brook trout exposed to municipal wastewaters before and after ozone treatment—Evidence of biological effects. Chemosphere 83: 564–571.

Lake Ontario Management Unit. 2007. 2006 annual report of the Lake Ontario management unit. Great Lake Fishery Commission. Ypsilanti, Michigan.

Laliberté, D. 2003. Évolution des teneurs en mercure et en BPC de quatre espèces de poissons du Saint-Laurent, 1976–1997. Ministère de l'Environnement, Direction du suivi de l'état de l'environnement, ENV/2003/0287, Québec.

La Violette, N. 2004. Les lacs fluviaux du Saint-Laurent: Hydrologie et modifications humaines. Nat. Can. 128: 98–104.

La Violette, N., D. Fournier, P. Dumont and Y. Mailhot. 2003. Caractérisation des communautés de poissons et développement d'un indice d'intégrité biotique pour le fleuve Saint-Laurent, 1995–1997. Faune et Parcs Québec, Direction de la recherche sur la faune, Québec.

Leclerc, E., Y. Mailhot, M. Mingelbier and L. Bernatchez. 2008. The landscape genetics of yellow perch (*Perca flavescens*) in a large fluvial ecosystem. Mol. Ecol. 17: 1702–1717.

Leclerc, J. 1984. Frayères et habitats potentiels de 11 espèces de poissons dans l'archipel de Montréal. Ministère du Loisir, de la Chasse et de la Pêche, Service Archipel, Montréal, Québec.

Leclerc, M. 1956. La pêche au Village-d'en-bas. Thèse de License ès Lettres (Civilisation). Université Laval, Québec.

Leclerc, P. 1987. Les perchaudes (*Perca flavescens*) du lac Saint-Pierre: biologie des populations et diagnose de l'intensité de l'exploitation sportive et commerciale. M.Sc. Thesis, Université du Québec à Montréal, Québec.

Lecomte, F., M. Mingelbier, P. Brodeur, P. Dumont, Y. Mailhot, N. Vachon, G. Richard and J. Morin. 2012. How human alterations and natural variations explain opposite responses in yellow perch stocks along the St. Lawrence River? Poster presented at the annual conference of the International Association for Great Lakes Research. Cornwall, Canada. 13–17 May 2012.

Leung, C., P. Magnan and B. Angers. 2011. Genetic evidence for sympatric populations of yellow perch (*Perca flavescens*) in Lake Saint-Pierre (Canada): the crucial first step in developing a fishery management plan. J. Aquat. Res. Dev. 2011, S6, http://dx.doi.org/10.4172/2155-9546.S6-001.

Magnan, P. 2002. Avis scientifique sur l'état du stock de perchaudes au lac Saint-Pierre, les indicateurs biologiques utilisés pour effectuer son suivi et la pertinence de protéger la période de fraye de façon partielle ou totale. Université du Québec à Trois-Rivières. Prepared for the Société de la faune et des parcs du Québec and the Ministère de l'Agriculture, des Pêcheries et de l'Alimentation du Québec, Trois-Rivières, Québec.

Magnan, P., P. Dumont, Y. Mailhot, F. Coulombe and L. Therrien. 2004. État du stock de perchaude du lac Saint-Pierre en 2003 et recommandations sur le niveau d'exploitation soutenable en 2004. Comité aviseur sur la gestion de la perchaude du lac Saint-Pierre, Université du Québec à Trois-Rivières, Trois-Rivières, Québec.

Magnan, P., Y. Mailhot and P. Dumont. 2008. État du stock de perchaude du lac Saint-Pierre en 2007 et efficacité du plan de gestion de 2005. Comité aviseur scientifique sur la gestion de la perchaude du lac Saint-Pierre, Université du Québec à Trois-Rivières et ministère des Ressources naturelles et de la Faune du Québec, Québec.

Magnan, P., P. Brodeur, N. Vachon, Y. Mailhot, P. Dumont and Y. Paradis. 2014. État des stocks de perchaude du lac Saint-Pierre et du tronçon Bécancour-Batiscan en 2011–2012 et bilan du plan de gestion de 2008. Comité aviseur scientifique sur la gestion de la perchaude du lac Saint-Pierre, Université du Québec à Trois-Rivières et ministère des Ressources naturelles et de la Faune du Québec.

Magnin, E. 1966. La pêche commerciale sur le fleuve Saint-Laurent et sa place dans l'ensemble des pêches commerciales de la province de Québec. Université de Montréal, Département des Sciences biologiques, Québec.

Mailhot, Y. 1997. Plan de gestion de la pêche à la perchaude et suivi scientifique de la population de la perchaude. pp. 141–146. *In*: M. Bernard and C. Groleau (eds.). Compte rendu du deuxième atelier sur les pêches commerciales. Duchesnay, Dec. 10–12 1996. Ministère de l'Environnement et de la Faune, Québec.

Mailhot, Y. 2001. Évaluation du taux annuel de mortalité totale des perchaudes du lac Saint Pierre et de son archipel en 1999. pp. 197–204. *In*: M. Bernard et C. Groleau (eds.). Compte rendu du cinquième atelier sur les pêches commerciales, Duchesnay, Jan. 18–20 2000. Faune et Parcs Québec, Québec.

Mailhot, Y., F. Axelsen, P. Dumont, H. Fournier, P. Lamoureux, C. Pomerleau and B. Portelance. 1987. Avis scientifique sur le statut de la population de la perchaude au lac Saint-Pierre. Ministère du Loisir, de la Chasse et de la Pêche du Québec et ministère de l'Agriculture, des Pêcheries et de l'Alimentation du Québec. Comité scientifique conjoint. Avis scientifique 87/3, Québec.

Mailhot, Y. and P. Dumont. 2003. Another yellow perch population decline in the 1990s: The Lake St. Pierre case study, St. Lawrence River, Quebec. pp. 135–136. *In*: T.P. Barry and J.A. Malison (eds.).

Proceedings of Percis III: The Third International Percid Fish Symposium, University of Wisconsin Sea Grant Institute, Madison, Wisconsin.

Makauskas, D. and D. Clapp. 2007. Status of yellow perch in Lake Michigan and yellow perch task group progress report. Report to the Lake Michigan Committee, Ypsilanti, Michigan.

Marcogliese, D.J., L. Gagnon-Brambilla, F. Gagné and A.D. Gendron. 2005. Joint effects of parasitism and pollution on oxidative stress biomarkers in yellow perch *Perca flavescens*. Dis. Aquat. Organ. 63: 77–84.

Massé, G. and J.-R. Mongeau. 1974. Répartition géographique des poissons, leur abondance relative et bathymétrie de la région du lac Saint-Pierre. Ministère du Tourisme, de la Chasse et de la Pêche, Service de l'aménagement de la faune, Tech. Rep. 06-01, Montréal, Québec.

Masson, S., Y. de Lafontaine, A.-M. Pelletier, G. Verreault, P. Brodeur, N. Vachon and H. Massé. 2013. Dispersion récente de la tanche au Québec. Nat. Can. 137: 55–61.

McCrimmon, H.R. 1972. La carpe au Canada. Office des Recherches sur les Pêcheries du Canada, Ottawa. Bull. 165.

Ménard, C., A.-M. Prudhomme, J. Bureau and M. Léveillé. 1993. Mixed-function oxygenase (MFO) enzymes as a tool for ecological field studies on fish of the St. Lawrence River. Canadian Conference for Fisheries Research, Trent University, Peterborough, Ontario, January 3–4 1993.

Mingelbier, M. and J. Morin. 2003. Hydrological delineation of the St. Lawrence River using water level and discharge variability. American Fisheries Society, 133rd Conference, Québec City, August 2003.

Mingelbier, M., P. Brodeur and J. Morin. 2008a. Spatialy explicit model predicting the spawning habitat and early stage mortality of northern pike (*Esox lucius*) in a large system: the St. Lawrence River between 1960 and 2000. Hydrobiologia 601: 55–69.

Mingelbier, M., Y. Reyjol, P. Dumont, Y. Mailhot, P. Brodeur, D. Deschamps, C. Côté and N. La Violette. 2008b. St. Lawrence River freshwater fish communities. 2nd edition. Monitoring the state of the St. Lawrence River. St. Lawrence Plan for a sustainable development, Québec and Canada. http://planstlaurent.qc.ca/fileadmin/site_documents/documents/PDFs_accessibles/A9R371B_FINAL_v1.1.pdf.

MLCP (Ministère du Loisir, de la Chasse et de la Pêche). 1984. L'importance de la plaine de débordement du lac Saint-Pierre pour la faune... et pour nous tous. Direction régionale de Trois-Rivières et Direction générale de la faune, Trois-Rivières et Québec.

Mongeau, J.-R. 1979. Recensement des poissons du lac Saint-François comtés de Huntingdon et Vaudreuil-Soulanges, pêche sportive et commerciale, ensemencements de Maskinongés, 1963 à 1977. Ministère du Tourisme, de la Chasse et de la Pêche, Tech. Rep. 06–25, Montréal, Québec.

Mongeau, J.-R. and G. Massé. 1976. Les poissons de la région de Montréal, la pêche sportive et commerciale, les ensemencements, les frayères, la contamination par le mercure et les PCB. Ministère du Tourisme, de la Chasse et de la Pêche. Tech. Rep. 06–13, Montréal, Québec.

Montpetit, C. 1897. Les poissons d'eau douce du Canada. C.-O. Beauchemin & Fils, Montréal, Québec.

Morin, J. and A. Bouchard. 2000. Les bases de la modélisation du tronçon Montréal/Trois-Rivières. Rapport scientifique SMC-Hydrométrie RS-100. Environnement Canada, Sainte-Foy, Québec.

Morin, J. and M. Leclerc. 1998. From pristine to present state: hydrology evolution of Lake Saint-François, St. Lawrence River. Can. J. Civ. Eng. 25: 864–879.

Morin, J., P. Boudreau, Y. Secretan and M. Leclerc. 2000. Pristine Lake Saint-François, St. Lawrence River: Hydrodynamic simulation and cumulative impact. J. Great Lakes Res. 26: 384–401.

Mortsh, L. and F.H. Quinn. 1996. Climate change scenarios for Great Lakes ecosystem studies. Limn. Oceanogr. 41: 903–911.

Munkittrick, K.R. and D.G. Dixon. 1989. Use of white sucker (*Catostomus commersoni*) populations to assess the health of aquatic ecosystems exposed to low-level contaminant stress. Can. J. Fish. Aquat. Sci. 46: 1455–1462.

Nelson, W.A. and C.H. Walburg. 1977. Population dynamics of yellow perch (*Perca flavescens*), sauger (*Stizostedion canadense*), and walleye (*S. vitreum vitreum*) in four main stem Missouri River reservoirs. J. Fish. Res. Board Can. 34: 1748–1763.

Ontario Ministry of Natural Resources. 2004. Regulatory guidelines for managing the yellow perch sport fishery in Ontario. Fisheries Section. Fish and Wildlife Branch, Ontario Ministry of Natural Resources. http://www.mnr.gov.on.ca/stdprodconsume/groups/lr/@mnr/@letsfish/documents/document/stel02_178925.pdf.

Ontario Ministry of Natural Resources. 2013. Lake Ontario Fish Communities and Fisheries: 2012 Annual Report of the Lake Ontario Management Unit. Ontario Ministry of Natural Resources, Picton, Ontario, Canada. http://www.glfc.org/lakecom/loc/mgmt_unit/LOA%2013.01.pdf.

Ouellet, V., M. Mingelbier, A. Saint-Hilaire and J. Morin. 2010. Frequency analysis as a tool for assessing adverse conditions during a massive fish kill in the St. Lawrence River, Canada. Water Qual. Res. J. Can. 45: 47–57.

Parrish, D.L. and F.J. Margraf. 1990. Interactions between white perch (*Morone americana*) and yellow perch (*Perca flavescens*) in Lake Erie as determined from feeding and growth. Can. J. Fish. Aquat. Sci. 47: 1779–1787.

Pluritec Ltée. 1982. Historique de la pêche commerciale en eaux douces depuis 1950. Rapport présenté au Ministère du Loisir, de la Chasse et de la Pêche du Québec, Québec.

Préfontaine, G. 1941. Perchaude, lac Saint-Louis. Ministère de la Chasse et de la Pêche, Montréal, Québec, Manuscript 1656.

R Development Core Team. 2007. R: A language and environment for statistical computing. R Foundation for statistical computing, Vienna, Austria.

Reyjol, Y., P. Brodeur, Y. Mailhot, M. Mingelbier and P. Dumont. 2010. Do native predators feed on non-native prey? The case of round goby in a fluvial piscivorous fish assemblage. J. Great Lakes Res. 36: 618–624.

Richard, G., D. Côté, M. Mingelbier, B. Jobin, J. Morin and P. Brodeur. 2011. Utilisation du sol dans la plaine inondable du lac Saint-Pierre (fleuve Saint-Laurent) durant les périodes 1950, 1964 et 1997 : interprétation de photos aériennes, numérisation et préparation d'une base de données géoréférencées, Québec, Tech. Rep. Prepared for the Ministère des Ressources naturelles et de la Faune and Environnement Canada.

Ricker, W.E. 1975. Computation and interpretation of biological statistics of fish populations. Fish. Res. Board Can. Bull. 191.

Rutherford, E.S., K.A. Rose, J.L. Forney, E.L. Mills, C.M. Mayer and L.G. Rudstam. 1999. Individual-based model simulations of a zebra mussel (*Dreissena polymorpha*) induced energy shunt on walleye (*Stizostedion vitreum*) and yellow perch (*Perca flavescens*) populations in Oneida Lake, NY. Can. J. Fish. Aquat. Sci. 56: 2148–2160.

Scott, W.B. and E.J. Crossman. 1973. Freshwater fishes of Canada. Fish. Res. Board Can. Bull. 184.

State of the St. Lawrence Monitoring Committee. 2008. Overview of the State of the St. Lawrence River 2008. St. Lawrence Plan. Environment Canada, Ministère du Développement durable, de l'Environnement et des Parcs du Québec, Ministère des Ressources naturelles et de la Faune du Québec, Fisheries and Oceans Canada, and Stratégies Saint-Laurent. http://planstlaurent.qc.ca/fileadmin/site_documents/documents/PDFs_accessibles/Portrait_global_2008_e_FINAL_v1.0.pdf.

Thibault, A. 1999. Groupe de travail sur la gestion de la pêche à la perchaude au lac Saint-Pierre. Rapport et recommandations. Présenté à Monsieur Guy Chevrette, Ministre des transports, Ministre délégué aux Affaires autochtones et Ministre responsable de la Faune et des Parcs. Trois-Rivières, Québec.

Thibault, A. 2004. Comité consultatif conjoint pour la gestion des stocks de poissons du lac Saint-Pierre. Bilan et recommandations. Pour le Ministère de l'Agriculture, des Pêcheries et de l'Alimentation et le Ministère des ressources naturelles, de la faune et des Parcs, Québec.

Thibault, A. 2008. Priorité à la perchaude. Comité consultatif conjoint pour la gestion des stocks de poissons du lac Saint-Pierre (CCCGP), Bilan et recommandations. Trois-Rivières, Québec.

Thorpe, J. 1977. Morphology, physiology, behaviour, and ecology of *Perca fluviatilis* L. and *Perca flavescens* Mitchill. J. Fish. Res. Board Can. 34: 1504–1514.

Tremblay, A. and P. Dumont. 1990. La pêche d'hiver dans la plaine du Saint-Laurent: portrait de l'activité et comparaison des techniques de pêche utilisées. Québec, Ministère du Loisir, de la Chasse et de la Pêche, Rep. 06-06, Montréal.

Vachon, N. 2002. Situation et évolution avec la qualité de l'eau des populations de doré jaune (*Stizostedion vitreum*), perchaude (*Perca flavescens*), grand brochet (*Esox lucius*) et achigan à petite bouche (*Micropterus dolomieui*) au lac Saint-François. Tech. rep. prepared for the Comité ZIP du Haut Saint-Laurent, Valleyfield, Québec.

Vachon, N. and P. Dumont. 2007. Examen comparé de l'état des stocks de poisson d'intérêt sportif dans cinq tronçons du fleuve Saint-Laurent à partir des pêches expérimentales effectuées de 1988 à 1997. Québec, Ministère des Ressources naturelles et de la Faune, Tech. Rep. 16–36, Longueuil.

Vachon, N., P. Dumont, P. Brodeur, C. Côté, Y. Mailhot, M. Mingelbier and Y. Paradis. 2013. Réseau de suivi ichtyologique: le lac Saint-François de 1996 à 2009. Québec, Ministère du développement durable, de l'Environnement, de la Faune et des Parcs, Longueuil.

Vanderploeg, H.A., T.F. Nalepa, D.J. Jude, E.J. Mills, K.T. Holeck, J.R. Liebig, I.A. Grigorovich and H. Ojaveer. 2002. Dispersal and emerging ecological impacts of Ponto-Caspian species in the Laurentian Great Lakes. Can. J. Fish. Aquat. Sci. 59: 1209–1228.

Van Guilder, M.A. and N.E. Seefelt. 2012. Double-crested cormorant (*Phalacrocorax auritus*) chick bioenergetics following round goby (*Neogobius melanostomus*) invasion and implementation of cormorant population control. J. Great Lakes Res. 39: 153–161.

Vincent, W.F. and J.J. Dodson. 1999. The St. Lawrence River, Canada—USA: the need for an ecosystem-level understanding of large rivers. Jpn. J. Limnol. 60: 29–50.

Vis, C., A. Cattaneo and C. Hudon. 2008. Shift from chlorophytes to cyanobacteria in benthic macroalgae along a gradient of nitrate depletion. J. Phycol. 44: 38–44.

Walters, C. and J.F. Kitchell. 2001. Cultivation/depensation effects on juvenile survival and recruitment: implications for the theory of fishing. Can. J. Fish. Aquat. Sci. 58: 39–50.

Wells, L. 1977. Changes in yellow perch (*Perca flavescens*) populations in Lake Michigan. J. Fish. Res. Board Can. 34: 1821–1829.

Willis, D.W. and C.S. Guy. 1991. Development and evaluation of a standard weight (Ws) equation for yellow perch. North Am. J. Fish. Manage. 11: 374–380.

[1] Ministère des Forêts, de la Faune et des Parcs, Direction de la gestion de la faune de la Mauricie et du Centre-du-Québec, 100, rue Laviolette, bureau 207, Trois-Rivières (QC) Canada, G9A 5S9 (Retired).
E-mail: yves.mailhot@tlb.sympatico.ca

[2] Ministère des Forêts, de la Faune et des Parcs, Direction de la gestion de la faune de l'Estrie, de Montréal, de la Montérégie et de Laval, 201 Place Charles-Le Moyne (4è étage), Longueuil (QC) Canada, J4K 2T5 (Retired).
E-mail: dumontpierre13@videotron.ca

[3] Ministère des Forêts, de la Faune et des Parcs, Direction de la faune aquatique, 880, Chemin Sainte-Foy (2è étage), Québec (QC) Canada, G1S 4X4.

[a] E-mail: yves.paradis@mffp.gouv.qc.ca

[b] E-mail: marc.mingelbier@mffp.gouv.qc.ca

[c] E-mail: frederic.lecomte.externe@mffp.gouv.qc.ca

[4] Ministère des Forêts, de la Faune et des Parcs, Direction de la gestion de la faune de la Mauricie et du Centre-du-Québec, 100, rue Laviolette, bureau 207, Trois-Rivières (QC) Canada, G9A 5S9.
E-mail: philippe.brodeur@mffp.gouv.qc.ca

[5] Ministère des Forêts, de la Faune et des Parcs, Direction de la gestion de la faune de l'Estrie, de Montréal, de la Montérégie et de Laval, 201, Place Charles-Le Moyne (4è étage), Longueuil (QC) Canada, J4K 2T5.
E-mail: nathalie.vachon@mffp.gouv.qc.ca

[6] Centre de recherche sur les interactions bassins versants - écosystèmes aquatiques (RIVE). Université du Québec à Trois-Rivières, C.P. 500, Trois-Rivières (QC) Canada, G9A 5H7.
E-mail: pierre.magnan@uqtr.ca

* Corresponding author

6

Insights into Percid Population and Community Biology and Ecology from a 70 year (1943 to 2013) Study of Perch *Perca fluviatilis* in Windermere, U.K.

Craig, John F.,[1] Fletcher, Janice M.[2,a] and Winfield, Ian J.[2,*]

ABSTRACT

The perch *Perca fluviatilis* L. is widely distributed throughout Europe, where it is a major component of many fish communities and frequently a target for commercial and recreational fisheries. Precipitated by a period of commercial exploitation in response to the disruption of marine fisheries during World War II, the lake perch population of Windermere, U.K., has been studied continuously from 1943 to the present. During this period, the lake has been subjected to environmental pressures including eutrophication, species introductions and climate change. The long-term effects of these and other perturbations, including disease, on the perch population and its place in the local fish community are reviewed.

Keywords: Climate change, community ecology, disease, eutrophication, population biology, species introductions, Windermere, U.K.

[1] Craig Consultancy, Whiteside, Dunscore, Dumfries, DG2 0UU, U.K.
Email: jfcraig@btconnect.com

[2] Lake Ecosystems Group, Centre for Ecology & Hydrology, Lancaster Environment Centre, Library Avenue, Bailrigg, Lancaster, Lancashire LA1 4AP, U.K.

[a] E-mail: jmf@ceh.ac.uk

* Corresponding author: ijw@ceh.ac.uk

6.1 Introduction

Many aspects of the biology of perch species *Perca* spp. can be productively studied by relatively short investigations, but the reproductive biology and longevity of these species are such that studies of their population biology and associated interactions require much longer-term observations. Even the three-year durations typical of many research studentships are too short for the collection of meaningful data in this context, leaving such studies to fisheries organisations, research institutes or university groups that can facilitate much longer programmes of research. Although the benefits of long-term ecological studies in an era of major environmental change are now appreciated by scientists and managers alike and inform much current freshwater environmental management (Maberly and Elliott 2012), such value was not so readily apparent a few decades ago when there was a paucity of long-term data sets in the environmental sciences (Elliott 1990). Furthermore, not all have subsequently survived, making those that do persist even more valuable.

The reviews (Elliott 1990; Maberly and Elliott 2012) cited above were both concerned primarily with long-term investigations carried out at Windermere (mere = lake) in the U.K., which include population and other studies of perch *Perca fluviatilis* L. which began in the late 1930s and have included a consistent annual population monitoring component since 1943 (Le Cren 2001). Having involved generations of scientists, this programme is now in its seventieth year and is without parallel elsewhere in the U.K. It also enjoys several aspects unique in a global context. Data gathered in earlier decades of this study made fundamental contributions to our understanding of basic perch biology and ecology. The lake and its perch population have also been subjected to a number of environmental pressures including disease, eutrophication, species introductions and climate change.

The approach taken in this review is to provide a brief introduction to the Windermere system and long-term study and then to describe the insights gained into percid population and community biology, including interactions with the wider environment.

6.2 Perch and Windermere

6.2.1 Perch

Perch is by far the most widespread and abundant representative of the genus *Perca* in European fresh waters. Its biology and ecology show great similarities with those of the yellow perch *Perca flavescens* (Mitchell) of North America, as evidenced by the detailed coverage of these two species by Thorpe (1977) and Craig (2000). Perch is widely distributed throughout Europe, where it is frequently a major component of fish communities in both running and particularly standing fresh waters. It is relatively tolerant of a variety of environmental conditions, including a wide range of water temperatures, and during its lifetime typically progresses through diets of zooplankton, macroinvertebrates and ultimately fishes, including conspecifics (Craig 2000). Unsurprisingly, this widespread and abundant species is a common target of commercial and recreational fisheries and in some locations is intensively exploited in lucrative fisheries.

6.2.2 Windermere

Windermere is England's largest natural lake and is situated (54°22'N, 2°56' W; altitude 39 m) in the English Lake District, U.K. It comprises a mesotrophic north basin (area 8.1 km^2, maximum depth 64 m) and a eutrophic south basin (area 6.7 km^2, maximum depth 44 m). The fish community is relatively simple with seven major populations comprising Arctic charr *Salvelinus alpinus* (L.), Atlantic salmon *Salmo salar* L., brown trout *Salmo trutta* L., European eel *Anguilla anguilla* (L.), perch, pike *Esox lucius* L. and in recent years roach *Rutilus rutilus* (L.), although nine other minor species are also present (Pickering 2001; Winfield et al. 2010; Winfield et al. 2011). Historically, Arctic charr and perch have dominated offshore (Frost 1977) and inshore (Le Cren 2001) habitats, respectively, although roach are now also abundant in the latter areas in particular (Winfield et al. 2010). The cultural eutrophication of the south basin has led periodically to low availability of dissolved oxygen, although these effects are limited to deep waters and so of no direct consequence to perch (Jones et al. 2008) and are now being actively managed. Like other lakes in Europe, Windermere has also experienced recent impacts from climate change (Maberly and Elliott 2012) and fish species introductions (Winfield et al. 2010).

6.3 Origins and History of the Long-term Study

Sustained scientific research at Windermere began in 1931 when the Freshwater Biological Association opened its first laboratory on the shore of the north basin, later moving to a more southerly location near the mid part of the lake. Perch were the subject of one of the first publications arising from research on the lake, with Allen (1935) describing their diet and seasonal migrations, but the outbreak of World War II in September 1939 changed the direction of research on this percid. Drawing on his experiences with fish traps in Africa, Worthington (1950) began to explore the feasibility of trapping perch in Windermere to enhance the supplies of food in a country subjected to a marine blockade. In addition, as recounted by Le Cren (2001) it was also considered that collecting extensive data on a perch population now subjected to an intensive fishery might increase scientific understanding of the 'overfishing problem' then challenging marine fisheries biologists around the world.

Thus began the long-term monitoring of perch in Windermere's two basins. Following a brief period of refinement of the trap design, subsequently to become known as the 'Windermere perch trap', and exploiting the new understanding of perch migrations produced by Allen (1935), a full-scale perch fishery began in the spring of 1941 using some 300 traps set on perch spawning grounds around the lake by about 30 volunteers (Le Cren 2001). The subsequent history of this commercial fishery, which started as a scientific study in 1943 and continues to the present, is described below; a key feature is that from the beginning detailed records were maintained for both catch and effort. The number of trapping areas was subsequently reduced as the programme evolved into a purely scientific activity, with the programme being taken over in 1989 by the Institute of Freshwater Ecology and subsequently by the Centre for Ecology & Hydrology. Four trapping locations (Green Tuft in the north basin and Chicken Rock, Rawlinson Nab and Lakeside in the south basin; extra counting stations around the lake were discontinued in 1977 as they showed similar trends to these four locations) have remained in use to the present day (Fig. 1), generating a perch data set of 70 years duration on catch-per-unit-effort (Fig. 2) and individual ages and other features. The Centre for Ecology & Hydrology also took over all other long-term scientific sampling programmes, including on Arctic charr and pike (Le Cren 2001), and initiated some new ones. This has led to the production of a broad suite of abiotic and biotic data (Maberly and Elliott 2012), with which the perch data can be analysed and interpreted in ways unforeseen by the study's founders. The long-term scientific

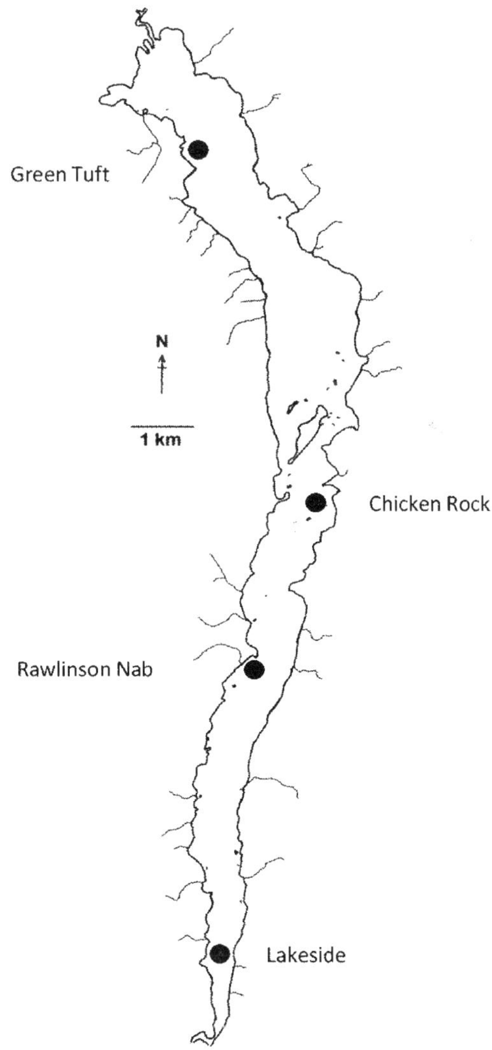

Fig. 1. Locations of the four main trapping sites (Green Tuft, Chicken Rock, Rawlinson Nab and Lakeside indicated by closed circles) used in the long-term study of perch in the two basins (north and south) of Windermere, U.K. The division between the two basins occurs in the area of islands just north of the trapping site of Chicken Rock.

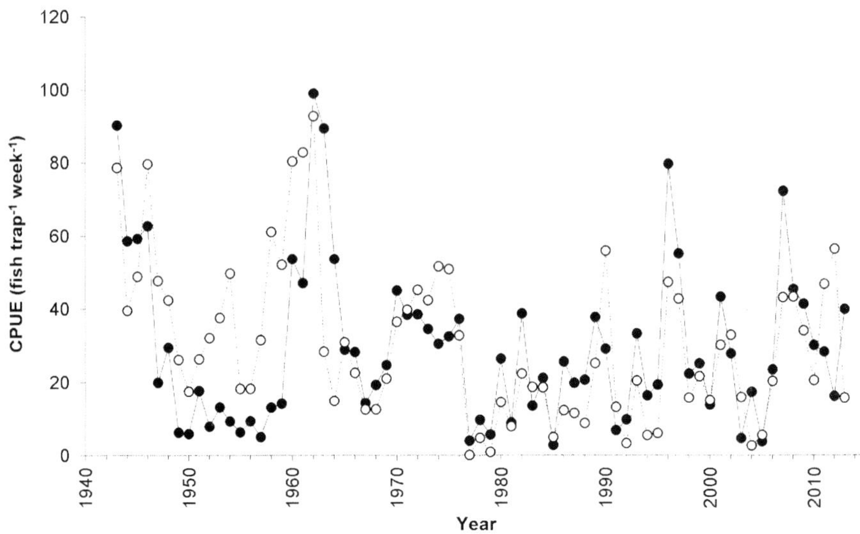

Fig. 2. Long-term variation in the annual catch-per-unit-effort (CPUE) of perch at the sites of Green Tuft in the north basin (closed circles, continuous line) and Lakeside in the south basin (open circles, broken line) of Windermere from 1943 to 2013. The effects of a massive disease outbreak in 1976 are clearly visible.

study of perch, which is now the only significant remover of fishes from Windermere, has also formed a framework against which other specific short-term perch studies, such as trophic ecology, could be undertaken.

6.4 Commercial and Recreational Fisheries

6.4.1 Commercial Fishery

Although a commercial net fishery for perch existed on Windermere before 1921 (Kipling 1972), this was limited and supplied only a relatively local market. This limitation was itself probably the result of a combination of limited infrastructure for transporting the catch and a general U.K. market preference for marine fishes. The outbreak of World War II clearly changed the latter and so the 1941-onwards trapping of perch from Windermere delivered a welcome catch, although its reliance on volunteer effort and the national food rationing programme meant that the operation was not truly commercial in the present use of the word. Similar trap fisheries were developed on a number of other water bodies, but none were as productive as the operation on Windermere and so few lasted more than a year or two (Le Cren 2001).

The scale of the perch fishery on Windermere was increased to a total of approximately 780 traps over the two basins in 1942. The catches in each successive year then declined as the population was 'overfished' in classic manner (Le Cren 2001). After the 1947 season, it was decided to restrict large-scale removal of perch to the south basin. By this time, even with rationing still in place, it was no longer economical for the catch to be sold for canning and so all subsequent trapping was carried out by scientists and the programme had evolved into the scientific study described above. No other commercial fisheries for perch, or any other species, have ever subsequently developed on Windermere.

6.4.2 Recreational Fishery

Recreational fishing for perch on Windermere has been even more limited than the above commercial activities. Historically, local anglers have much preferred fishing for Arctic charr or brown trout rather than for perch. Although visiting anglers became more common in the post-war years as a result of increased leisure time and improved travel opportunities, their visits to Windermere have tended to focus on brown trout or pike and in later years, to a lesser extent, on roach. Moreover, with the exception of some fishing for brown trout, as elsewhere in the U.K. such recreational angling is practised almost exclusively as catch-and-release. As a result, recreational fishing has no direct impacts on the fish populations of the lake, although some past live-baiting practices appear to have led to a number of undesirable and unauthorized species introductions (Winfield and Durie 2004).

6.5 Age, Growth and Mortality

Growth and mortality of perch, as in many other teleosts, can show marked fluctuations both temporally and spatially. It is often difficult to measure accurately these fluctuations although they are the cornerstone to understanding population dynamics, the effects of perturbations such as fishing and disease, and the management and conservation of the perch resource.

6.5.1 Age

Scales were initially collected to age trapped Windermere perch but by 1942 opercular bones were found to be better for age and growth studies (Le Cren 1947, 2001). In addition the total length and sex of the fish were determined. Fish were not weighed on a regular basis until 1969. In years when the catches were very high a stratified subsample (always >50% of the catch) was aged. These data have provided the basic information for studying the population dynamics of Windermere perch. For example estimates were made of year-class strengths from 1941 to 1964 (Le Cren et al. 1977) and 1967 to 1975 (Craig et al. 1979).

6.5.2 Growth

Le Cren (1958) provided an analysis of perch growth based on fish caught between 1942 and 1956 in the north basin of Windermere and Craig (1980) provided similar analysis on perch year classes from 1955 to 1972. Le Cren (1958) found that perch growth did not follow a von Bertalanffy (1938) growth curve although later he found that exceptionally big perch caught at the same time did follow this relationship (Le Cren 1992). In later years the growth patterns of the perch changed and the von

Bertalanffy growth model fitted the data well (Craig 1980). A general trend over the above time period was a gradual increase in total length with age, and an increase in the von Bertalanffy coefficient *K* and L_∞ values for both sexes.

Differences in growth have been noted in different year-classes of perch (Craig 1978a). For example the 1959 year-class grew slowly achieving in theory 95% of its growth in 25 years compared to the 1968 year-class with 95% of its growth being completed in 7 years. The availability of perch fry as food was considered to be the major factor in determining differences in growth, metabolism and ultimate age in these year-classes. In a similar fashion Le Cren (1992) found that accelerated growth at a mean age of 4 years for a period of *c*. 4 years occurred when there was an abundance of young perch to cannibalise on. Thus diet has been suggested as being important for perch growth, e.g., perch grow faster on a diet of fishes than on invertebrates (Craig 2000). The negative effect of density causing competition for available food and thus limiting growth has been shown for Windermere perch (Craig 1980). Temperature has also been shown to be a major factor in controlling growth and Le Cren (1958) found that two thirds of yearly Windermere perch growth could be explained by temperature >14°C.

More recently the mean length of adult perch was found to be best predicted by phosphorus concentration, summer temperature and densities of different age classes of perch through a complex combination of competition- and predation-based interactions. For example, higher nutrient levels, temperature and densities of small perch as prey increased food availability and reduced competition resulting in increased mean length and decreased skewness (Ohlberger et al. 2013).

6.5.3 Mortality

Various estimates of mortality in perch in Windermere have been calculated throughout the history of this long-term study (Le Cren et al. 1977; Craig et al. 1979; Langangen et al. 2011).

Craig (1978a, 1980, 1985) and Craig and Fletcher (1984) found evidence for the fundamental relationship between growth and mortality rates in fishes, i.e., faster growing fishes had a higher mortality rate (physiological ageing). This was mainly based on comparisons between different year classes of Windermere perch.

A very high adult mortality in 1976 caused by disease (see below) released predation on juveniles resulting in an increase of juvenile biomass by nearly as much as the adult biomass decreased (Ohlberger et al. 2011a). This was termed 'stage-specific biomass overcompensation'. In addition, age-specific adult fecundity and length at age were found to be higher after the disease outbreak than before. Ohlberger et al. (2011a) suggested that the high mortality in adults meant that the survivors were released from competition and energy could be directed to somatic and reproductive growth. Pre-disease seasonal allocation of resources in adult perch to somatic and gonadal development has been described by Craig (1977). The energy that is available for growth can be used to increase the energy content of the soma or the development of the gonad in mature fish. As the perch get older a greater proportion of the available energy is directed towards the gonads.

6.6 Reproduction and Recruitment

Perch congregate at spawning and males spend more time in these congregations than females. The trapping of perch in Windermere was based on knowledge of these congregations and the bias towards males in the catch was acknowledged. The latter has led to several problems with calculating statistics about the perch populations (Bagenal 1972; Le Cren 2001). The traps thus catch mature fish and historically these fish are aged 2+ years or older.

Variations in reproduction (net reproduction rate) and recruitment are probably determined by fecundity, growth, mortality, age of maturity and biomass of the adult stock as discussed by Craig and Kipling (1983). The number of eggs laid in a particular year, however, appears to be unrelated to the resulting year-class strength (Craig and Kipling 1983) as environmental factors are likely to control the latter. In the pre-disease period, changes in absolute fecundity were noted in perch of the same size, a 46% increase from samples taken in 1944 and 1960 to those examined in 1979 to 1981 (Craig and Kipling 1983). Ohlberger et al. (2011a) using 60 years of data suggest that age-specific fecundity (and size) were higher after than before the disease outbreak although they used gonad mass not fecundity data.

Throughout the period of this study, there have been publications on the population dynamics of Windermere perch in particular calculations of recruitment, year-class strengths and stock biomass and their relationships with external factors (Le Cren et al. 1977; Craig et al. 1979; Jones et al. 1982; Craig 1980, 1982; Craig and Kipling 1983; Mills and Hurley 1990; Winfield et al. 1998; Paxton et al. 2004; Ohlberger et al. 2013). All these studies indicate, with different approaches and sophistication, the relationship between stock and recruitment and the importance of external factors in particular temperature in the first year of life, cannibalism and other predation such as by pike (Winfield et al. 2012). The data collected on perch over such a long period of time are unique and have been used in elucidating stock and recruitment relationships.

6.7 Food and Feeding

Detailed studies on the food and feeding of perch have been sporadic (Allen 1935; Frost 1946; Smyly 1952; McCormack 1970; Craig 1978b; Guma'a 1978) and nothing has been published since Craig (1978b) on adults and Guma'a (1978) on juveniles. Guma'a (1978) observed algal cells and ciliates in larval perch and then noted that larger members of the plankton were selected as the perch grew in length (prey size was gape-limited); fishes appeared in the diet of perch when they were about 40 mm in length. Smyly (1952) had also found previously that the prey size increased as the young perch grew. Zooplankton was the main prey from May to September although some benthic organisms were found in the stomachs of fish when they reached an age of about 1 month. Smyly (1952) noted cannibalism by 0+ year perch on their siblings during July. Craig (1978b) studied the food, feeding and consumption of adult perch. With minor exceptions Windermere perch fed from November to April on benthic prey (including cannibalism on young perch), from May to June on planktonic and benthic prey and from July to October on zooplankton (mainly *Daphnia hyalina* var. *galeata*, *Leptodora kindti* and *Bythotrephes longimanus*) and perch fry. Year classes of perch were strong during this study period (1973 to 1976) (Craig et al. 1979) and thus larval perch were abundant for the larger perch to cannibalise them. McCormick (1970) found that young perch were not very common in the diet of older perch. During the period of her study (1963 to 1967), however, year classes of perch were poor (Le Cren et al. 1977). Both McCormick (1970) and Craig (1978b) found that perch fed on similar types of food over a large size range, e.g., perch fry and zooplankton were found in the stomachs of perch ranging in mass from 32 to 311 g in July (Craig 1978b). Cannibalism by Windermere perch has been significant in their population dynamics influencing not only mortality but also growth. Le Cren (1992) found that large perch in Windermere accelerated their growth through piscivory, i.e., through eating their own young.

6.8 Arrival and Impacts of Disease

In the mid 1970s, the dynamics of the perch population of Windermere were subjected to a major natural experiment by the arrival of a 'perch disease'. Although secondary infections were described in detail by Pickering and Willoughby (1977), the pathogenic organism has never itself been unambiguously identified. Le Cren (2001) records that it has been suggested to be a form of furunculosis (*Aeromonas* spp.), although he also considers the possibility that secondary infection with *Saprolegnia* spp. may have been the ultimate cause of death.

Whatever its specific pathology, it is clear that the disease's outbreak was unexpected, sudden and massive. Extensive sampling of the newly-diseased population by Bucke et al. (1979) revealed a wide variety of fungal and bacterial infections, with the evidence again suggesting that these were not the primary agents of mortality. Infections were distributed through all age groups of perch during the course of one year, although adults appeared to be particularly vulnerable at spawning time. A dramatic increase in mortality occurred in 1976, with Bucke et al. (1979) estimating that over 98% of the adult perch population (over 1,000,000 fish) died in that year. The effects on perch catch-per-unit-effort and by implication on population abundance are clearly visible in Fig. 2.

This remarkable event occurred approximately 33 years into the long-term study, facilitating subsequent detailed assessments of its effects on perch population biology through a series of before-and-after studies. Most such effects have been remarkable in their clarity and some have radiated out from the perch population to other components of the Windermere ecosystem. Unsurprisingly given the magnitude of its associated mortality, the disease's arrival produced major changes in the structure of the perch population which persisted for many years. Most notably this included a very clear example of stage-specific biomass compensation by juveniles as noted above (Ohlberger et al. 2011a), facilitated by the pathogen-induced decrease in adult biomass leading to a reduction in cannibalism and so facilitating a corresponding increase in juvenile biomass (Fig. 3). These impacts of disease on recruitment processes have also been shown to be a significant contributor to statistical (Paxton et al. 2004) and multi-stage delay differential equation (Brown et al. 2005) models of the dynamics of the perch population.

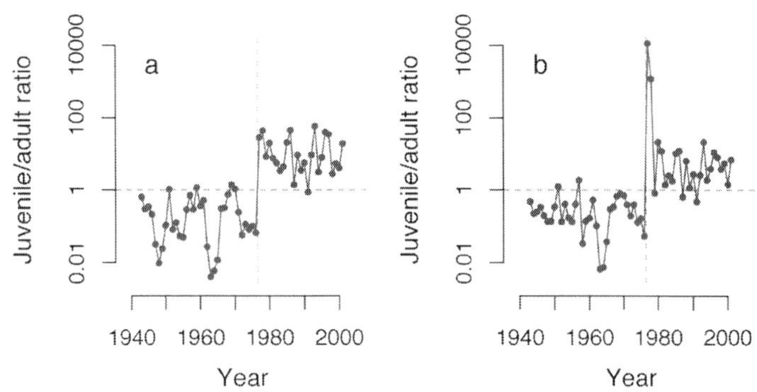

Fig. 3. Time series of juvenile: adult biomass in the (a) north and (b) south basins of Windermere, indicating a shift in the biomass distribution from being dominated by adults (<1) to being dominated by juveniles (>1) after the disease outbreak in 1976 (vertical dashed line). Redrawn with permission from Ohlberger et al. (2011a).

Study of the Windermere perch population has also contributed to wider understanding of the impact of disease-induced adaptive evolution, which is critical to inform management of the ecological and economic effects of infectious diseases around the world. Studies of pathogen-induced trait changes are extremely rare, in part because they must be of very long duration. Examination of trends in the maturation pattern of Windermere perch has revealed that the pathogen induced a phenotypic change from a previously increasing to a subsequently decreasing size at maturation which was independent of cohort-specific growth rates (Ohlberger et al. 2011b). This aspect of life-history chronology is of great importance in many animals, particularly those exploited by size-selective activities such as fishing. Remarkably, the Windermere data have demonstrated that the evolutionary rates of phenotypic change (0.28 to 0.68 Darwins) imposed by the pathogen are high and comparable with those observed in populations exposed to intense harvesting. Moreover, the age-size truncating effect of the disease has also been found to affect how phenological shifts (see below) translate into changes in population abundance. Specifically, it has been shown that disease-induced size truncation of the perch population has amplified its sensitivity to climate-change-induced changes in predator-prey asynchrony with respect to the lake's zooplankton by reducing intraspecific density regulation due to competition and cannibalism (Ohlberger et al. 2014). Again, although mediated specifically by disease impact in the case of the Windermere perch, the underlying mechanisms are also likely to apply to generic size-selective harvesting and so make a significant contribution to our understanding of the circumstances in which exploited populations may collapse.

Finally, the impact of disease on the perch population has also had ramifications for other components of Windermere's food web through its selection against larger perch (Edeline et al. 2008). Before the arrival of the disease in 1976, predation pressure from pike selected against small individuals and dominated selective forces acting on the perch population as a whole. After this time, however, pathogen-induced selection overrode predator-induced selection and drove a rapid change to smaller, slower-growing perch. In turn, these effects made perch easier prey for pike and weaker competitors against juvenile pike, ultimately increasing juvenile pike survival and total pike numbers. This study by Edeline et al. (2008) clearly emphasises the importance of considering the relative strengths and directions of multiple selective pressures in attempts to understand community functioning.

6.9 Effects of Eutrophication

Eutrophication has been defined as excessive plant growth resulting from nutrient enrichment by human activity (Smith and Schindler 2009) and is widespread in fresh (and other) waters around the world (Schindler 2012). Consequently, its effects on percids and other freshwater fishes have a long history of study and management as recently reviewed by Winfield (2015). While some fish populations in some water bodies may initially increase in abundance in response to limited eutrophication, its effects usually quickly become negative and result in reduced population abundance or changes in relative species composition of the fish community. These effects result from a range of potential specific mechanisms within the overall eutrophication syndrome. These may include a reduction in water transparency, decrease in dissolved oxygen, loss of macrophytes (the increased plant growth is often in the form of algae or cyanobacteria which are not actually plants, but which are often referred to as 'blue-green algae'), and shifts in competitive balances within the fish community. In temperate fresh waters, a very common general pattern along a developing eutrophication gradient is an early decline in salmonids and coregonids, a temporary increase in percids and ultimately an increase in cyprinids (Mehner et al. 2005).

Windermere has a long and well documented history of eutrophication, with relatively little impact on its north basin but more significant effects in its south basin which receives a much greater loading of nutrients (Pickering 2001). Advanced eutrophication can have adverse impacts on percids, but there is no indication that this has happened in Windermere. In the early years of the perch study, concentrations of soluble reactive phosphorus (the key nutrient in the system) were similar in the two basins and always less than 5 mg m^{-3} (Winfield et al. 2008). These levels began to increase in both basins in the early 1960s, but much more so in the south basin which by the 1970s reached values in excess of 15 mg m^{-3}. The situation continued to deteriorate such that the south basin reached a peak of 28.0 mg m^{-3} in 1990, almost three times greater than the peak of 10.4 mg m^{-3} observed in the north basin in 2001. The introduction of phosphate stripping in 1992 produced some improvement in conditions, but even by the mid 2000s levels of soluble reactive phosphorus remained much higher than in the 1940s.

Despite this considerable history of eutrophication, the perch population of Windermere has shown a remarkable degree of resilience in both basins over the long-term study as illustrated by the patterns of catch-per-unit-effort shown in Fig. 2. Abundance has certainly varied greatly over this period, but it has not shown any marked or prolonged decline which may be attributed to eutrophication as has frequently been observed for percids elsewhere (Leach et al. 1977). This lack of an obvious negative impact is probably due to the degree of eutrophication experienced at Windermere actually being relatively limited in a global context. For example, water transparency as assessed by a Secchi disc has consistently been measured in metres (I.J.W., unpublished data), significant decreases in dissolved oxygen have been limited to deep waters rarely frequented by perch (Jones et al. 2008) and macrophytes have persisted (I.J.W., unpublished observations). The role of eutrophication in shifting competitive balances within the Windermere fish community will be discussed below.

Although eutrophication has not had any catastrophic impacts on the perch population of Windermere, population models developed by Edeline et al. (2008) and Ohlberger et al. (2013) both retained phosphorus concentration as a significant explanatory factor for observed variations in perch body size. Although the direct effects of water nutrient concentrations are at least two (primary producers and primary consumers) or three (primary producers, primary consumers and secondary consumers) trophic steps away from carnivorous or piscivorous perch, respectively, the duration and breadth of the Windermere study are such that such complex effects of eutrophication can be detected, quantified and used in sophisticated explorations of the Windermere ecosystem.

6.10 Species Introductions and Community Ecology

With a total of only 16 species (Winfield et al. 2011) of which just seven are abundant as described above, the fish community of Windermere is relatively simple despite it being England's largest lake. This depauperate condition is in accordance with a general European pattern of decreasing freshwater fish diversity in the north and west of the continent (Brucet et al. 2013), further exacerbated by a similar pattern within the U.K. itself which arises in large part from the effects of the last glaciation (Davies et al. 2004).

As a result of the above biogeographical patterns, for thousands of years the perch population of Windermere has existed in a simple fish community dominated by salmonids, itself as the sole percid, and pike as the sole esocid. Indeed, the pre-1990s research reviewed above focussed almost exclusively on only the local populations of Arctic charr, perch and pike as reviewed in detail by Mills and Hurley (1990) and Le Cren (2001). Predator-prey interactions between pike and perch formed much of this earlier work and indeed continue to the present through studies such as those of Edeline et al. (2008), Langangen et al. (2011), Ohlberger et al. (2011b) and Winfield et al. (2012) discussed elsewhere in this review. In contrast, opportunities for direct interactions between perch and Arctic charr are very limited in Windermere because the latter species only frequents the littoral zone during brief spawning periods. Consequently, for most of the lake's history, competitive interspecific interactions involving perch have been largely limited to potential limited interactions with juvenile pike as considered by Edeline et al. (2008).

This relatively simple situation began to change in the 1990s with a remarkable increase in the abundance of the roach in first the south and then the north basins of the lake. This cyprinid is of particular concern because of the potential for strong competitive interactions with perch through shared zooplanktonic prey which has been clearly demonstrated elsewhere in Europe (Svanbäck et al. 2008). Roach is not native to most of the water bodies of the English Lake District, but it has recently appeared in several of the larger lakes apparently as the result of live-baiting practices (now banned) by anglers fishing for pike (Winfield and Durie 2004). Remarkably for the specific case of Windermere, this angler-mediated introduction of roach occurred in late 1890s (Winfield et al. 2011). Roach subsequently remained a scarce and spatially restricted component of the fish community throughout almost all of the extensive research reviewed by Le Cren (2001), despite the period of marked eutrophication in the lake's south basin described above which elsewhere has been observed to favour roach populations (Lappalainen et al. 2001). It was only in the 1990s, when the water temperature of Windermere increased, that the introduced roach numbers started to develop. Periodic sampling of the fish communities of the two basins between 1995 and 2008 showed that by the early 2000s roach had become significant components of the fish communities of offshore surface and particularly inshore habitats (Winfield et al. 2010).

Despite obvious concerns of competitive impacts on the perch population by the introduced roach population, it is empirically evident from Fig. 2 that there has been no dramatic decline in perch in either basin of Windermere during the post-1990 increase in the roach population. Although roach has indeed increased markedly and is now an important component of the fish community, it is not dominant in numerical terms (Winfield et al. 2010). Such dominance may occur in the future if local warming trends persist, but at present the perch population shows no signs at the population level of significant competitive impact from roach.

Finally in the context of the contemporary fish community in which the perch population of Windermere now exists, it is also relevant to note that roach is not the only introduced species giving rise to local concerns. The related common bream [*Abramis brama* (L.)] has also increased in the lake in recent years, although it remains relatively scarce (Winfield et al. 2011). In addition, the ruffe [*Gymnocephalus cernuus* (L.)] has recently been introduced to the smaller lake of Rydal Water just a short distance upstream of Windermere (Winfield et al. 2010) and this species is known to have the potential for competitive interactions with perch (Dieterich et al. 2004). Clearly, after a long historical period of existence in a relatively simple fish community, the perch population of Windermere may face a much more complex situation in the years to come.

6.11 Effects of Climate Change

The effects of water temperature on perch biology and ecology have been a recurring theme in the many studies cited above of this species in Windermere. Le Cren (2001) also amassed many of these investigations into a detailed but highly accessible overview, noting for example the marked effects of temperature on growth and year-class strength. The subsequent population modelling investigations cited above have almost all included temperature as a factor, with for example that of Paxton et al. (2004) examining the significance of water temperatures at different times of the year and finding that higher temperatures in the late summer have a particularly important positive influence on perch recruitment. In addition, Le Cren (1965) and Kipling (1976) have documented effects of water temperature on the spawning activity and spawning success of perch in the lake. Clearly, sustained changes in water temperature have the potential to result in major changes to the ecology of perch in Windermere through effects on reproduction and growth at the levels of the individual and the population.

Climate change is of course considered to be the major potential source of such changes and George (2010) has documented a range of impacts on water temperature, lake ice phenology, the supply and recycling of nitrogen and phosphorus, the flux of dissolved organic carbon and the growth and seasonal succession of phytoplankton at a wide range of European lakes. Moreover, Windermere is amongst these water bodies and Winfield et al. (2008) have shown that its annual average temperature has showed a significant and sustained increase since the early 1990s. An associated phenological effect on perch was evident as early as 2003 in terms of a shift of spawning activity to earlier in the spring by approximately 13 days (Winfield et al. 2004). An extensive meta-analysis of rates of phenological change in marine, freshwater and terrestrial environments by Thackeray et al. (2010) put this shift into a wider context, showing that in fresh waters, changes were generally less pronounced for vertebrates such as fishes than for invertebrates such as zooplankton, and so, raising the possibility of increasing trophic level asynchrony. A more specific multi-trophic-level phenological study of phytoplankton and zooplankton abundance, and perch spawning in Windermere by Thackeray et al. (2013) has shown that zooplankton have advanced their dynamics most and perch least rapidly during the spring months (Fig. 4), resulting in altered food-web synchrony in both basins. Such developing trophic mismatch between perch and its zooplankton food resources may ultimately affect under-yearling survival and recruitment to the adult stage, for which evidence from Windermere has recently been demonstrated (Ohlberger et al. 2014).

Clearly, climate change is having direct impacts on the perch population of Windermere and will inevitably have a significant influence on its future dynamics. In addition, as discussed above, it also appears to have a direct bearing on the local population dynamics of roach, the potential competitor. Although introduced in the late 1890s (Winfield et al. 2011), this warm-water cyprinid remained extremely scarce (despite an extensive period of eutrophication) until the early 1990s, when Windermere's temperature began to increase. As anticipated for elsewhere in the northern parts of the European temperate zone (Lehtonen 1996), it seems likely that there is a causal relationship between lake warming and roach abundance. Such a thermal mechanism may also explain the recent limited increase of common bream in Windermere described by Winfield et al. (2011). As in many other European lakes (Jeppesen et al. 2012), perch, salmonids and cyprinids face very different futures in Windermere which is increasingly being affected by climate change.

6.12 Concluding Remarks

The long-term study of perch in Windermere is both remarkable and unique. It originated during the dark days of a world war, survived a period of uncertainty and organisational change towards the end of the twentieth century, and is now the focus of extended international and interdisciplinary collaborations. This review alone cites 47 publications arising directly out of the perch programme and it is by no means

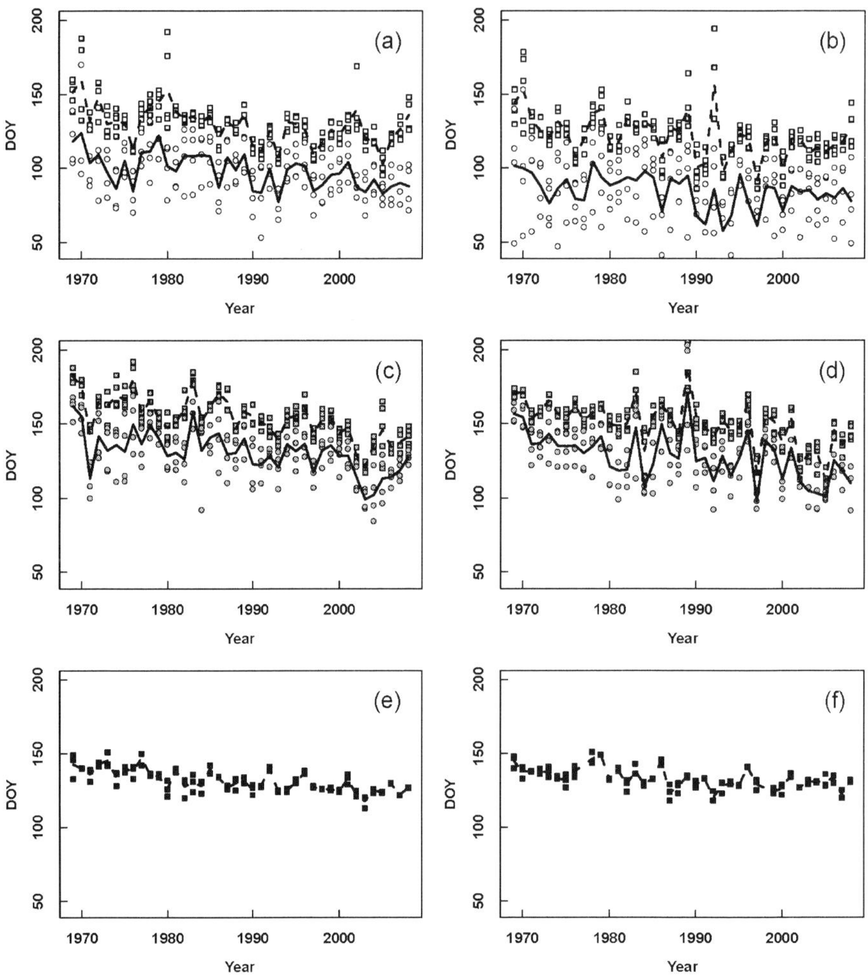

Fig. 4. Long-term changes in the seasonal timing of phytoplankton (a, b) and zooplankton (c, d) spring population growth, and perch spawning (e, f). Data are shown for the north (a, c, e) and south (b, d, f) basins of Windermere. Points show the original phenological metric data and lines show average seasonal timings for distinct metric classes. In plots a–d, solid lines show the mean seasonal timing of onset-type metrics (circles), and dashed lines show the mean seasonal timing of peak/middle-type metrics (square symbols). For the perch data, only peak/middle-type metrics were calculated. Redrawn with permission from Thackeray et al. (2013).

exhaustive. Without doubt, research at Windermere has made major contributions to understanding perch biology and ecosystem-based studies and management of fresh waters.

It is particularly pleasing that the research programme has certainly met, and in many ways surpassed, the original vision described by Worthington (1950) of increasing scientific understanding of the 'overfishing problem' then (and still now) challenging fisheries biologists around the world. In closing, it is fitting to follow Le

Cren (2001) and quote a statement that he originally made in his earlier review of Le Cren (1987), i.e., "Each period in our studies of fish in Windermere has produced new events, interpretations, and ideas which have caused us to modify previous conclusions. Long-term studies have a special value in fishery research." In our view, this conclusion still remains relevant and continuation of the Windermere study has much to offer to future generations. It has been a privilege for us to play a role in it.

6.13 Acknowledgements

Although the emphasis and views expressed here are our own, they have formed as a result of our involvement over several decades with many colleagues in a variety of research projects revolving around the long-term research programmes of Windermere. We are grateful to the many individuals who have participated in the Windermere data collection and to the Freshwater Biological Association for their historical role in the production of these invaluable long-term data. As an inevitable consequence of the nature of this long-term study many of these colleagues are no longer with us. For J.M.F. and I.J.W., this work has been funded primarily by the Natural Environment Research Council of the U.K.

6.14 References

Allen, K.R. 1935. The food and migration of the perch (*Perca fluviatilis*) in Windermere. J. Anim. Ecol. 4: 264–273.

Bagenal, T.B. 1972. The variability in numbers of perch, *Perca fluviatilis* L. caught in traps. Freshwater Biol. 2: 27–36.

Bertalanffy, L. von. 1938. A quantitative theory of organic growth. Human Biol. 10: 181–213.

Brown, A.R., A.M. Riddle, I.J. Winfield, J.M. Fletcher and J.B. James. 2005. Predicting the effects of endocrine disrupting chemicals on healthy and disease impacted populations of perch (*Perca fluviatilis*). Ecol. Modell. 189: 377–395.

Brucet, S., S. Pédron, T. Mehner, T.L. Lauridsen, C. Argillier, I.J. Winfield, P. Volta, M. Emmrich, T. Hesthagen, K. Holmgren, L. Benejam, F. Kelly, T. Krause, A. Palm, M. Rask and E. Jeppesen. 2013. Fish diversity in European lakes: geographical predictors dominate over anthropogenic pressures. Freshwater Biol. 58: 1779–1793.

Bucke, D., G.D. Cawley, J.F. Craig, A.D. Pickering and L.G. Willoughby. 1979. Further studies of an epizootic of perch, *Perca fluviatilis* L., of uncertain aetiology. J. Fish. Dis. 2: 297–311.

Craig, J.F. 1977. The body composition of adult perch, *Perca fluviatilis* in Windermere, with reference to seasonal changes and reproduction. J. Anim. Ecol. 46: 617–632.

Craig, J.F. 1978a. A note on ageing in fish with special reference to the perch, *Perca fluviatilis* L. Verh. Internat. Verein. Limnol. 20: 2060–2064.

Craig, J.F. 1978b. A study of the food and feeding of perch, *Perca fluviatilis* L., in Windermere. Freshwater Biol. 8: 59–68.

Craig, J.F. 1980. Growth and production of the 1955 to 1972 cohorts of perch, *Perca fluviatilis* L. in Windermere. J. Anim. Ecol. 49: 291–315.

Craig, J.F. 1982. Population dynamics of Windermere perch. FBA Annual Report 50: 49–59.

Craig, J.F. 1985. Aging in fish. Can. J. Zool. 63: 1–8.

Craig, J.F. 2000. Percid Fishes: Systematics, Ecology and Exploitation. Blackwell Sciences, Oxford, U.K.

Craig, J.F. and J.M. Fletcher. 1984. Growth and mortality of zebra fish, *Brachydaniorerio* (Hamilton Buchanan), maintained at two temperatures and on two diets. J. Fish Biol. 25: 43–55.

Craig, J.F. and C. Kipling. 1983. Reproduction effort versus the environment; case histories of Windermere perch, *Perca fluviatilis* L., and pike, *Esox lucius* L. J. Fish. Biol. 22: 713–727.

Craig, J.F., C. Kipling, E.D. Le Cren and J.C. McCormack. 1979. Estimates of the numbers, biomass and year-class strengths of perch (*Perca fluviatilis* L.) in Windermere from 1967 to 1977 and some comparison with earlier years. J. Anim. Ecol. 48: 315–325.

Davies, C., J. Shelley, P. Harding, I. McLean, R. Gardiner and G. Peirson. 2004. The distribution of freshwater fishes in Britain. pp. 19–30. *In*: C. Davies, J. Shelley, P. Harding, I. McLean, R. Gardiner and G. Peirson (eds.). Freshwater Fishes in Britain—The Species and their Distribution. Harvey Books, Colchester, U.K.

Dieterich, A., D. Baumgärtner and R. Eckmann. 2004. Competition for food between Eurasian perch (*Perca fluviatilis* L.) and ruffe (*Gymnocephalus cernuus* (L.)) over different substrate types. Ecol. Freshwater Fish 13: 236–244.

Edeline, E., T.B. Ari, L.A Vøllestad, I.J. Winfield, J.M. Fletcher, J.B. James and N. Chr. Stenseth. 2008. Antagonistic selection from predators and pathogens alters food-web structure. Proc. Natl. Acad. Sci. U.S.A. 105: 19792–19796.

Elliott, J.M. 1990. The need for long-term investigations in ecology and the contribution of the Freshwater Biological Association. Freshwater Biol. 23: 1–5.

Frost, W.E. 1946. On the food relationships of the fish in Windermere. Biologisches Jaarboek 13: 216–231.

Frost, W.E. 1977. The food of charr, *Salvelinus willoughbii* Gunther, in Windermere. J. Fish Biol. 11: 531–547.

George, G. (ed.). 2010. The Impact of Climate Change on European Lakes. Springer, The Netherlands.

Guma'a, S.A. 1978. The food and feeding habits of young perch, *Perca fluviatilis*, in Windermere. Freshwater Biol. 8: 177–187.

Jeppesen, E., T. Mehner, I.J. Winfield, K. Kangur, J. Sarvala, D. Gerdeaux, M. Rask, H.J. Malmquist, K. Holmgren, P. Volta, S. Romo, R. Eckmann, A. Sandström, S. Blanco, A. Kangur, H. RagnarssonStabo, M. Tarvainen, A.-M. Ventelä, M. Søndergaard, T.L. Lauridsen and M. Meerhoff. 2012. Impacts of climate warming on the long-term dynamics of key fish species in 24 European lakes. Hydrobiologia 694: 1–39.

Jones, R., J.F. Craig and C. Kipling. 1982. Recruitment as a function of an environmental and a biological variable—a freshwater example. ICES CM/1982/L: 42.

Jones, I.D., I.J. Winfield and F. Carse. 2008. Assessment of long-term changes in habitat availability for Arctic charr (*Salvelinus alpinus*) in a temperate lake using oxygen profiles and hydroacoustic surveys. Freshwater Biol. 53: 393–402.

Kipling, C. 1972. The commercial fisheries of Windermere. Trans. Cumberland Westmorland Antiquarian Archeological Soc. 72: 156–204.

Kipling, C. 1976. Year-class strengths of perch and pike in Windermere. Rep. Freshwater Biol. Assoc. 44: 68–75.

Langangen, O., E. Edeline, J. Ohlberger, I.J. Winfield, J.M. Fletcher, J.B. James, N. Chr. Stenseth and L.A. Vøllestad. 2011. Six decades of pike and perch population dynamics in Windermere. Fish. Res. 109: 131–139.

Lappalainen, A., M. Rask, H. Koponen and S. Vesala. 2001. Relative abundance, diet and growth of perch (*Perca fluviatilis*) and roach (*Rutilus rutilus*) at Tvärminne, north Baltic Sea, in 1975 and 1997: responses to eutrophication? Boreal Environ. Res. 6: 107–118.

Leach, J.H., M.G. Johnson, J.R.M. Kelso, J. Hartmann, W. Nümann and B. Entz. 1977. Responses of percid fishes and their habitats to eutrophication. J. Fish. Res. Board. Can. 34: 1959–1963.

Le Cren, E.D. 1947. The determination of the age and growth of the perch (*Perca fluviatilis*) from the opercular bone. J. Anim. Ecol. 16: 188–204.

Le Cren, E.D. 1958. Observations on the growth of perch (*Perca fluviatilis* L.) over twenty-two years with special reference to the effects of temperature and changes in population density. J. Anim. Ecol. 27: 287–334.

Le Cren, E.D. 1965. Some factors regulating the size of populations of freshwater fishes. Mitt.—Int. Ver. Theor. Angew. Limnol. 13: 88–105.

Le Cren, E.D. 1987. Perch (*Perca fluviatilis*) and pike (*Esox lucius*) in Windermere from 1940 to 1985; studies in population dynamics. Can. J. Fish. Aquat. Sci. 44 Supplement II: 216–228.

Le Cren, E.D. 1992. Exceptionally big individual perch (*Perca fluviatilis*) and their growth. J. Fish Biol. 40: 599–625.

Le Cren, D. 2001. The Windermere perch and pike project: an historical review. Freshwater Forum 15: 3–34.

Le Cren, E.D., C. Kipling and J.C. McCormack. 1977. A study on the numbers, biomass and year-class strangths of perch (*Perca fluviatilis* L.) in Windermere from 1941 to 1966. J. Anim. Ecol. 46: 261–307.

Lehtonen, H. 1996. Potential effects of global warming on northern European freshwater fish and fisheries. Fish. Manage. Ecol. 3: 59–71.

Maberly, S.C. and J.A. Elliott. 2012. Insights from long-term studies in the Windermere catchment: external stressors, internal interactions and the structure and function of lake ecosystems. Freshwater Biol. 57: 233–243.

McCormack, J.C. 1970. Observations on the food of perch (*Perca fluviatilis* L.) in Windermere. J. Anim. Ecol. 39: 255–267.

Mehner, T., M. Diekmann, U. Brämick and R. Lemcke. 2005. Composition of fish communities in German lakes as related to lake morphology, trophic state, shore structure and human-use intensity. Freshwater Biol. 50: 70–85.

Mills, C.A. and M.A. Hurley. 1990. Long-term studies on the Windermere populations of perch (*Perca fluviatilis*), pike (*Esox lucius*) and Arctic charr (*Salvelinus alpinus*). Freshwater Biol. 23: 119–136.

Ohlberger, J., Ø. Langangen, E. Edeline, D. Claessen, I.J. Winfield, N. Chr. Stenseth and L.A. Vøllestad. 2011a. Stage-specific biomass overcompensation by juveniles in response to increased adult mortality in a wild fish population. Ecology 92: 2175–2182.

Ohlberger, J., Ø. Langangen, E. Edeline, E.M. Olsen, I.J. Winfield, J.M. Fletcher, J.B. James, N. Chr. Stenseth and L.A. Vøllestad. 2011b. Pathogen-induced rapid evolution in a vertebrate life-history trait. Proc. R. Soc. B 278: 35–41.

Ohlberger, J., J. Otero, E. Edeline, I.J. Winfield, N. Chr. Stenseth and L.A. Vøllestad. 2013. Biotic and abiotic effects on cohort size distributions in fish. Oikos 122: 835–844.

Ohlberger, J., S.J. Thackeray, I.J. Winfield, S.C. Maberly and L.A. Vøllestad. 2014. When phenology matters: age-size truncation alters population response to trophic mismatch. Proc. R. Soc. B 281: 20140938.

Paxton, C.G.M., I.J. Winfield, J.M. Fletcher, D.G. George and D.P. Hewitt. 2004. Biotic and abiotic influences on the recruitment of perch (*Perca fluviatilis*) in Windermere, U. K. J. Fish Biol. 65: 1622–1642.

Pickering, A.D. and L.G. Willoughby. 1977. Epidermal lesions and fungal infection on the perch, *Perca fluviatilis* L., in Windermere. J. Fish Biol. 11: 349–354.

Pickering, A.D. 2001. Windermere: Restoring the Health of England's Largest Lake. Freshwater Biological Association Special Publication No. 11. Freshwater Biological Association, Ambleside, U.K.

Schindler, D.W. 2012. The dilemma of controlling cultural eutrophication of lakes. Proc. R. Soc. B.DOI:10.1098/rspb.2012.1032.

Smith, V.H. and D.W. Schindler. 2009. Eutrophication science: where do we go from here? Trends Ecol. Evol. 24: 201–207.

Smyly, W.J.P. 1952. Observations on the food of the fry of perch (*Perca fluviatilis* L.) in Windermere. Proc. Zool. Soc. London 122: 407–416.

Svanbäck, R., P. Eklöv, R. Fransson and K. Holmgren. 2008. Intraspecific competition drives multiple species resource polymorphism in fish communities. Oikos 117: 114–124.

Thackeray, S.J., T.H. Sparks, M. Frederiksen, S. Burthe, P.J. Bacon, J. Bell, M.S. Botham, T.M. Brereton, P.W. Bright, L. Carvalho, T. Clutton-Brock, A. Dawson, M. Edwards, J.M. Elliott, R. Harrington, D. Johns, I.D. Jones, J.T. Jones, D.I. Leech, D.B. Roy, W.A. Scott, M. Smith, R.J. Smithers, I.J. Winfield and S. Wanless. 2010. Trophic level asynchrony in rates of phenological change for marine, freshwater and terrestrial environments. Glob. Change Biol. 16: 3304–3313.

Thackeray, S.J., P.A. Henrys, H. Feuchtmayr, I.D. Jones, S.C. Maberly and I.J. Winfield. 2013. Food web de-synchronisation in England's largest lake: an assessment based upon multiple phenological metrics. Glob. Change Biol. 19: 3568–3580.

Thorpe, J. 1977. Synopsis of biological data on the perch, *Perca fluvitilis* (Linnaeus 1758) and *Perca flavescens* (Mitchill 1814). FAO Fisheries Synopsis 113.

Winfield, I.J. and N.C. Durie. 2004. Fish introductions and their management in the English Lake District. Fish. Manage. Ecol. 11: 1–7.

Winfield, I.J., J.M. Fletcher, D.P. Hewitt and J.B. James. 2004. Long-term trends in the timing of the spawning season of Eurasian perch (*Perca fluviatilis*) in the north basin of Windermere, U.K. pp. 95–96. *In*: T.P. Barry and J.A. Malison (eds.). Proceedings of Percis III: The Third International Percid Fish Symposium. University of Wisconsin Sea Grant Institute, Madison, Wisconsin, U.S.A.

Winfield, I.J., J.M. Fletcher and J.B. James. 2008. The Arctic charr (*Salvelinus alpinus*) populations of Windermere, U.K.: population trends associated with eutrophication, climate change and increased abundance of roach (*Rutilus rutilus*). Environ. Biol. Fish. 83: 25–35.

Winfield, I.J., J.M. Fletcher and J.B. James. 2010. An overview of fish species introductions to the English Lake District, UK, an area of outstanding conservation and fisheries importance. J. Appl. Ichthyol. 26(Supplement 2): 60–65.

Winfield, I.J., J.M. Fletcher and J.B. James. 2011. Invasive fish species in the largest lakes of Scotland, Northern Ireland, Wales and England: the collective U.K. experience. Hydrobiologia 660: 93–103.

Winfield, I.J., J.M. Fletcher and J.B. James. 2012. Long-term changes in the diet of pike (*Esox lucius*), the top aquatic predator in a changing Windermere. Freshwater Biol. 57: 373–383.

Winfield, I.J. (2015). Eutrophication and freshwater fisheries. *In*: J.F. Craig (ed.). Freshwater Fisheries Ecology. Wiley-Blackwell, Oxford, U.K.

Worthington, E.B. 1950. An experiment with populations of fish in Windermere, 1939–48. Proc. Zool. Soc. Lond. 120: 113–149.

7

Reproductive Biology and Environmental Determinism of Perch Reproductive Cycle

Pascal Fontaine,* Abdulbaset Abdulfatah and Fabrice Teletchea

ABSTRACT

Perch are annual spawners whose spawning season occurs when both water temperature and photoperiod are increasing during spring. In the northern hemisphere, according to latitude the spawning season is observed from March in southern countries to June in northern countries. The reproductive cycle is initially induced by the photoperiod decrease after the summer solstice in late June, however major morpho-anatomical changes are chiefly recorded in late August-September. At that period, testicles are well developed, maximum gonado-somatic indexes are recorded and males are often spermiating. For females, oogenesis is a much longer process than spermatogenesis and vitellogenesis continues over autumn and early winter and requires a wintering period to achieve full oocyte development. Spermatogenesis and oogenesis are mainly determined by annual photoperiod and temperature variations, which explain the timing of the main steps of the reproductive cycle, and can be regulated by other abiotic and biotic factors that modulate reproductive performances.

Keywords: Reproductive cycle, spermatogenesis, oogenesis, spawning, environmental determinism, modulating factors

Unité de Recherche Animal et Fonctionnalités des Produits Animaux, Université de Lorraine, 2 avenue de la Forêt de Haye, B.P. 172, 54505 Vandoeuvre-lès-Nancy, France.

* Corresponding author: p.fontaine@univ-lorraine.fr

7.1 Introduction

Yellow perch and European perch (*Perca flavescens* and *P. fluviatilis*, respectively) are much appreciated for human consumption, particularly in inland areas. They are consumed in various ways (fillet and whole fish) and size (small fried fish as well as large fish) (Tamazouzt et al. 1993). As a result, perch have been exploited in several water bodies for a long time, particularly in the north (Estonia, Finland, Sweden), the east (Poland, Russia) and in Western Europe for European perch and in North America (Canada and northern United States) for yellow perch. Perch harvest strongly increased at the beginning of the 1980s before stabilizing at around 25,000 tons per year ever since (www.fao.org). More recently, to respond to the increasing demand of the human consumption market, perch aquaculture has started to develop based on intensive rearing models and recirculated water technologies (Fontaine et al. 2008; Toner and Rougeot 2008). Nevertheless, these species, particularly the European perch, are also reared in much more extensive systems, namely polyculture in ponds, for the restocking markets and sport fisheries. In this kind of system, perch are introduced as a pelagic piscivorous species and represent a low percentage of the overall production (<1%). Whatever the activity considered (management of wild populations for sustainable fisheries, intensive or extensive aquaculture), an accurate knowledge of (1) the natural reproductive cycle; (2) the environmental determinism controlling these cycles; and (3) the factors regulating reproductive performance, are all required to improve the management of perch stocks in natural ecosystems or in polyculture ponds and to improve productivity of perch aquaculture.

Based on morpho-anatomical, histological and endocrine indicators, the first part of this chapter describes the reproductive cycle of perch under natural conditions based on field studies. The second part is a synthesis of the research carried out in the past 20 years on the role of variations of temperature and photoperiod in the control of reproductive cycles, especially as they affect each sex differentially. Finally, in the wider context of the regulation of reproductive performance (multifactorial approaches), the third part analyzes our current knowledge of the role of modulating factors on perch reproductive performance, including environmental (other than photoperiod and temperature), nutritional, and population-level (linked to the characteristics of breeders) factors. This chapter is focused on yellow perch and European perch, but evidence suggests that the Balkhash Perch *P. schrenkii* has the same reproductive biology (see Chapter 3).

7.2 Description of the Annual Reproductive Cycle

7.2.1 Gonads

In males, the two testes are located in the peritoneal cavity between the intestine and the swim bladder. They are linked together by a septum posteriorly and join the basal part of the urinary bladder. In females, two ovaries are present in immature fish, but during gonad maturation, they fuse into a single organ. The ovaries have the same position as the testes within the ventral cavity. The oviduct and the urinary bladder merge to form a urogenital sinus (Craig 2000). Note that only two cases of hermaphroditism have been described in European perch. Chevey (1922) observed an individual that had released sperm during the spawning season, but during dissection the gonad appeared to be an ovary. Such observation has been confirmed by Jellyman (1976) who found a gonad consisting of three different parts: one containing oocytes and located dorsally in the ventral cavity, another composed of two lobes below the first one and containing spermatids and spermatozoa only, and finally one not entirely differentiated and composed of connective tissues only.

7.2.2 Sexual Maturity and Sexual Dimorphism

In percids, sexual maturity occurs earlier in males than in females (Willemsen 1977; Mann 1978). In Western Europe (France, northern Italy, United Kingdom), sexual maturity in European perch is generally acquired after one or two summers for males (age 0+ and 1+), thus at a size of 7 to 12 cm, and between two to four years for females, thus at a length of 12 to 20 cm (Alm 1954; Craig 1974; Papageorgiou 1977; Treasurer 1981; Jamet et al. 1990; Flesch 1994; Jamet and Desmolles 1994). Actually, in natural conditions, the age at sexual maturity varies according to sex, habitats (thermal regimes) and fish communities (densities of perch, their prey, and interspecific competitors). An environment where food is abundant and fish biomass is low allows for more rapid growth and earlier sexual maturity for perch (Houthuijzen et al. 1993). Obviously, the effect of food availability on perch growth is tightly linked to interspecific competition, for example in European perch with cyprinids such as roach *Rutilusrutilus* (Persson 1990).

The effect of water temperature on sexual maturity is also well established. For instance, in a water body where water was artificially heated by a nuclear power plant (the Forsmark Reservoir in Sweden, Sandström et al. 1995), sexual maturity was reached earlier compared to fish sampled from a reference area (SW Bothnian Sea, normal temperature regime). In yellow perch, Malison et al. (1986) have found that the size of breeders is more important than environmental factors or age for determining the onset of the first reproductive cycle in both males and females. In rearing conditions (with optimal environmental conditions), the age at sexual maturity can be very early (Malison et al. 1986; Ben Ammar et al. 2012). In such conditions, in yellow perch, vitellogenesis and spermatogenesis start for the first time at a size of 85–90 mm (5–10 g) corresponding to an age of 3–4 months (Malison et al. 1986).

The identification of sex in percids is tricky based only on morphological characters, except during the spawning season, during which mature females display a swollen and rounded belly, and males release their milt easily (Flesch 1994). In European perch during the spawning season, Collette and Banarescu (1977) noticed that males are brighter than females. Jamet (1991) observed the presence of nuptial tubercles on the fish's head and Lindesjöö (1994) found that the structure of the skin is different between sexes. During prespawning (February) and spawning (May–June) periods, the epidermis and dermis of males is thicker than that of females (higher number of layers of epithelial cells, larger cells). Noteworthy, in these species, there is a sexual dimorphism of growth that is observed both in natural (Craig 2000) and rearing (Malison et al. 1982; Fontaine et al. 1997) conditions. This sexual dimorphism is linked to early sexual differentiation (Mezhnin 1978; Zelenkov 1982; Malison et al.

1986) and might be due to certain sexual steroids (estradiol, E_2) circulated at elevated concentrations in the blood. Such sexual differentiation starts earlier in females (Kayes et al. 1982).

7.2.3 Spawning Period and Fecundity of Breeders

Yellow and European perch are both early spring spawners that do not provide any parental care to their offspring (Teletchea et al. 2009). A statistical analysis of 29 biological traits related to the function of reproduction have shown that perch have reproductive traits close to those exhibited by other percids (*Sander lucioperca, Sander vitreus, Gymonocephaluscernuus*), some cyprinids (*Rutilusrutilus, Aspiusaspius, Chondrostomanasus, Phoxinusphoxinus*) and esocids (*Esoxlucius, Esoxmasquinongy*). In natural conditions, perch reproduce once a year during the spring (March–June) just after a wintering period of 160 days at a temperature below 10°C (Hokanson 1977; Craig 2000). Spawning occurs when both water temperature and photoperiod (the duration of day daylight) are increasing, yet this varies strongly with geographical region; generally, more northern populations spawn later and at lower temperatures. In the Northern Hemisphere (some populations have been introduced in the Southern Hemisphere like in New-Zealand, Jellyman 1980), the spawning season of European perch lasts from March-April in France and in Greece (Papageorgiou 1977; Flesh 1994; Sulistyo et al. 1998) to May–June in England and Finland (Guma'a 1978; Urho 1996). Obviously, this global effect of latitude depends strongly on the characteristics of the water body considered (deep and cold lakes vs. rivers or plain ponds) and altitude (effect on the thermal regimes). The water body characteristics (size, thermal regime) can also influence the duration of the spawning period. For instance, in Lake Geneva (Franco-Swiss Lake), the spawning season lasts from April to June (Gillet et al. 1995). The authors also found that spawning time and depth at which the spawns were observed in the lake were closely related to the size of females. Larger females tended to spawn later than smaller females (Gillet and Dubois 2007). Consequently, in large lakes, several successive peaks of high densities of spawns (ribbons of eggs) are frequently observed (Jones 1982). In consequence, because of the large geographical distribution of European perch and the diversity of its habitats, spawning can occur at various periods and temperatures in spring. According to most authors, spawning can occur between 7 and 20°C (Sulistyo 1998). For yellow perch (in lakes Mendota, Michigan and Minnesota), similar observations (spawning takes place from the end of February to the beginning of July, in water having temperatures ranging from 3 to 18°C) have been reported by Hokanson (1977) and Heidinger and Kayes (1986). Nevertheless, this species mostly spawns from mid-April up to the end of May at a water temperature between 8.9 to 12.2°C (Scott and Crossman 1973).

For all perch, including the Balkhash perch (see Chapter 3), the egg masses display features that are similar to egg strings observed in some anurans (Amphibia). Eggs are tightly grouped and make a gelatinous ribbon with zig-zag folds. Translucent and yellowish eggs are tied together by a mucilaginous sheath. The eggs of perch have a thick membrane made of an external sticky envelope, a large medium space composed of fine fibres arranged radially and an internal layer called *zonaradiata* (Scott and Crossman 1973; Craig 2000). In European perch, the relative fecundity (number of eggs per g of female) varies strongly according to the diversity of habitats and the environmental conditions encountered; fecundities varying from 33 to 202 eggs per gram of female weight have been described in the literature (Table 1). Thorpe (1977) considered that a fecundity of 210 eggs per gram of female weight is very high for this species. In yellow perch, the relative fecundity varies between 79 to 233 eggs per gram of female weight (Heidinger and Kayes 1986). By comparison to other temperate freshwater fish species, perch have a high fecundity, as revealed by the important gonadosomatic index (GSI) [(weight of gonads/total weight of females) x 100, %)] found during the spawning season (20–30%) (Treasurer and Holiday 1981; Heidinger and Kayes 1986; Jamet and Desmolles 1994; Sulistyo et al. 1998).

Table 1. Relative fecundity (number of eggs per gram of female weight) of female European perch, *Perca fluviatilis*, from different habitats and localities.

Locality and country	Female size (mm)	Relative fecundity (number of eggs.g^{-1})	References
Lake Pounui, New-Zealand	140, 422	33, 141	Jellyman 1980
Loch Davan, Scotland	176, 323	78, 134	Treasurer 1981
Reservoir, Forsmark, Sweden	170, 250	126, 137	Sandström et al. 1995
Lake Agios Vasiolios, Greece	160, 202	133, 202	Papageorgiou 1977
Lake Aydat, France		182	Jamet and Desmolles 1994

The number of eggs within a ribbon is correlated to its length and breadth (Dalimier et al. 1982; Heidinger and Kayes 1986; Gillet et al. 1995). In yellow perch, before water hardening, the diameter of fertilized eggs varies between 1.62 and 2.09 mm, and between 1.87 and 2.81 mm after the swelling of the egg membrane (Mansueti 1964). Actually, when in contact with water, the gelatinous membrane increases in volume and reaches a thickness of 500 μm (Malservisi and Magnin 1968). In European perch, the diameter of fertilized eggs is usually smaller (0.86–1.62 mm) (Jellyman 1980; Treasurer 1983).

In male perch, the number of spermatozoa per millilitre of milt (sperm) is very important, varying from 40.1^9 mL^{-1} to 76.1^9 mL^{-1} (Piironen and Hyvärinen 1983; Ciereszko and Dabrowski 1993). During spawning, males release their spermatozoa in five seconds, which make a white cloud around the egg ribbon (Hergenrader 1969; Treasurer 1981). Spermatozoa are viable only for a few minutes after their release in the water, probably less than two minutes (Piironen and Hyvärinen 1983). Concerning the fine structure of spermatozoa, it is uniflagellated and possesses an ovoid head (length: 1.87 ± 0.16 μm; 1.78 ± 0.14 μm, the central part of the body has only one mitochondrion and a tail region and lacks an acrosome). As the flagellum inserts mediolaterally on the nucleus, the spermatozoon is asymmetrically shaped (Lahnsteiner et al. 1995).

7.2.4 Gonadogenesis, Spermatogenesis and Ovogenesis

The morpho-anatomical and histological development of ovaries in females (Mansueti 1964; Treasurer and Holiday 1981; Sulistyo et al. 1998) and of testes in males (Turner 1919; Sulistyo et al. 2000) are well known in perch.

Females

Oogenesis displays a group-synchronous development of oocytes, which means that two batches of oocytes could be encountered in an ovary at a given time: a first batch of primary oocytes (reserves for new oogenic cycles) and a second one in development (on-going cycle). When focusing only on this second batch of oocytes, the oocyte development is synchronous (Marza 1938). Hereafter, based on the studies of Sulistyo et al. (1998) and the different defined oocyte development stages (Table 2), the reproductive cycle of European perch females is described from a population of perch (size: 19–24 cm; origin: pond of 620 ha, Domaine de Lindre, Moselle, France) during April 1995 until April 1996 (Fig. 1).

In summer (July–August), a period of sexual quiescence, gonads were weakly developed. Thus, by the end of July females displayed a very small ovary mass (GSI < 1%) with ovaries containing only primary oocytes with a diameter (OD) of 100–140 µm (Fig. 2). By mid-August, some oocytes in early stages of endogenous vitellogenesis (accumulation of yolk vesicles) were identified and showed a steady and regular increase of ovary mass until December (GSI (%) = 0.073 x ND + 0.648, where ND = number of days from August 1, $r^2 = 0.99$, Sulistyo 1998). The increase of ovary mass was lower during winter (January–February, <4°C, Fig. 1). The increase of GSI from July to December was very closely correlated to the steady and regular increase in oocyte diameter (OD = 430 µm at the end of September, OD = 750 µm at the end of December). Similar observations were made by Treasurer and Holliday (1981) when studying populations of European perch in Scottish lakes. By December, oocytes were at the beginning of exogenous vitellogenesis. From this period, oocyte

Table 2. Different stages of ovary and oocyte development in female European perch, *Perca fluviatilis*, during an annual reproductive cycle (Rinchard and Kestemont 1996).

	Stages of ovary development	Oocyte stages within the ovary	Description of the most advanced oocytes
(1)	Previtellogenic	Previtellogenic oocytes	Oocytes with free vacuoles in the cytoplasm
(2)	Beginning of the endogenous vitellogenesis	Previtellogenic oocytes in endogenous vitellogenesis	Apparition of yolk vesicles forming 2 to 3 rings in the periphery of the cytoplasm (beginning of the endogenous vitellogenesis)
(3)	End of the endogenous vitellogenesis	Previtellogenic oocytes and oocytes having achieved their endogenous vitellogenesis	Oocytes full of yolk vesicles. Follicle and cellular layers differentiated (end of the endogenous vitellogenesis)
(4)	Exogenous vitellogenesis	Previtellogenic oocytes and oocytes at different stages of exogenous vitellogenesis	Oocytes accumulating yolk globules and yolk vesicles at the periphery of the cytoplasm
(5)	Final maturation	Previtellogenic oocytes and oocytes in final maturation	Appearance of the micropyle and migration of the germinal vesicle towards the micropyle
(6)	Post-spawning	Previtellogenic oocytes and pre- and post-ovulatory follicles	Pre- and post-ovulatory follicles hypertrophied, degeneration of the yolk substance

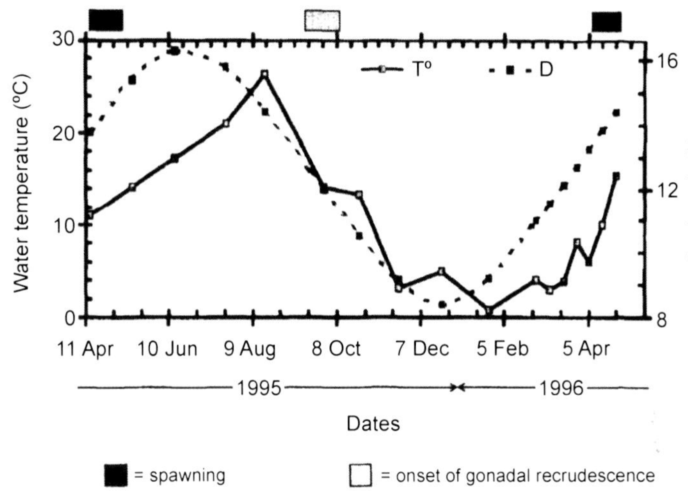

Fig. 1. Variations of water temperature (T) and day length (D) in Lindre pond (Moselle, France) during the period from April 1995 to April 1996 (Sulistyo et al. 1998).

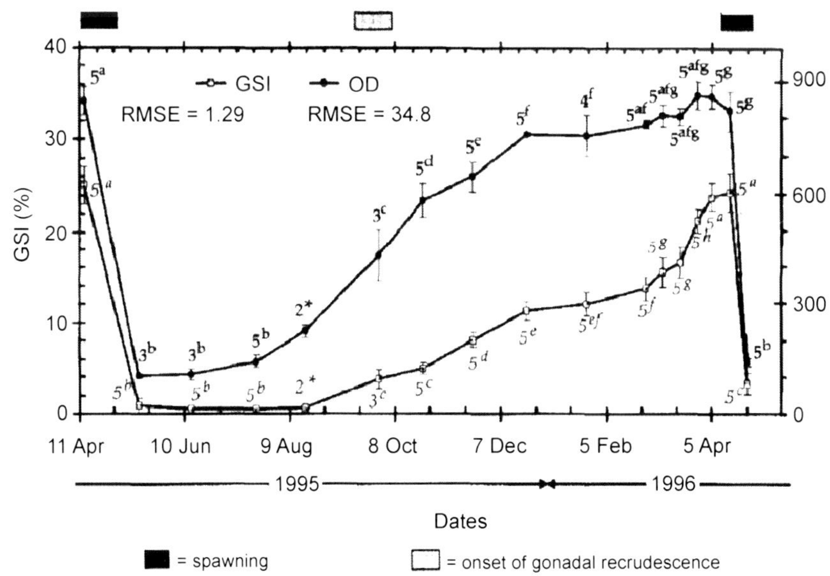

Fig. 2. Variations of the gonadosomatic index (GSI) and oocyte diameter (OD) in females of European perch, *Perca fluviatilis*, during an annual reproductive cycle (April 1995–April 1996). Values represent mean ± standard deviations. Numbers indicate sample size and letters indicate significant differences ($p < 0.05$). RMSE: Root mean square errors (Sulistyo et al. 1998).

size did not significantly increase, except for a few days prior to spawning, at the end of March, when oocytes achieved their maximum size (OD = 860 µm), similar to that observed the year before (OD = 850 µm).

Histological analysis demonstrated that the gelatinous ribbon was already present around the oocytes at the end of March, well before the oocyte maturation (migration of the nucleus from a central position to a peripheral position, breakdown of the nuclear envelope, Fig. 3) is observed in the pre-ovulatory period (Migaud et al. 2003a). In fact, during the pre-ovulatory period, the migration of the nucleus (germinal vesicle) seems

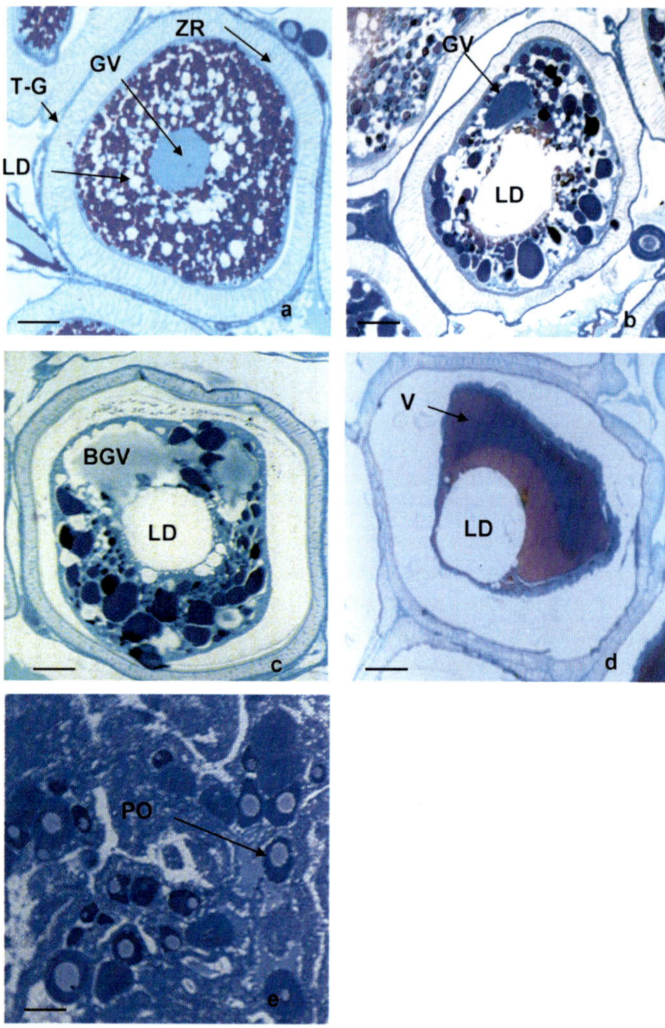

Fig. 3. Illustrations of histological stages determined in pre-ovulatory period (March-April 1999) in European perch *Perca fluviatilis* females. (a) Oocytes at the end of exogenous vitellogenesis (April 7); (b) Mature oocyte with germinal vesicle migrated to the periphery (April 14); (c) Mature oocyte with the breakdown of the nuclear envelope (April 21); (d) Oocyte ovulated (April 23); (e) Ovary in post-spawning period (April 26). Scale = 100 µm. GV: germinal vesicle; LD: lipid droplet; T-G: theca and granulosa; ZR: zonaradiata (chorion with microvilosities); BGV: breakdown of the germinal vesicle; V: Vitellus; PO: primary oocyte (Migaud et al. 2003a).

to operate during several weeks, whereas the breakdown of the nuclear envelope of the germinal vesicle and the ovulation of oocytes are rapid phenomena that are tricky to observe (Fig. 4, Sulistyo 1998; Migaud et al. 2003a). An *in vitro* study by Goetz and Theofan (1979) using yellow perch, highlighted that the change from the nuclear migration stage to the breakdown of the germinal vesicle could be achieved in four days confirming the rapidity of the cell changes. This rapid change has also been observed in other percids, such as walleye *Sander vitreus* (Malison and Held 1996).

In the northeast of France, spawning occurs over a two to three week period in April during which both water temperature and photoperiod increase. In adult female perch, important morpho-anatomical changes are noticed during oogenesis (Figs. 5 and 6). First, the weight of viscera is proportionally much higher during summer than values recorded during the rest of the year. The viscerosomatic index (VSI) [(weight of viscera/total weight of female) x 100, %] increases during the post-spawning period (April–June) from 2 to 6% (maximal value), reflecting an active feeding of breeders and an accumulation of mesenteric adipose tissue (Makarova 1973; Kirillov and Akhremenko 1982; Sulistyo 1998). These reserves decrease rapidly after the start of autumn and the development of the ovary mass. The hepatosomatic index (HSI) [(weight of liver/total weight of the individual) x 100, %] in females is minimal during summer (1%) and regularly increases during autumn to reach a maximum during winter. Such variations have also been observed in the same species by Makarova (1973) and in walleye (Henderson et al. 1996). It seems that during winter, the liver is very active (hepatic synthesis, stocking and allocation of lipids to gonads) linked to the development of gonads.

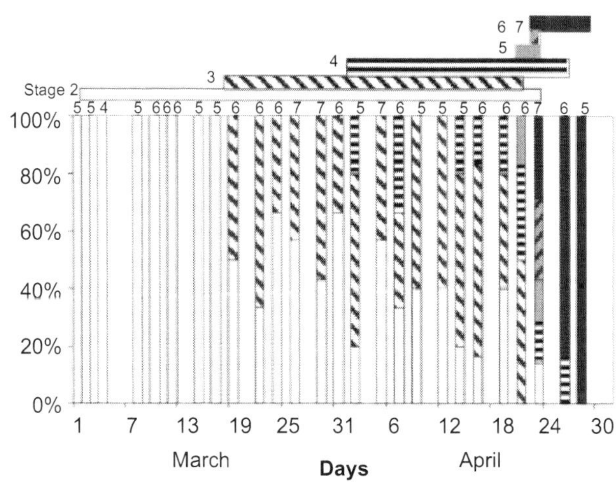

Fig. 4. Oogenesis stages (% of abundance) observed in female European perch, *Perca fluviatilis*, during the pre-ovulatory period in natural conditions (1 March–30 April 1999). Numbers above each bar represent sample size. Stage 2: ovaries with vitellogenic oocytes, stage 3: pre-migrating germinal vesicle, stage 4: oocytes with peripheral germinal vesicle, stage 5: germinal vesicle break down; stage 6: ovulation, stage 7: spawning (empty ovary) (Migaud et al. 2003a).

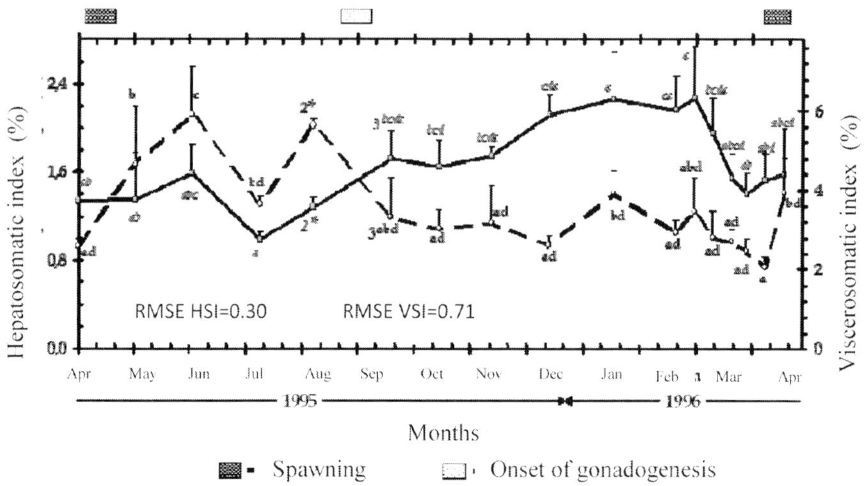

Fig. 5. Variations of the hepatosomatic (solid line) and viscerosomatic (dashed line) indexes in female European perch, *Perca fluviatilis*, during an annual reproductive cycle (April 1995–April 1996). Values represent mean ± standard deviations (n = 5, except when a number is indicated near the value) and letters indicate significant differences ($p < 0.05$). RMSE: Root mean square errors (Sulistyo 1998).

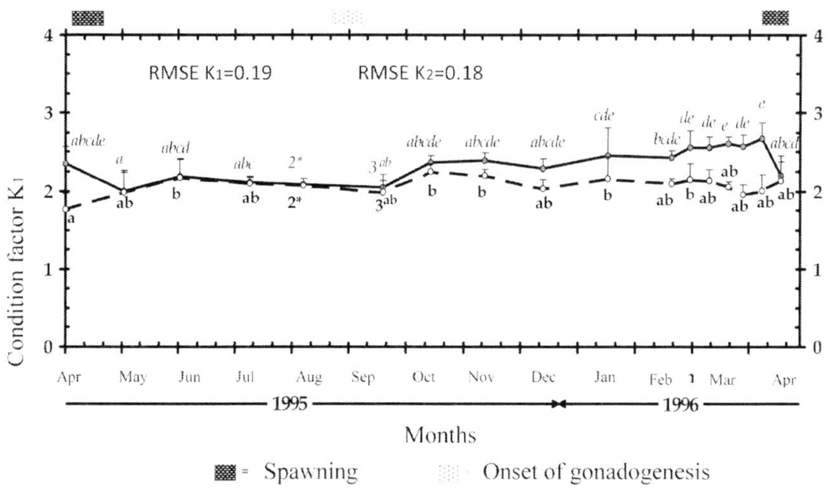

Fig. 6. Variations of condition factors K_1 (solid line) and K_2 (dashed line) in female European perch, *Perca fluviatilis*, during an annual reproductive cycle (April 1995–April 1996). Values represent mean ± standard deviations (n = 5, except when a number is indicated near the value) and letters indicate significant differences ($p < 0.05$). RMSE: Root mean square errors (Sulistyo 1998).

The important energetic investment of female perch towards reproduction is well highlighted by comparing the evolution of the condition indexes K_1 [(total weight of the female/standard length^3) x 100, %] and K_2 [((total weight of the female—weight of the gonads)/standard length^3) x 100, %] (Fig. 6). These two indexes are similar and have a value close to 2 in summer, whereas just before the spawning period (beginning of April) a difference of 0.7 is observed ($K_1 = 2.5$ vs. $K_2 = 1.8$). The annual variations observed are similar to those described by Le Cren (1951) and Jamet and Desmolles (1994) in other hydrosystems.

In fish, reproductive cycles are synchronized by annual variations of environmental factors, chiefly temperature and photoperiod for temperate species such as perch (Migaud et al. 2010; Taranger et al. 2010). After the perception and analysis of environmental stimuli, the hypothalamic-pituitary axis controls oogenesis in females via the secretion of gonadotropin hormones (FSH: follicle-stimulating hormone, LH: luteinizing hormone) and the regulation and production of sexual steroids (testosterone, T; 17β-estradiol, E_2; 17-α,20-β-dihydroxy-4-pregnen-3-one, 17,20β-P) by the ovary and of vitellogenin by the liver. In female perch (Figs. 7 and 8), the plasma concentrations of T (< 0.5 ng.mL^{-1}), E_2 (< 0.5 ng.mL^{-1}), 17,20β-P (<0.4 ng.mL^{-1}) and PPP (< 0.2 μg·mL^{-1}) (plasma protein phosphorus, indirect measure of vitellogenin) are low during summer, after which the levels of E_2 and PPP rise significantly in September (Fig. 7). At the end of September, a large increase in the level of E_2 (2.5–2.7 ng.mL^{-1}) is observed followed by an increase of the level of T (16 ng·mL^{-1}) one month later. Thereafter, plasma levels of T, E_2 and PPP (2 μg·mL^{-1}) remain high until the spawning, which indicates active vitellogenesis, and peaks of concentrations of T (26 ng·mL^{-1}) and E_2 (5 ng·mL^{-1}) are respectively observed at the end of March and beginning of April. Plasma T concentration decreases sharply during the spawning period, whereas plasma E_2 decreases after the spawning period (Figs. 7 and 8).

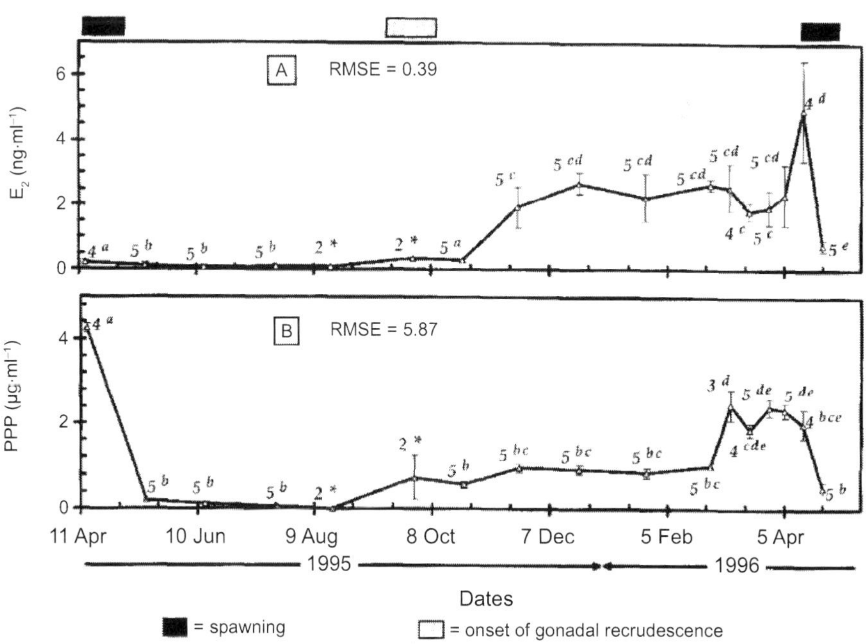

Fig. 7. Variations of plasma levels of 17β-estradiol (E_2) and plasma protein phosphorus (PPP) in female European perch, *Perca fluviatilis*, during an annual reproductive cycle (April 1995–April 1996). Values represent mean ± standard deviations. Numbers indicate sample size and letters indicate significant differences ($p < 0.05$). RMSE: Root mean square errors. *indicates that the value is not analyzed statistically (Sulistyo et al. 1998).

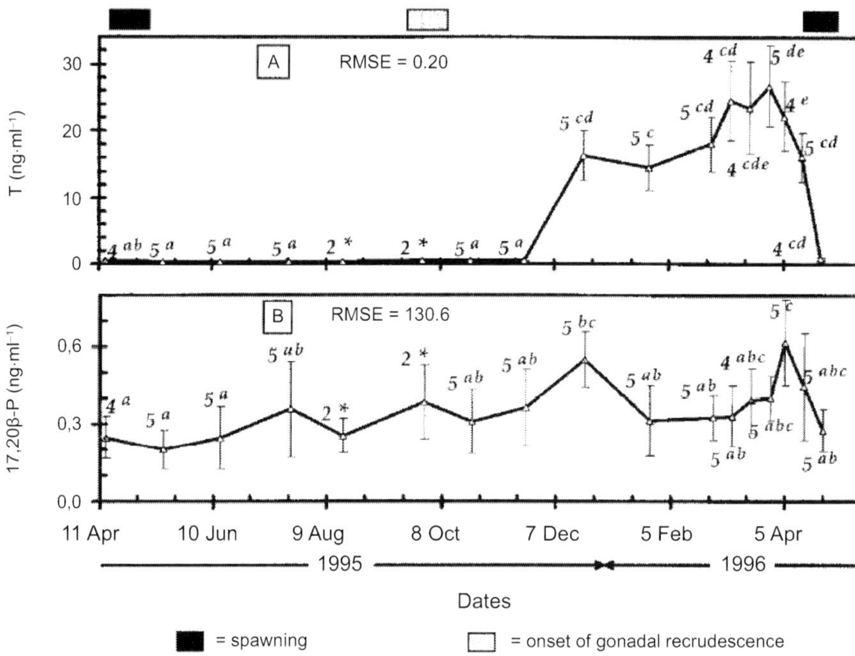

Fig. 8. Plasma levels of testosterone (T) and 17,20β-P in European perch *Perca fluviatilis* females during an annual reproductive cycle (April 1995–April 1996). Values represent mean ± standard deviations. Numbers indicate sample size and letters indicate significant differences ($p < 0.05$). RMSE: Root mean square errors. *indicates that the value is not analyzed statistically (Sulistyo et al. 1998).

Males

In males of both yellow perch and European perch, the GSI remains stable during summer (Turner 1919; Craig 1974; Treasurer and Holiday 1981; Tanasichuk and Mackay 1989; Hayes and Taylor 1994). Sulistyo et al. (2000) have reported GSI values lower than 0.3% in male European perch captured in June, two months after the last spawning period occurred in April (Fig. 9).

In September, a strong increase of testis mass is observed in males of European perch and the GSI reaches its maximal value (8–10%) at the beginning of autumn (Le Cren 1951; Treasurer and Holiday 1981; Fontaine et al. 1996; Sulistyo et al. 2000). Note that in yellow perch, the dynamics is similar, yet the maximal values of GSI are often observed later in November–January (Tanasichuk and Mackay 1989; Hayes and Taylor 1994). This strong gonadal development in autumn reflects an intensification of the production of spermatogonia and spermatocytes (Figs. 10 and 11). Thereafter, the GSI remains high and constant until the spawning period and then decreases just after (Tanasichuk and Mackay 1989; Dabrowski et al. 1994; Dabrowski and Ciereszko 1996) (Fig. 9). A slight decrease or increase in GSI can occur prior to the spawning season in both European and yellow perch (GSI = 4–6%) (Craig 1974; Hayes and Taylor 1994; Sulistyo et al. 2000).

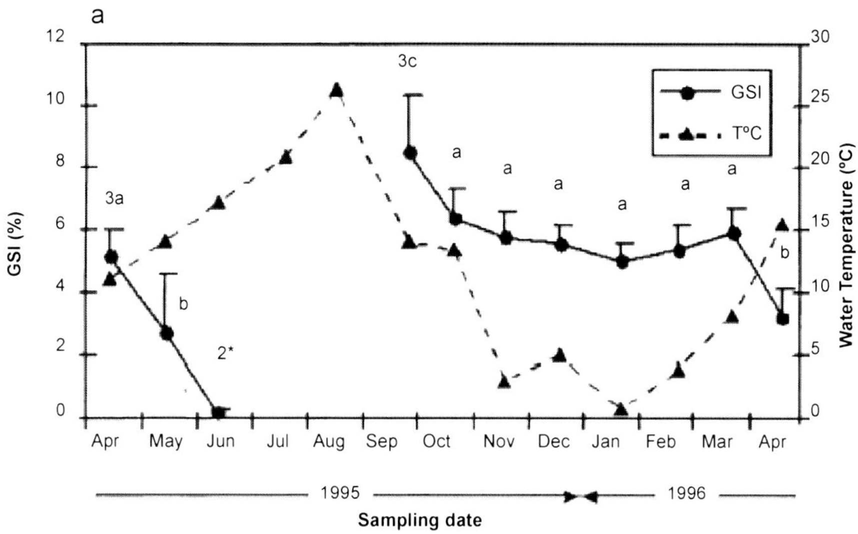

Fig. 9. Variations of the water temperature in Lindre pond (Moselle, France) and gonadosomatic index (GSI) in male European perch, *Perca fluviatilis*, during an annual reproductive cycle (April 1995–April 1996). Values represent mean ± standard deviations (n = 5, except when a number is indicated near the value) and letters indicate significant differences ($p < 0.05$). * indicates that the value is not analyzed statistically. RMSE (root mean square error) = 1.06 (Sulistyo et al. 2000).

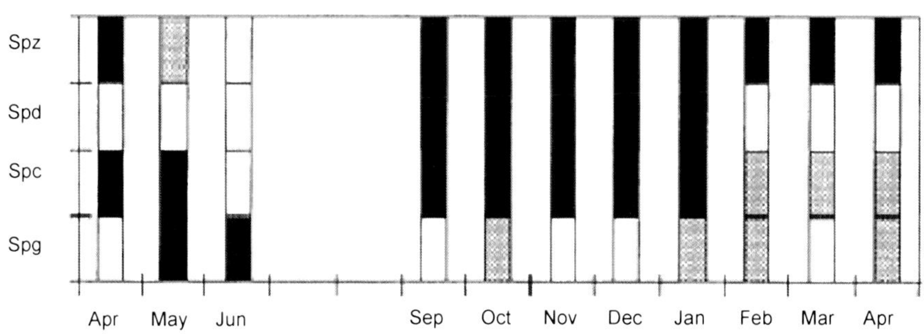

Fig. 10. Stages of spermatogenesis observed in male European perch *Perca fluviatilis* during an annual reproductive cycle (April 1995–April 1996). Black boxes indicate the most abundant stages, the grey boxes the less abundant stages and the white boxes their absence. Spg: spermatogonia; Spc: spermatocyte; Spd: spermatid; Spz: spermatozoa (Sulistyo et al. 2000).

Spermatozoa and spermatocytes were very abundant in testes in April during the reproductive season (Fig. 11A). In May, about one month after spawning, testes contained some residual spermatozoa surrounded by empty spaces, as well as stem cells, spermatogonia and spermatocytes (Fig. 11B). Males remained spermiating in May. Only spermatogonia were found in males caught in June (Figs. 10 and 11C). Then in September, during the period of testis growth, gonads contained spermatocytes, spermatids and spermatozoa in equal proportions (Fig. 11D). At that time, spermatogonia were observed in the periphery of the tissue wall. In December,

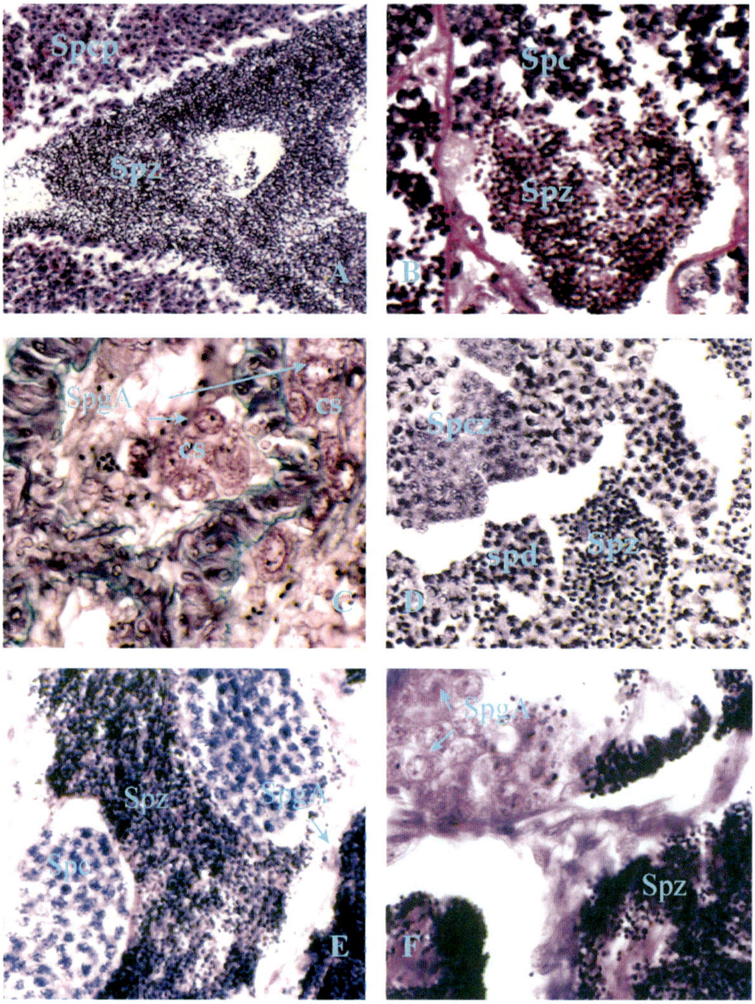

Fig. 11. Sequence of testis development stages in male European perch, *Perca fluviatilis*, during an annual reproductive cycle (April 1995–April 1996). Testes of males sampled in mid-April (A, x 100), mid-May (B, x 250), mid-June (C, x 250), late September (D, x 250), late December (E, x 250) and late February (F, x 250). Spg: spermatogonia; Spc: spermatocyte (z: stage zygoten; p: stage pachyten); Spd: spermatid; Spz: spermatozoa; Cs: Sertoli cells (Sulistyo et al. 2000).

spermatozoa increased in numbers (Fig. 11E). From December onward, all males were spermiating with a small amount of milt flowing out during fish handling. Two months before spawning, in February, spermatozoa were dominant (Fig. 11F) and few spermatids were observed.

In fish, testosterone (T) and 11-ketotestosterone (11KT) have been identified as major sex steroids involved in spermatogenesis and spermiation during the reproductive cycle (Billard et al. 1990; Borg 1994). The role of these androgens in early spermatogenesis and spermiogenesis has been confirmed by Sulistyo et al. (2000) (Fig. 12). From the spawning season in April to November, plasma levels of T and 11KT were low (< 0.5 ng·mL^{-1}) in European perch males. Plasma T levels increased significantly in December and reached peak levels (12–13 ng·mL^{-1}), then decreased in February and increased again until spawning in April (6–7 ng·mL^{-1}). The first increase is thought to regulate male germ cell differentiation and the pre-spawning peak may stimulate secondary sexual behaviours, increase the pituitary GTH levels or serve as a precursor for the production of other steroids like 11KT (Crim et al. 1981; Kobayashi et al. 1989). Plasma levels of 11KT increased significantly during autumn in yellow perch (1.2–1.5 ng·mL^{-1}, Dabrowski et al. 1996) or winter in European perch (4–5 ng·mL^{-1}, Sulistyo et al. 2000) and then decreased slowly (Fig. 12). The high levels of 11KT maintained until spawning may have a major role in keeping spermatozoa alive during the prolonged period of sperm storage within the testes.

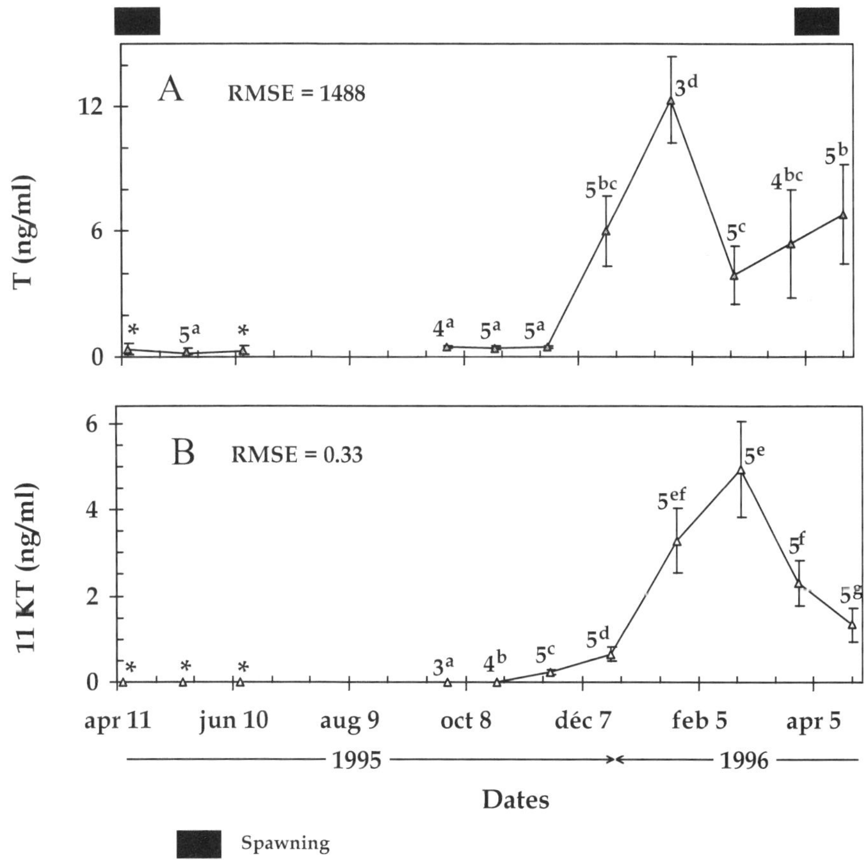

Fig. 12. Changes in plasma levels (mean ± standard deviation) of testosterone (T) and 11-ketotestosterone (11KT) in male European perch, *Perca fluviatilis*, during an annual reproductive cycle (April 1995–April 1996) in Lindre Pond (Moselle, France). Values represent mean ± standard deviations. Numbers indicate sample size and the numbers with the same superscript were not significantly different ($p < 0.05$). *indicates that the value is not analyzed statistically. RMSE = root mean square error (Sulistyo et al. 2000).

7.3 Environmental Control of the Reproductive Cycle

In order to develop perch aquaculture, research has been conducted since the 1990s to better understand the environmental control of the perch reproductive cycle. This research focuses mainly on two themes:

- The initiation of the reproductive cycle in order to optimize growth performance, perch displaying a precocious sexual maturity (well before reaching commercial size), which interferes negatively with the objectives of aquaculture;
- The induction and acquisition of fertilized eggs outside the natural spawning season (out-of-season spawning) by manipulating environmental factors in order to significantly increase the availability of larvae throughout the year.

Temperate fish species are annual spawners that mainly rely on annual cues (temperature and photoperiod) to synchronize the three main phases of their reproductive cycle: induction (initiation of oogenesis or spermatogenesis), gonadogenesis and spawning (including oocyte maturation, ovulation and eggs expulsion) (Wang et al. 2010). Consequently, manipulations of temperature and photoperiod can be used to manage the perch reproductive cycle with two purposes: inhibition of puberty or production of out-of-season spawning (Taranger et al. 2010). Compared to other freshwater species (salmonids, cyprinids), percids have a more complex environmental control due to the respective roles of temperature and photoperiod, which can vary according to the timing at which variations are applied during the reproductive cycle (Wang et al. 2010). This environmental control over perch reproduction is detailed in the following sections, and mainly concerns females because the first environmental manipulations have shown that it is relatively easy to obtain spermiating males (Tamazouzt et al. 1994).

7.3.1 Induction or Inhibition of the Reproductive Cycle

In fish species with marked seasonality of breeding activity such as European and yellow perch, the reproductive cycle is controlled and synchronized by seasonal environmental changes in relation to local climatic conditions (Taranger et al. 2010). The photoperiod is considered the main factor that synchronizes the reproductive cycle in temperate species (Bromage et al. 2001; Migaud et al. 2010). However, in percids, the reproductive cycle is induced by both decreasing temperature and photoperiod (Wang et al. 2010).

The role of the initial values and variations of both temperature and photoperiod have been extensively studied in European perch (Tamazouzt et al. 1994). Early work demonstrated that the application of artificial photothermal variations induced the development of gonads in both males and females that were initially sexually quiescent (GSI: 0.5–0.6%). Subsequent studies demonstrated that a decrease of temperature only or a decrease of photoperiod only could also induce a reproductive cycle, like the simultaneous decrease of both factors. In fact, a reproductive cycle can be induced by these three ways, but the quality of the gonadogenesis is highly variable and depends strongly on the level of the other factor. Obtaining an optimal answer of breeders (induction rate of 100%, normal gonadal growth) requires (1) applying a decrease of both temperature and photoperiod, (2) initiating the decrease of photoperiod before the decrease of temperature (Wang et al. 2006), (3) decreasing the photoperiod significantly from 2 to 8 hours (Wang et al. 2006; Abdulfatah et al. 2011) and (4) progressively decreasing water temperature (Wang et al. 2006). A much smaller decrease (1 hour) of photoperiod is not sufficient for normal reproductive cycle induction in females of European perch, especially if the decrease of temperature is rapid (Abdulfatah et al. 2011). As mentioned earlier, a decrease of photoperiod alone can induce a reproductive cycle, wherever the temperature is maintained constant and high (22°C, optimal temperature for growth), slightly decreased (18°C), strongly decreased (14°C) or very strongly decreased (6°C) (Abdulfatah et al. 2013). Nevertheless, maintaining a high temperature (22°C) or a slight decrease (18°C) strongly alters ovarian development in females (lower GSI, less advanced oocyte stages) (Abdulfatah et al. 2013). Finally, a decrease of temperature alone without a decrease of photoperiod can also induce a reproductive cycle, but it is only possible if the applied photoperiod presents a long photoperiod. Thus, under a constant photoperiod (L:D 12:12), a decrease of water temperature alone from 21 to 6°C could induce a reproductive cycle (Migaud et al. 2002). Nevertheless, it remains that the decrease of the photoperiod displays a stronger inducing effect compared to a decrease of temperature. Keeping breeders at a high temperature (22°C) does not preclude them from starting a reproductive cycle when the photoperiod is sharply decreased (4–8 hours) (Abdulfatah et al. 2013), whereas maintaining a constant photoperiod with a long photoperiod (L:D 16:8, L:D 17:7 or L:D 24:0) inhibits breeders to start a reproductive cycle even if temperature is decreased from 22°C to 6°C (Migaud et al. 2003b, 2004; Abdulfatah et al. 2011, 2013). Inhibition of the reproductive cycle in European perch due to inappropriate photoperiods, like constant photothermal conditions, is related to lower sex-steroid levels caused by a lack of gonadotropin stimulation (Milla et al. 2009). In addition, inappropriate variations of photoperiod (increase) during the pre-inductive period (i.e., before the application of an inductive program) could also inhibit the onset of a reproductive cycle in breeders (Fontaine et al. 2006). The quality of the gonadogenesis to different photothermal conditions applied is thus the result of a trade-off between inducing (decrease of the length of days and temperature) and inhibiting effects (maintaining constant photoperiod with a long photoperiod). For induction, the decrease of the duration of the photoperiod can be qualified as the main inducing factor, whereas the decrease of temperature should be considered as a secondary inducing factor.

7.3.2 Gonadogenesis and Spawning

Once the reproductive cycle has been induced, gonadogenesis, and particularly oogenesis in females, requires specific environmental conditions to allow its progress and the acquisition of good quality offspring. Thus, in accordance with observations made under natural conditions (Hokanson 1977), controlled experiments demonstrated that European perch requires a long vernalization period (prolonged exposure to cold temperature) for a normal gonadogenesis, particularly during the progress of vitellogenesis in females (Migaud et al. 2002). For European perch, compared to the reproductive cycle under natural conditions (Sulistyo et al. 1998, 2000), it was demonstrated that a vernalization period (6°C) of five months is necessary to complete vitellogenesis in females (GSI: 15%), whereas a period of three months is not sufficient (GSI: 10–12%) (Migaud et al. 2002). The temperature applied during this vernalization period is very important. A recent study has shown that maintaining elevated temperatures (14 and 18°C) resulted in a decrease of the GSI (Fig. 13), in oocyte diameter (Fig. 14) and in the early appearance of atretic oocytes (Fig. 15) (Abdulfatah 2010). The decrease of the ovary mass and the appearance of atretic oocytes occur earlier when the duration of the photoperiod is long. In conclusion, the proper progress of oogenesis requires a cold temperature and a short photoperiod.

Once vitellogenesis is achieved, such oogenesis stage is observed in females in natural conditions when captured in February–March (GSI: 15–20%), spawning is induced by an increase of the water temperature from 6 to 14°C (Abdulfatah et al. 2012). At this stage, an increase of day length does not seem to play a key role in the control of the spawning time. Finally, it was determined that female perch spawn chiefly during the first part of the photoperiod, between dawn and early afternoon (Migaud et al. 2006). The daily variations of light intensity synchronize reproductive behaviours.

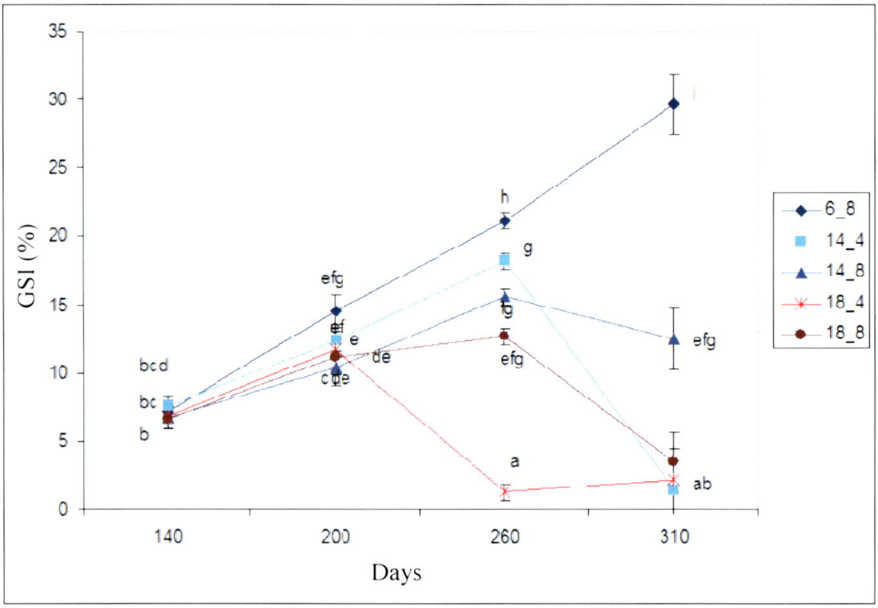

Fig. 13. Effects of photo-thermal treatments applied from the end of the initial induction period (day 140) up to the spawning period on the gonadosomatic index (GSI) of female European perch, *Perca fluviatilis*. Each treatment is presented as **a**_**b** in which **a** indicates the temperature (6, 14 or 18°C) and **b** indicates the photophase decrease (minus 4 or 8 hours from the initial photoperiod L:D 16:8 at day 1). Values represent mean ± standard deviations (n = 5). The values with the same superscript were not significantly different ($p < 0.05$) (Abdulfatah 2010).

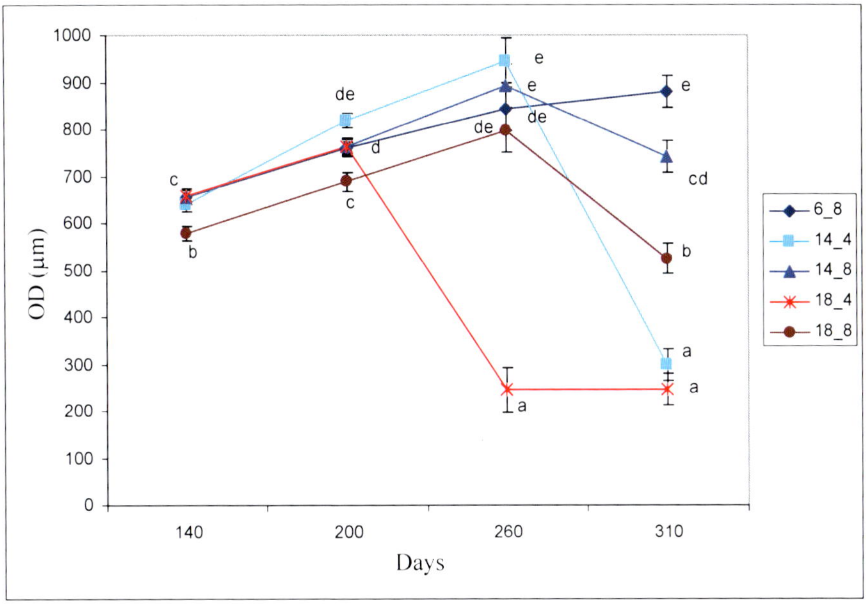

Fig. 14. Effects of photo-thermal treatments applied from the end of the initial induction period (day 140) up to the spawning period on the oocyte diameter (OD) of female European perch, *Perca fluviatilis*. Each treatment is presented as **a**_**b** in which **a** indicates the temperature (6, 14 or 18°C) and **b** indicates the photophase decrease (minus 4 or 8 hours from the initial photoperiod L:D 16:8 at day 1). Values represent mean ± standard deviations (n = 5). The values with the same superscript were not significantly different ($p < 0.05$) (Abdulfatah 2010).

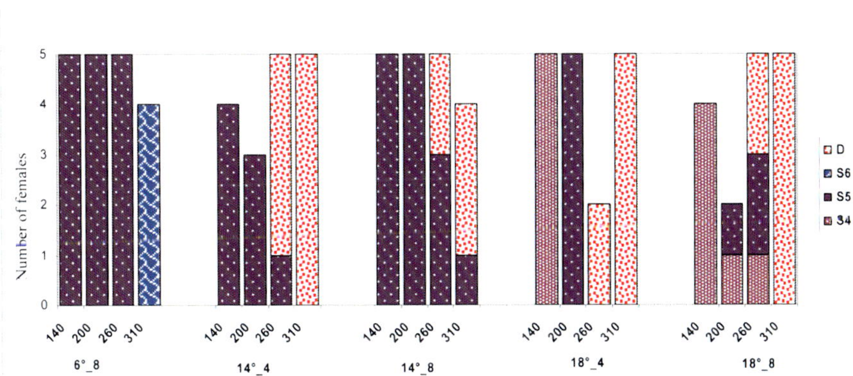

Fig. 15. Effects of photo-thermal treatments applied from the end of the initial induction period (day 140) up to the spawning period on the oogenesis stages of female European perch, *Perca fluviatilis*. Each treatment is presented as **a**°_**b** in which **a**° indicates the temperature (6, 14 or 18°C) and **b** indicates the photophase decrease (minus 4 or 8 hours from the initial photoperiod L:D 16:8 at day 1). S4: early vitellogenesis stage, S5: advanced vitellogenesis stage, S6: last maturation stage, D: atretic oocyte (Abdulfatah 2010).

7.4 Regulation of Reproductive Performance

As described above, the reproductive cycle of perch is controlled by annual variations of photoperiod and water temperature, the importance of which varies according to the phases of the cycle considered (induction, gonadogenesis, spawning). Thus, artificial photothermal programs have been developed and are used in rearing conditions to obtain out-of-season spawning several times per year (based on different batches of breeders reared in dedicated rooms specifically equipped). Thus, perch farms can produce fish on the basis of 4 to 12 spawnings a year. Nowadays, the main bottleneck for the hatchery-nursery activities is the regulation of reproductive performance, which still remains too variable.

An inappropriate photoperiod could alter gonadogenesis and consequently reduce the number of spawns obtained, delay the spawning period and reduce the fecundity of females and fertilization and hatching rates (Migaud et al. 2003b; Migaud et al. 2006). In yellow perch, females subjected to a condensed photothermal regime (4 months) applied by the end of August spawned one month earlier than control (simulated natural temperature and photoperiod). Moreover, egg quality was reduced with a fertilization rate of only 36.5% compared to 81.7% in controls (Ciereszko et al. 1997).

The determinism of the quality of gametes, eggs and larvae is multifactorial. Three main types of factors can be distinguished: environmental factors including physical and chemical characteristics of water and biological environment, nutritional factors linked to the feeding of breeders and population factors linked to the characteristics of the breeders. Even though numerous studies have been carried out on reared species in the last decades, very few have focused on percids. Nevertheless, a synthesis of current knowledge is proposed below.

The influence of environmental factors other than water temperature and photoperiod on perch reproduction has received little attention. Wang et al. (2006) reported no effect of the light spectrum (white light vs. red light) on the induction of oogenesis in European perch. Nevertheless, European perch are very sensitive to stress (see below), thus it is likely that adverse environmental conditions (high turbidity for instance) could have negative effects on their reproduction.

Concerning nutrition, percids are "capital breeders" (Henderson et al. 1996), which means that the development of gonads is largely based on the reserves accumulated during sexual quiescence. Consequently, the initial nutritional state is a key to the success of the reproductive cycle. The quantity and quality of reserves (perivisceral adipose tissue) seem to be very important. Nevertheless, Wang et al. (2006), when comparing the response of European perch females with different fat indices (5.3 vs. 1.3%), did not observe any significant effect on their responses nine weeks after the induction of a reproductive cycle. The quantity and quality of initial nutritional reserves had no effect on the first step of a reproductive cycle (induction), but could influence gonadogenesis (in particular oogenesis) in later steps (vitellogenesis). Besides, the type of food distributed (pellets vs. live prey such as forage fish) during the wintering period modifies the GSI in females (Castets 2011). At the end of the wintering period, female perch showed higher GSI when they were fed pellets (21%) compared to females fed live prey (17%). At the qualitative level, studies examining the essential role of lipid reserves on embryonic survival have mainly focused on the role of fatty acids. In walleye (*Sander vitreus*), a deficit in FA n-3 can stop embryonic development and result in a low survival rate (Czesny and Dabrowki 1998). More recently, Henrotte et al. (2010a) have demonstrated that the optimal ratio of DHA/EPA/ARA for the feeding of breeders of European perch should be 3/2/2. This ratio ensures the acquisition of good quality eggs and larvae. The composition of fatty acids also affects some characteristics of milt in male breeders. For example, the ratio n-3/n-6 in the feed modifies the composition of male semen (composition of unsaturated fatty acids), without modifying sperm quality either in terms of sperm volume and density, spermatozoa mobility and velocity, or sperm osmolality (Henrotte et al. 2010b).

High levels of cortisol (200 ng·mL^{-1}) are observed after fish handling (Wang et al. 2004a and b; Jentoft et al. 2005) compared to the low basal levels (<5 ng.mL^{-1}) measured before handling. This high sensitivity to handling and the weakness of the non-specific immune system (lysozyme activity: 10–20 IU?mL^{-1}, complement activity: 10–15 IU•mL^{-1}), compared to other species (Tort et al. 1998), could explain the high mortalities observed in breeders during the breeding season (Wang et al. 2003).

Other population factors could also strongly affect reproductive performance, such as the geographical origin of breeders. By comparing the reproductive performance of breeders coming from the Meuse River (Belgium) and Lake Geneva (France) under the same photothermal regime, Castets (2011) observed a delay of spawning events by one month between the two populations, which suggests that the spawning event is partially a heritable character. The broodstock originating from Lake Geneva spawned later and at higher temperatures than the broodstock originating from the Meuse River (12–14°C vs. 6–8°C). This result was observed for wild and domesticated breeders of both geographical origins.

Studies have also been carried out to identify cellular, biochemical and molecular indicators of oocyte quality. At the cellular level, it has been established that the fragmentation of oil droplets is correlated with the quality of oocytes (Zarski et al. 2011). More fragmented droplets are associated with a lower survival of oocytes. At the biochemical level, the activity of certain enzymes could be also good indicators. For instance, the reproductive performance of European perch (hatching rate, resistance of larvae to an osmotic stress) is linked to the activity level of cathepsin L, a lysosomal protease, assayed in eggs 7 days after fertilization (Kestemont et al. 1999). Finally, at the molecular level, proteins involved in cell response to oxidative stress, as well as in energy metabolism, heat shock proteins and vitellogenins appear of particular interest and may constitute molecular markers predictive of the reproductive performance in European perch (Castets et al. 2012). Further investigations are required to validate the potential of these biomarkers as indicators of reproductive performance in perch.

7.5 Conclusions

The reproductive biology and the environmental determinism of perch reproductive cycle are now well understood, especially for the European perch. The three *Perca* species are "early spring spawners without parental care" whose reproductive cycles are determined by annual day length and temperature variations. Efficient artificial photothermal programmes (100% of spawning) have been used by fish farmers to obtain several out-of-season spawns per year. This knowledge has contributed to the development of a European percid culture using indoor recirculated aquaculture systems (RAS). Future research should focus on a better understanding of the multifactorial regulation of reproductive performance.

7.6 References

Abdulfatah, A. 2010. Study of the environmental determinism of the reproductive cycle of the European perch (*Perca fluviatilis*). PhD thesis, I.N.P.L., Nancy, France 182 p.

Abdulfatah, A., P. Fontaine, P. Kestemont, J.-N. Gardeur and M. Marie. 2011. Effects of photothermal kinetic and amplitude of photoperiod decrease on the induction of the reproduction cycle in female European perch *Perca fluviatilis*. Aquaculture 322-323: 169–176.

Abdulfatah, A., P. Fontaine and M. Marie. 2012. Effect of final temperature and photoperiod kinetics on female European perch spawning induction. AQUA 2012: Global aquaculture—Securing our future, 1–5 September, Prague, Czech Republic.

Abdulfatah, A., P. Fontaine, P. Kestemont, S. Milla and M. Marie. 2013. Effects of the thermal threshold and the timing of temperature reduction on the initiation of reproduction cycle in female of European perch *Perca fluviatilis*. Aquaculture 376-379: 90–96.

Alm, G. 1954. Maturity, mortality and growth of Perch, *Perca fluviatilis* L., grown in ponds. Rep. Inst. Freshw. Res. Drottningholm. 35: 11–20.

Ben Ammar, I., Y. Le Doré, A. Iuretig, F. Teletchea, B. Schaerlinger, S. Milla and F. Fontaine. 2012. Age effect on puberty onset of European perch reared under RAS conditions. AQUA 2012, Global Aquaculture—securing our future, September 1–5, Prague, Czech Republic.

Billard, R., F. Le Gac and M. Loir. 1990. Hormonal control of sperm production in teleost fish. pp. 329–335. *In*: A. Epple, C.G. Scanes and M.H. Stetson (eds.). Progress in Comparative Endocrinology. Willey-Liss, New-York.

Borg, B. 1994. Androgens in teleost fishes. Comp. Biochem. Physiol. 109C: 219–245.

Castets, M.-D. 2011. Function of reproduction and regulation of quality in European perch *Perca fluviatilis*. PhD. Thesis INPL, Ecole doctorale RP2E, Nancy, France 299 p.

Castets, M.-D., B. Schaerlinger, F. Silvestre, J.-N. Gardeur, M. Dieu, C. Corbier, P. Kestemont and P. Fontaine. 2012. Combined analysis of *Perca fluviatilis* reproductive performance and oocyte proteomic profile. Theriogenology 78(2): 432–442.

Chevey, P. 1922. Observation sur une perche hermaphrodite (*Perca fluviatilis* Linn.). Bull. Soc. Zool. Fr. 47: 60–64.

Ciereszko, A. and K. Dabrowski. 1993. Estimation of sperm concentration of rainbow trout, whitefish and yellow perch using a spectrophotometric technique. Aquaculture 109: 367–373.

Ciereszko, R.E., K. Dabrowski and A. Ciereszko. 1997. Effects of temperature and photoperiod on reproduction of female yellow perch *Perca flavescens*: plasma concentrations of steroid hormone, spontaneous and induced ovulation, and quality of eggs. J. World Aquac. Soc. 28: 344–356.

Collette, B.B. and T. Banarescu. 1977. Systematics and zoogeography of the fishes of the family Percidae. J. Fish. Res. Board Can. 34: 1450–1463.

Craig, J.F. 1974. Population dynamics of perch, *Perca fluviatilis* L. in Slepton Ley, Devon. Freshw. Biol. 4: 433–444.

Craig, J.F. 2000. Percid Fishes: Systematics, Ecology and Exploitation. Blackwell Science, Oxford.

Crim, L.W., R.E. Peter and R. Billard. 1981. Onset of gonadotropic hormone accumulation in the immature trout pituitary gland in response to estrogen of aromatizable androgen steroid hormones. Gen. Comp. Endocrinol. 37: 192–196.

Czesny, S. and K. Dabrowski. 1998. The effect of egg fatty acid concentrations on embryo viability in wild and domesticated walleye (*Stizostedionvitreum*). Aquat. Living Resour. 2 (6): 371–378.

Dabrowski, K., A. Ciereszko, L. Ramseyer, D. Culver and P. Kestemont. 1994. Effects of hormonal treatment on induced spermiation and ovulation in the yellow perch (*Perca flavescens*). Aquaculture 120: 171–180.

Dabrowki, K. and A. Ciereszko. 1996. The dynamics of gonad growth and ascorbate status in yellow perch, *Perca flavescens* (Mitchill). Aquac. Res. 27: 539–542.

Dabrowski, K., R.E. Ciereszko, A. Ciereszko, G.P. Toth, S.A. Christ, D. El-Saidy and J.S. Ottobre. 1996. Reproductive physiology of yellow perch (*Perca flavescens*): environmental and endocrinological cues. J. Appl. Ichthyol. 12: 139–148.

Dalimier, N., J.-C. Philippart and J. Voss. 1982. Etude éco-éthologique de la reproduction de la perche (*Perca fluviatilis* L.): observations en plongée dans une carrière innondée. Cah. Ethol. Appl. 2: 37–50.

Flesch, A. 1994. Biology of the European perch (*Perca fluviatilis*) in the Mirgenbach Reservoir (Cattenom, Moselle). PhD thesis, University of Metz, France 241p.

Fontaine, P., L. Tamazouzt and B. Capdeville. 1996. Growth of the European perch *Perca fluviatilis* L. reared in floating cages and in water recirculated system: first results. J. Appl. Ichthyol. 12 (3-4): 181–184.

Fontaine, P., J.N. Gardeur, P. Kestemont and A. Georges. 1997. Influence of feeding level on growth, intraspecific weight variability and sexual growth dimorphism of European perch *Perca fluviatilis* L. reared in a recirculation system. Aquaculture 157(1-2): 1–9.

Fontaine, P., C. Pereira, N. Wang and M. Marie. 2006. Influence of pre-inductive photoperiod variations on European perch *Perca fluviatilis* broodstock response to an inductive photothermal program. Aquaculture 255: 410–416.

Fontaine, P., P. Kestemont, F. Teletchea and N. Wang. 2008. Percid Fish Culture: From Research to Production. Presses universitaires de Namur 150 p.

Gillet, C., J.-P. Dubois and S. Bonnet. 1995. Influence of temperature and size of females on the timing of spawning of perch *Perca fluviatilis* in Lake Genova from 1984 to 1993. Env. Biol. Fish. 42: 355–363.

Gillet, C. and J.-P. Dubois. 2007. Effect of water temperature and size of females on the timing of spawning of perch *Perca fluviatilis* L. in Lake Geneva from 1984 to 2003. J. Fish Biol. 70: 1001–1014.

Goetz, F.W. and G. Theofan. 1979. *In vitro* stimulation of germinal vesicle breakdown and ovulation of yellow perch (*Perca flavescens*) oocytes, effects of 17α-hydroxy-20β-dihydroprogesteron and prostaglandins. Gen. Comp. Endocrinol. 37: 273–285.

Guma'a, S.A. 1978. The effects of temperature on the development and mortality of eggs of perch, *Perca fluviatilis*. Freshwater Biol. 8: 221–227.

Hayes, D.B. and W.W. Taylor. 1994. Changes in the composition of somatic and gonadal tissues of yellow perch following white sucker removal. Trans. Am. Fish. Soc. 123: 204–216.

Heidinger, R. and T.D. Kayes. 1986. Yellow perch. pp. 103–113. *In*: R.R. Stickney (ed.). Culture of Non-Salmonid Freshwater Fish: Yellow Perch. CRC Press, Boca Raton, FL.

Henderson, B.A., J.L. Wong and S.J. Nepszy. 1996. Reproduction of walleye in Lake Erie: allocation of energy. Can. J. Fish. Aquat. Sci. 53: 127–133.

Henrotte, E., R. Mandiki, T.P. Agbohessi, M. Vandecan, C. Mélard and P. Kestemont. 2010a. Egg and larval quality, and egg fatty acid composition of European perch breeders (*Perca fluviatilis*) fed different dietary DHA/EPA/AA ratios. Aquac. Res. 41: 51–63.

Henrotte, E., V. Kaspar, M. Rodina, M. Psenicka, O. Linhart and P. Kestemont. 2010b. Dietary n-3/n-6 ratio affects the biochemical composition of European perch (*Perca fluviatilis*) semen but not indicators of sperm quality. Aquac. Res. 41: 31–38.

Hergenrader, G.L. 1969. Spawning behavior of *Perca flavescens* in aquaria. Copeia 4: 839–841.

Hokanson, K.E.F. 1977. Temperature requirements of some percids and adaptations to the seasonal temperature cycle. J. Fish. Res. Board Can. 34: 1524–1550.

Houthuijzen, R.P., J.J.G.M. Backx and A.D. Buijse. 1993. Exceptionally rapid growth and early maturation of perch in a freshwater lake recently converted from an estuary. J. Fish Biol. 43: 320–324.

Jamet, J.L. 1991. Importance de la faune ichtyologique dans le lac d'Aydat, milieu eutrophe de la zone tempérée Nord : ses relations trophiques avec les autres composants de l'écosystème. Thèse Université Blaise Pascal, Clermond-Ferrand, France 263 pp.

Jamet, J.L., C. Caravaglia, R. Dal Molin and D. Sargos. 1990. Fécondité, croissance et régime alimentaire de la perche adulte (*Perca fluviatilis* L.) du lac Monate (Italie du Nord). Riv. Idrobiol. 29: 599–615.

Jamet, J.L. and F. Desmolles. 1994. Growth, reproduction and condition of roach (*Rutilusrutilus* L.), perch (*Perca fluviatilis* L.) and ruffe (*Gymnocephaluscernuus* L.) in eutrophic lake Aydat (France). Int. Rev. Ges. Hydrobiol. 79: 305–322.

Jellyman, D. 1976. Hermaphrodite European perch, *Perca fluviatilis* L. N. Zeal. J. Mar. Freshw. Res. 10: 721–723.

Jellyman, D. 1980. Age, growth, and reproduction of perch, *Perca fluviatilis* L. in Lake Pounui. N. Zeal. J. Mar. Freshw. Res. 14: 391–400.

Jentoft, S., A. Aastveit, P. Torjesen and Ø. Andersen. 2005. Effects of stress on growth, cortisol and glucose levels in non-domesticated European perch (*Perca fluviatilis*) and domesticated rainbow trout (*Oncorhynchus mykiss*). Comp. Biochem. Physiol. A 141: 353–358.

Jones, D.H. 1982. The spawning of perch (*Perca fluviatilis* L.) in Loch Leven, Kinross, Scotland. Fish. Mgmt. 13(4): 139–151.

Kayes, T.B., C.B. Best and J.A. Malison. 1982. Hormonal manipulation of sex in yellow perch (*Perca flavescens*). Ann. Zool. 22: 955.

Kestemont, P., J. Cooremans, A. Abi-Ayad and C. Mélard. 1999. Cathepsin L in eggs and larvae of perch *Perca fluviatilis*: variations with developmental stage and spawning period. Fish Physiol. Biochem. 21: 59–64.

Kirillov, A.F. and A.K. Akhremenko. 1982. Seasonal changes in energy substrate concentrations in the tissues of the Vilyui Reservoir perch, *Perca fluviatilis*. J. Ichthyol. 13: 129–133.

Kobayashi, M., K. Aida and I. Hanyu. 1989. Induction of gonadotropin surge by steroid hormone implantation in ovariectomized and sexually regressed female goldfish. Gen. Comp. Endocrinol. 73: 469–476.

Lahnsteiner, F., B. Berger, T. Weismann and R. Patzner. 1995. Fine structure and motility of spermatozoa and composition of the seminal plasma in the perch. J. Fish Biol. 47: 492–508.

Le Cren, E.D. 1951. The length-weight relationship and seasonal cycle in gonad weight and condition in the perch (*Perca fluviatilis*). J. Anim. Ecol. 20: 201–219.

Lindesjöö, E. 1994. Temporal variation and sexual dimorphism of the skin of perch *Perca fluviatilis* L.: A morphological study. J. Appl. Ichthyol. 10: 154–166.

Makarova, N.P. 1973. Seasonal changes in some of the physiological characteristics of the perch (*Perca fluviatilis* L.) of Ivan'kovo Reservoir. J. Ichthyol. 13: 742–752.

Malison, J.A., C.D. Best and T. Kayes. 1982. Hormonal control of growth and sexual size dimorphism in yellow perch (*Perca flavescens*). Am. Zool. 22: 955.

Malison, J.A., T.B. Kayes, C.D. Best, C.H. Amundson and B. Wentworth. 1986. Sexual differentiation and use of hormones to control sex in yellow perch (*Perca flavescens*). Can. J. Fish. Aquat. Sci. 43(1): 26–35.

Malison, J.A. and J.A. Held. 1996. Reproduction and spawning in walleye (*Stizostedionvitreum*). J. Appl. Ichthyol. 12(3-4): 153–156.

Malservisi, A. and E. Magnin. 1968. Changements cycliques annuels se produisant dans les ovaires de *Perca fluviatilis flavescens* (Mitchill) de la region de Montréal. Naturaliste Can. 95: 929–945.

Mann, R.H.K. 1978. Observations on the biology of the perch, *Perca fluviatilis*, in the River Stour, Dorset. Freshw. Biol. 8: 229–239.

Mansueti, A.J. 1964. Early development of the yellow perch, *Perca flavescens*. Chesapeake Sci. 5: 46–66.

Marza, V.D. 1938. Histophysiologie de l'ovogenèse. Hermann, Paris.

Mezhnin, F.I. 1978. Development of the sex cells in the early ontogeny of the common perch, *Perca fluviatilis*. J. Ichthyol. 18(1): 71–86.

Migaud, H., P. Fontaine, I. Sulistyo, P. Kestemont and J.-N. Gardeur. 2002. Induction of out-of-season spawning in European perch *Perca fluviatilis*: effects of cooling and chilling periods on female gametogenesis and spawning. Aquaculture 205: 253–267.

Migaud, H., R. Mandiki, J.-N. Gardeur, A. Fostier, P. Kestemont and P. Fontaine. 2003a. Synthesis of sex steroids in final oocyte maturation and induced ovulation in female European perch, *Perca fluviatilis*. Aquat. Living Resour. 16 (4): 380–388.

Migaud, H., R. Mandiki, J.-N. Gardeur, P. Kestemont, N. Bromage and P. Fontaine. 2003b. Influence of photoperiod regimes on the European perch gonadogenesis, spawning and eggs and larvae quality. Fish Physiol. Biochem. 28: 395–397.

Migaud, H., P. Fontaine, P. Kestemont, N. Wang and J. Brun-Bellut. 2004. Influence of photoperiod on the onset of gonadogenesis in European perch *Perca fluviatilis*. Aquaculture 241: 561–574.

Migaud, H., N. Wang, J.-N. Gardeur and P. Fontaine. 2006. Influence of photoperiod on reproductive performances in European perch *Perca fluviatilis*. Aquaculture 252: 385–393.

Migaud, H., A. Davie and J.F. Taylor. 2010. Current knowledge on the photoneuroendocrine regulation of reproduction in temperate fish species. J. Fish Biol. 76: 27–68.

Milla, S., R. Mandiki, P. Hubermont, C. Rougeot, C. Mélard and P. Kestemont. 2009. Ovarian steroidogenesis inhibition by constant photothermal conditions is caused by a lack of gonadotropin stimulation in European perch. Gen. Comp. Endocrinol. 163: 242–250.

Papageorgiou, N.K. 1977. Fecundity and reproduction of perch (*Perca fluviatilis* L.) in Lake AgiosVasilios, Greece. Freshw. Biol. 7: 559–565.

Persson, L. 1990. A field experiment on the effects of interspecific competition from roach *Rutilusrutilus* (L.) on age at maturity and gonad size in perch, *Perca fluviatilis* (L.). J. Fish Biol. 37: 899–906.

Piironen, J. and H. Hyvärinen. 1983. Composition of the milt of some teleost fishes. J. Fish Biol. 22: 351–361.

Rinchard, J. and P. Kestemont. 1996. Comparative study of reproductive biology in single- and multiple-spawner cyprinid fish. I. Morphological and histological features. J. Fish Biol. 49: 883–894.

Sandström, O., E. Neuman and G. Thoresson. 1995. Effects of temperature on life history variables in perch. J. Fish Biol. 47: 652–670.

Scott, W.B. and E.J. Crossman. 1973. Freshwater fishes of Canada. Fish. Res. Board Can. 184: 966.

Sulistyo, I. 1998. Contribution to the study and the control of the reproductive cycle of the European perch *Perca fluviatilis* L. PhD Thesis, University Henri Poincaré, Nancy, France 144 p.

Sulistyo, I., J. Rinchard, P. Fontaine, J.-N. Gardeur, B. Capdeville and P. Kestemont. 1998. Reproductive cycle and plasma levels of sex steroids in female European perch *Perca fluviatilis*. Aquat. Living Resour. 11(2): 101–110.

Sulistyo, I., P. Fontaine, J. Rinchard, J.-N. Gardeur, H. Migaud, B. Capdeville and P. Kestemont. 2000. Reproductive cycle and plasma levels of sex steroids in male European perch *Perca fluviatilis*. Aquat. Living Resour. 13(2): 80–106.

Tamazouzt, L., J.P. Dubois and P. Fontaine. 1993. Production et Marché actuels de la perche *Perca fluviatilis* L. en Europe. La Pisciculture Française 113: 4–8.

Tamazouzt, L., P. Fontaine and D. Terver. 1994. Décalage de la période de reproduction de la perche (*Perca fluviatilis*) en eau recyclée. Ichtyophysiologica Acta 7: 29–40.

Tanasichuk, R.W. and W.C. Mackay. 1989. Quantitative and qualitative characteristics of somatic and gonadal growth of yellow perch (*Perca flavescens*) from Lac Ste Anne, Alberta. Can. J. Fish. Aquat. Sci. 46: 989–994.

Taranger, G.L., M. Carillo, R.W. Schulz, P. Fontaine, S. Zanuy, A. Felip, F.A. Weltzien, S. Dufour, Ǿ. Karlsen, B. Norberg, E. Andersson and T. Hansen. 2010. Control of puberty in farmed fish. Gen. Comp. Endocrinol. 165: 483–515.

Teletchea, F., A. Fostier, E. Kamler, J.-N. Gardeur, P.-Y. Le Bail, B. Jalabert and P. Fontaine. 2009. Comparative analysis of reproductive traits in 65 freshwater fish species: application to the domestication of new species. Rev. Fish Biol. Fish. 19: 403–430.

Thorpe, J. 1977. Synopsis of biological data on the perch Perca fluviatilis and Perca flavescens. F.A.O. Fish Synopsis 113: 138.

Toner, D. and C. Rougeot. 2008. Farming of European perch. Aquaculture Explained 24: 78.

Tort, L., F. Padros, J. Rotlant and S. Crespo. 1998. Winter syndrome in the gilthead sea bream Sparusaurata. Immunological and histopathological features. Fish Shellfish Immunol. 8: 37–47.

Treasurer, J.W. 1981. Somes aspects of the reproductive biology of perch *Perca fluviatilis* L. fecundity, maturation and spawning behaviour. J. Fish Biol. 18: 729–740.

Treasurer, J.W. 1983. Estimates of egg and viable embryo production in a lacustrine perch, *Perca fluviatilis*. Env. Biol. Fish. 8: 3–16.

Treasurer, J.W. and F.G.T. Hollliday. 1981. Some aspects of the reproductive biology of perch *Perca fluviatilis* L.: A histological description of the reproductive cycle. J. Fish Biol. 18: 359–376.

Turner, C.L. 1919. The seasonal cycle in spermaty of the perch. J. Morphol. 32: 681–711.

Urho, L. 1996. Habitat shifts of perch larvae as survival strategy. Ann. Zool Fennici 33(3-4): 329–340.

Wang, N., H. Migaud, L. Acerete, J.-N. Gardeur, L. Tort and P. Fontaine. 2003. Mortality and non-specific immune response of European perch, *Perca fluviatilis*, during the spawning season. Fish Physiol. Biochem. 28: 523–524.

Wang, N., P. Fontaine, J.-N. Gardeur and M. Marie. 2004a. Temperature effect on the delay of cortisol release after an acute stress in European perch (*Perca fluviatilis*). The 5th International Symposium of Fish Endocrinology, September 5–9, 2004, Castellon, Spain.

Wang, N., P. Fontaine, L. Tort, M. Marie and J.-N. Gardeur. 2004b. Effect of wintering condition on European perch (*Perca fluviatilis*), stress and non-specific immune responses. The 5th International Symposium of Fish Endocrinology, September 5–9, 2004, Castellon, Spain.

Wang, N., J.-N. Gardeur, E. Henrotte, M. Marie, P. Kestemont and P. Fontaine. 2006. Determinism of the induction of the reproductive cycle in female European Perch, *Perca fluviatilis*: effects of environmental cues and modulating factors. Aquaculture 261: 706–714.

Wang, N., F. Teletchea, P. Kestemont, S. Milla and P. Fontaine. 2010. Photothermal control of the reproductive cycle in temperate fishes. Rev. Aquaculture 2: 209–222.

Willemsen, J. 1977. Population dynamics of percids in Lake Ijssel and some smaller lakes in The Netherlands. J. Fish. Res. Board Can. 30(10): 1710–1719.

Zarski, D., K. Palińska, K. Targońska, Z. Bokor, L. Kotrik, S. Krejszeff, K. Kupren, A. Horváth, B. Urbányi and D. Kucharczyk. 2011. Oocyte quality indicators in European perch, *Perca fluviatilis* L., during reproduction under controlled conditions. Aquaculture 313: 84–91.

Zelenkov, V.M. 1982. Early gametogenesis and sex differentiation in the perch, *Perca fluviatilis*. J. Ichthyol. 21(2): 124–130.

8

Biology and Ecology of Perch Parasites

Jasminca Behrmann-Godel[1,*] and Alexander Brinker[2]

ABSTRACT

The European perch *Perca fluviatilis* (L.) and the yellow perch *Perca flavescens* (M.) are generalist feeders occupying a number of different habitats during their life stages and are exposed to infection by a variety of ecto- and endoparasites. We provide a short general overview of parasite ecology and introduce major macroparasite groups including examples of relevant species infecting European perch and yellow perch in their respective distribution ranges. We describe the ways in which specific perch parasites impair host fitness by measures like growth, fecundity or mortality. Effects of parasites on adult and juvenile perch are discussed separately and accompanied with specific examples from field data and laboratory experiments. The important role of both perch species as model systems is discussed especially for the research areas of parasite host ecology and evolution. Thereby we incorporate basic studies on both the European and yellow perch although almost no comparative studies for both species exist which could be a major aim for future research. We also review some long-term studies on the European perch populations of Lake Constance as an instructive case study.

Keywords: European perch, yellow perch, parasite host co-evolution, parasite life cycles, parasite community ecology, trophic linkage, fitness constraints, bioindicators, *Perca fluviatilis*, *Perca flavescens*

[1] Limnological Institute, University of Konstanz, Mainaustrasse 252, 78457 Konstanz, Germany.
[2] Fisheries Research Station Baden-Württemberg, Argenweg 50/1, 88085 Langenargen, Germany. E-mail: Alexander.Brinker@lazbw.bwl.de
* Corresponding author: Jasminca.Behrmann@uni-konstanz.de

8.1 Introduction

Almost every major group of organisms including viruses, bacteria, protists, higher metazoans, fungi, plants and mammals has parasitic members and there are indications that species with at least one parasitic phase in their life history may actually outnumber non-parasites. Some parasites develop directly on or in their host, while others have extremely complex life cycles involving a sequence of obligate hosts infected at different stages of development. For a glossary of important parasitological terms used in this chapter, see the Infobox. From a parasite's point of view, the host can be seen as its environment, providing a number of distinct habitats for colonization (Bush et al. 1977). The many different ways in which parasites infect their hosts are mostly specific to parasite species and/or life cycle stage. Infection may be direct as in most trematode cercariae, which penetrate the host's skin. However parasites can also be transmitted actively by vectors or indirectly via the food chain—for example

Infobox: Parasitological terminology

Definition of Parasitism:

Parasites are organisms that benefit at the expense of another organism belonging to another species, called the **host**, mostly by trophic exploitation.

Life strategies of parasites:

Endoparasites live inside the body of the host, **ectoparasites** on the surface. **Microparasites** are unicellular organisms such as bacteria and protozoa. Viruses may also be called microparasites. **Macroparasites** are eumetazoa including flatworms, nematodes, arthropods and several others. Parasites that depend on a host for development are called **obligate parasites**. In contrast, a **facultative parasite** can complete its life cycle without infecting a host. Organisms that are not typically parasitic but can become so under certain specific conditions (e.g., immune deficiency of the host) are called **opportunistic parasites**.

Parasite life cycles:

A **direct (monoxenous) life cycle** can be found in parasites that exploit only one host for development. An **indirect or complex (heteroxenous) life cycle** is realized when parasites depend on several hosts for development. Sexual maturity and reproduction is always realized in the **definitive (final, primary) host**. One or several **intermediate (secondary) hosts** are needed for preceding stages of development, which may involve asexual reproduction or metamorphosis to the next developmental stage. If the intermediate or definitive host is used as a carrier to the next host it is typically called a **vector**. In terrestrial systems most vectors are blood-feeding arthropods, for example insects such as mosquitoes. They transmit the parasite (e.g., *Plasmodium* spp., which causes malaria) to the next host of the parasite's life cycle. Typical vectors in the aquatic environment (especially for fish) are blood-feeding arthropods or annelids such as leeches. Some parasites may infect a host but do not undergo development. These so-called **paratenic (transport)** hosts can be used for dispersal or to reach the next trophic level, which often raises the parasite's chance of being transmitted to the next host in the life cycle.

Population and community concepts (after, Bush et al. 1997):

All parasites of a given species in a single host individual are defined as the **infrapopulation**. All infrapopulations of a given parasite species in a given host species in an ecosystem comprise the **component population.** Finally, all component populations of a given parasite species in different host species of its life cycle make up the **supra population.** Similarly, the sum of the infrapopulations of all parasite species within a single host individual is known as the **infracommunity**. All infracommunities of all parasite species in a given host species in an ecosystem make up the **component community**. Finally the sum of all component communities in all life cycle hosts comprises the **supra community**.

as a result of predation on the infective stage of the parasite itself or when a predatory host consumes an infected prey organism.

The major benefits of parasitism compared with a free-living lifestyle are shelter, access to resources, the relatively stable environment provided by the host, energy savings though not having to forage and the ease of transportation to new areas (Combes 2001). However there are also disadvantages for parasites, most notably the challenge of finding and successfully infecting host organisms (Combes 2001). Many parasites have developed fascinating adaptations to overcome this problem, for example manipulating host behavior or increasing host conspicuousness in order to improve the likelihood of successful transmission to the next host (Moore 2002). A further challenge is presented by the immune responses mounted by hosts in an effort to repel parasitic invaders (Combes 2001). However, the strength and efficacy of the host immune response can be highly stage dependent. Generally, parasites tend to provoke stronger immune responses in an intermediate host than in the definitive host (Ewald 1995). Immune responses are often dependent on the feeding strategy of the parasite and its exact location in or on the host's body (Jones 2001; Alvarez-Pellitero 2008). For example, skin-penetrating cercariae generally provoke a strong host reaction whereas the immune response to parasites residing in the gut, such as most adult tapeworms, is often rather weak (Woo 1992). Several parasite species are able to manipulate host immunity, for example by down-regulating the strength of the immune response (Goater et al. 2014).

Despite the localized pathology that can be inflicted on individual hosts even in established parasite-host systems, the typically aggregate nature of parasite distribution within a host population usually means that only a fraction of hosts are severely affected. Negative effects such as reduced fecundity or survival tend to be observed only in individual hosts where infection intensity is unusually high, and not, as a rule, at the population level.

During recent decades, anthropogenic environmental change, in particular rises in temperature, environmental pollution and introductions of new species, have been shown to trigger negative changes in parasite-host interactions. Even small changes may be deleterious, and the cumulative effect of multiple environmental stresses can result in a significant negative impact on immune function and animal health. For example, a rise in temperature can prolong the period of parasite transmission and affect the abundance and virulence of particular pathogens and parasites (Vidal-Martínez et al. 2010; Marcogliese and Pietrock 2011). Increased pathogenicity of opportunistic parasites mediated by increased virulence or a prolonged season can result in a serious threat to fish populations. Furthermore, rising temperatures may also favor incursions by alien parasites, which either enter the system directly as soon as temperature limitation disappears or arrive via co-introduction with non-native intermediate hosts (Marcogliese 2001; Poulin 2006; Paull and Johnson 2011). Introduced parasites may cause epizootic outbreaks in naïve host populations, which lack adaptations to reduce pathogenicity or to defend against the invaders (Marcogliese 2001; Britton et al. 2011; Behrmann-Godel et al. 2014).

The extensive distribution of European and yellow perch and the wide range of habitat and food sources utilized mean the number of described parasites for these two species is correspondingly high. Information regarding Balkhash perch

(*Perca schrenkii*) parasites is provided in Chapter 3 (Section 3.6). Therefore, the information given in the following section is restricted to European (*Perca fluviatilis*) and yellow perch (*Perca flavescens*). Several parasite species lists for European and yellow perch exist in the literature (e.g., Craig 2000; Morozinska-Gogol 2008). From these it can be seen that the biodiversity of perch parasites including myxozoan, protozoan and metazoan parasites described today includes about 160 species for yellow perch and 147 for European perch, 19 of which are shared by both species (Craig 2000). An additional high species number of parasitic bacteria and viruses affecting perch can be assumed. Instead of repeating these lists, we aim to provide specific information about the biology and ecology of the major parasite groups, illustrated with specific examples of parasite species infecting perch and their impact on host fish population. Throughout the chapter we will concentrate on protozoans, myxozoans and macroparasites (hereafter called parasites) and exclude viruses, bacteria and fungi. We will begin with general information about the broad spectrum of parasites and illustrate the ways in which ecological factors enhance and limit the number of parasite species infecting host species in general and perch specifically. We will follow by focusing on some specific parasite species and illustrate their negative impacts on adult and larval European perch. Finally we review the role of perch parasites in providing insights into the ecology and evolution of host-parasite interactions.

8.2 Parasites—The Hidden Biodiversity On and Within Perch

The definition of ecology as the relationship between an organism and its biotic and abiotic environment means parasitism has to be seen as an ecological concept (Goater et al. 2014). In addition to the wider environmental factors also faced by free-living organisms, parasites must also deal with challenges arising from their habitat being another organism. Environmental factors influencing the host's ecology impact simultaneously on the parasite, which also has to deal with factors such as continuous attack from the host immune system (Buchmann et al. 2012; Dezfuli et al. 2014). As with free-living organisms, parasites have basic needs for space, nutrition and shelter and requirements for reproduction and survival, which combine to form a specific niche, and restrict the parasite to particular habitats on or within the host organism. Likewise, the distribution ranges of both free-living and parasitic organisms are limited by factors such as dispersal ability, tolerance to environmental variability and interaction with other species. For parasites, these factors translate to the availability of suitable life cycle hosts, the geographical limitations on transmission, an ability to adapt to host immune reactions and interspecific competition with other parasites attempting to exploit the same host individual. These requirements and limitations have a number of observable consequences: (1) the geographic occurrence of parasites is highly variable, (2) parasites tend to be specialists, with evolutionary adaptations that allow them to explore a single or very restricted range of host species, and (3) the behavioral, morphological and physiological adaptations of parasites also restrict them to specific sites, tissues or organs on or within hosts (see Poulin 1998; Goater et al. 2014). This last point is especially helpful for parasitologists, fish farmers and aquarium hobbyists, in taking advantage to locate and identify particular parasite species on or in an infected host. Figure 1 shows such a typical "distribution map" for parasite

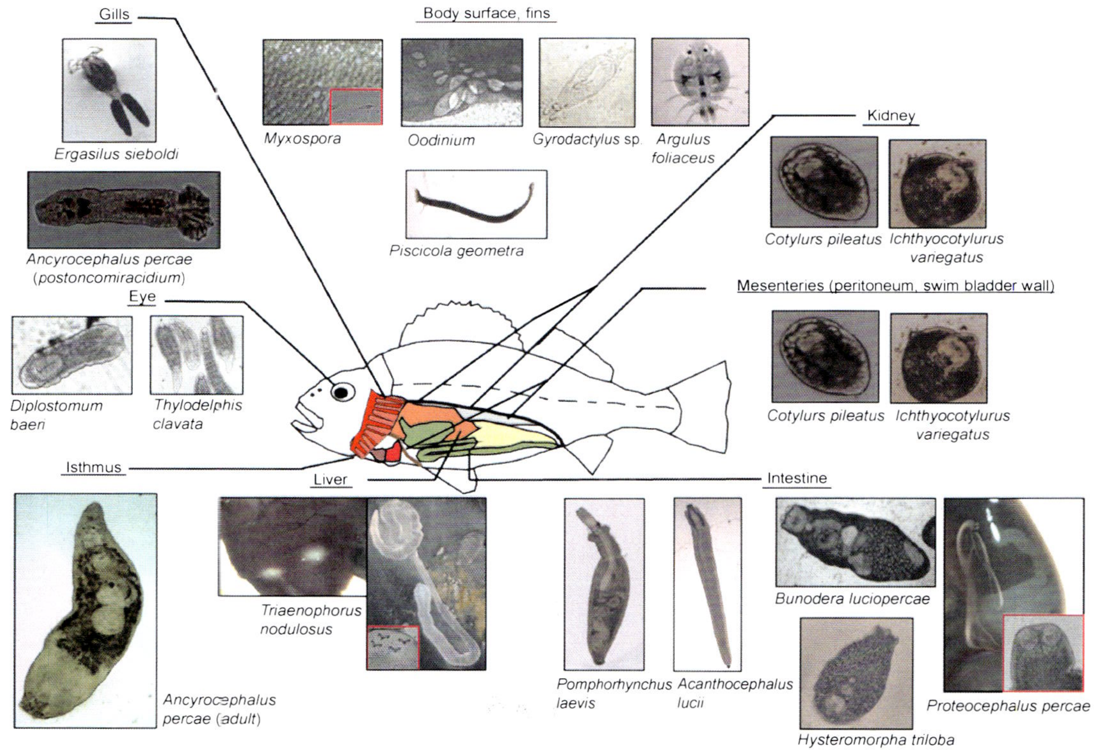

Fig. 1. Component parasite community from European perch in Lake Constance, Germany, indicating the site specificity on or in their host. Each parasite species occupies a specific and predictable habitat (organ or tissue).

species occurring in a parasite component community (see Infobox for definition) of European perch from Lake Constance in Germany. However, only a subset of these species will contribute to the infracommunity (see Infobox for definition) present in an individual host.

Goater et al. (2014) describe parasite community composition and species richness in terms of supply and screens (after Combes 2001). 'Supply' refers to the availability of parasite species from a global pool, the contents of which determine which species are theoretically available at a regional level. 'Screens' are biotic and abiotic factors that restrict or exclude specific parasites from particular localities and thereby control the composition of the local pool and ultimately the composition of the component- and infracommunities within the local pool (Fig. 2). Abiotic factors acting as screens may include pH, salinity, drought, acid rain or host extinctions. Biotic features might be the feeding habits of hosts, host immunology, species interactions or phylogeny, including recent and contemporary coevolution between parasites and hosts and host specificity. Thus the parasite species found in any infra- or component community are usually only a subset of those available within the local parasite pool, and these are in turn always only a subset of a theoretical maximum of parasites available in the global parasite pool.

Thus parasite species richness and abundance vary greatly, not only between host species but also among host populations of the same species in different habitats or geographic locations (Johnson et al. 2004). Additional factors influencing parasite species richness for a particular host are its physical size (body length or mass), geographical range and population density. Increases in host size, geographical range or density have been shown to correlate positively with parasite species richness

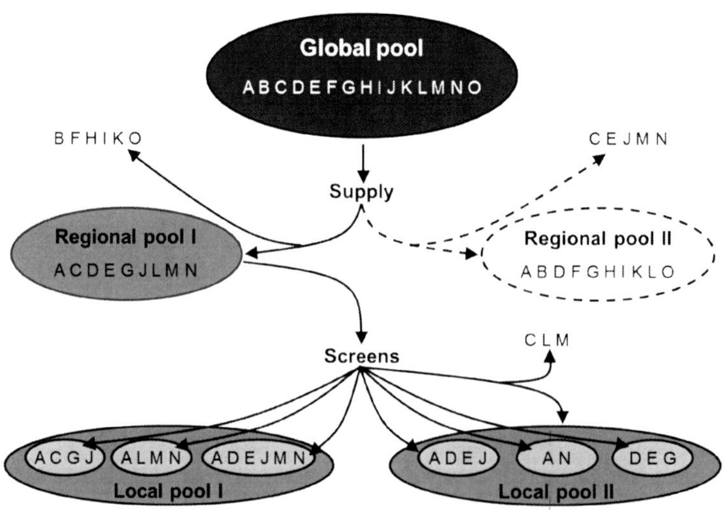

Fig. 2. The relationship between parasites present in the global pool and two different local pools. Letters represent individual parasite species available in the different pools. Supply dictates which species are available in the two different regional pools I and II, while screens determine which are available in the local pools. Light grey ellipses within the local pools represent component communities of parasites (see text for a more detailed explanation). Figure was redrawn from Goater et al. (2014).

(Poulin and Morand 2004). However the commonality of these rules cannot be taken for granted, given the lack of information concerning most parasite host systems and a number of confounding results (Goater et al. 2014).

Both European and yellow perch are distributed in a broad geographic range including fresh, brackish and salt-water habitats. Both species are generalist feeders, taking many different food taxa from a variety of trophic levels (including algae, zooplankton, benthic invertebrates and fish), and shift between habitats (pelagic to littoral zones) during ontogenetic development (Coles 1981; Wang and Eckmann 1994; Wang and Appenzeller 1998; Craig 2000). Furthermore, both species can occur in very dense populations, in particular during spawning migrations (Craig 2000). Thus despite potential uncertainties, we would expect parasite species richness for European and yellow perch to be high on a global as well as a local scale. Additionally, we would expect to find them hosting parasites with a broad range of infection strategies including direct infection and indirect routes such as predator-prey interactions and transmission by consumption of intermediate hosts from a range of trophic levels. These expectations have been borne out in many investigations, documenting species rich parasite component communities for each perch species (Andrews 1979; Carney and Dick 1999, 2000a; Morozinska-Gogol 2008) as shown in Fig. 1 and Table 1.

Besides a few very common parasite species that appear in almost all component communities of yellow and European perch (Carney and Dick 1999), the precise composition of the local parasite pool available at a particular sampling location within a given ecosystem remains highly unpredictable. Thus the parasite communities of perch have been found to vary considerably between ecosystems in close geographical proximity to one another, between sampling years at the same location (Poulin and Valtonen 2002; Johnson et al. 2004) and between contemporary host populations within the same ecosystem (Fig. 3).

Parasite identification is not a trivial task and specialists are often required for accurate species determination. The taxonomic determination of non-model organisms can prove especially difficult and time-consuming, increasingly so when working with the early developmental stages of host fish. The study of parasite infections in fish fry presents particular challenges. The young age of the hosts mean that any infection observed must be very recent, and the parasites themselves are therefore in early stages of development (Kuchta et al. 2009) and may lack important morphological characteristics required for identification. Classically, this problem has been addressed with life-cycle studies including experimental infections. However, such studies are laborious and usually limited to small numbers of congeneric parasites. Molecular identification techniques based on sequences of nuclear genes such as ribosomal rRNA and the mitochondrial cytochrome oxidase gene can be helpful in otherwise tricky species diagnoses (Zehnder and Mariaux 1999; Scholz et al. 2007; Sonnenberg et al. 2007; Locke et al. 2010a,b). A combination of classical morphological analysis and genetic surveys of various parasite life-cycle stages sampled from a range of hosts has proved helpful in identifying early developmental stages of parasites in perch fry and in elucidating the typical parasite succession during perch ontogeny (Behrmann-Godel 2013) (Table 2). In addition, genetic surveys have provided new insights into parasite transmission pathways between several fish species and their parasites within the study area (Behrmann-Godel 2013).

Table 1. Examples of component communities of macroparasites for European perch *Perca fluviatilis* sampled in Lake Constance, Germany (oligotrophic lake, surface area 536 m^2, mean depth 90 m) and yellow perch *Perca flavescens* from Dauphin Lake, Manitoba (eutrophic to mesotrophic lake, surface area approximately 520 m^2, mean depth 3.5 m) (Data from Carney and Dick 2000a). Prev = prevalence (percent of infected fish); Mean int. = mean intensity (mean number of parasites per infected host individual).

***Perca fluviatilis* Lake Constance 2008 (n = 255)**					***Perca flavescens*, Dauphin Lake 1993 (n = 102)**				
Parasite	Route of infection	Site of infection	Prev [%]	Mean int.	Parasite	Route of infection	Site of infection	Prev [%]	Mean int.
Monogenea									
*Ancyrocephalus percae**	Direct	Gills, isthmus	67	23	*Urocleidus adspectus*	Direct	Gills	26	8
Gyrodactylus gasterostei	Direct	Body surface	7	6					
Digenea/Trematoda									
Bunodera luciopercae	Cladoceran/ amphipod ingestion	Intestine	99	155	*Crepidostomum cooperi*	Benthic insect larv. ingestion	Intestine	37	54
Cotylurus pileatus	Cercarial penetration	Mesenteries	92	33	*Centrovarium lobotes*	Fish ingestion	Intestine	9	6
Ichthyocotylurus variegatus	Cercarial penetration	Mesenteries	79	48	*Clinostomum* spp.	Cercarial penetration	Flesh	2	2
Hysteromorpha triloba	Cercarial penetration	Mesenteries	30	4	*Apophallus* spp.	Cercarial penetration	Flesch and fins	33	5
Diplostomum spp.	Cercarial penetration	Eye	99	31	*Diplostomum* spp.	Cercarial penetration	Eye	22	4
Tylodelphys clavata	Cercarial penetration	Vitreous humor	100	193	*Neochasmus* spp.	Cercarial penetration	Eye and flesh	25	3
Phyllodistomum pseudofolium	Cercarial penetration	Urinary bladder	1	3					
Cestoda									
Triaenophorus nodulosus	Copepod ingestion	Liver	93	5	*Triaenophorus nodulosus*	Copepod ingestion	Liver	9	1

Proteocephalus percae	Copepod ingestion	Intestine	88	18	*Proteocephalus pearsi*	Copepod ingestion	Intestine	24	26
Eubothrium crassum	Copepod ingestion	Intestine	29	4	*Bothriocephalus cuspidatus*	Copepod ingestion	Intestine	12	21
					Ligula intestinalis	Copepod ingestion	Body cavity	1	1
Acanthocephala									
Pomphorhynchus laevis	Amphipod ingestion	Intestine	1	3	*Pomphorhynchus bulbocolli*	Amphipod ingestion	Intestine	4	2
Acanthocephalus lucii	Amphipod ingestion	Intestine	7	9					
Acanthocephalus anguillae	Amphipod ingestion	Intestine	0.4	5					
Nematoda									
Raphidascaris acus	Insect larvae, fish ingest.	Liver and mesentery	1	1	*Raphidascaris acus*	Insect larvae fish ingest.	Liver mesentery	66	13
Angullicoloides crassus	Copepod ingestion	Swim bladder	0.4	1	*Spinitectus gracilis*	Insect larvae ingestion	Intestine	27	4
Hirudinea									
Piscicola geometra	Direct	Body surface	1	1	*Myzobdella moorei*	Direct	Fins and body surface	25	3
Arthropoda/Crustacea									
Ergasilus sieboldi	Direct	Gills	13	3	*Ergasilus luciopercarum*	Direct	Gills	14	3
Argulus foliaceus	Direct	Body surface	7	2					

**A. percae* is an invasive parasite first documented in Lake Constance in 2012 in a study of examining n = 539 adult European perch (data from Behrmann-Godel et al. 2014).

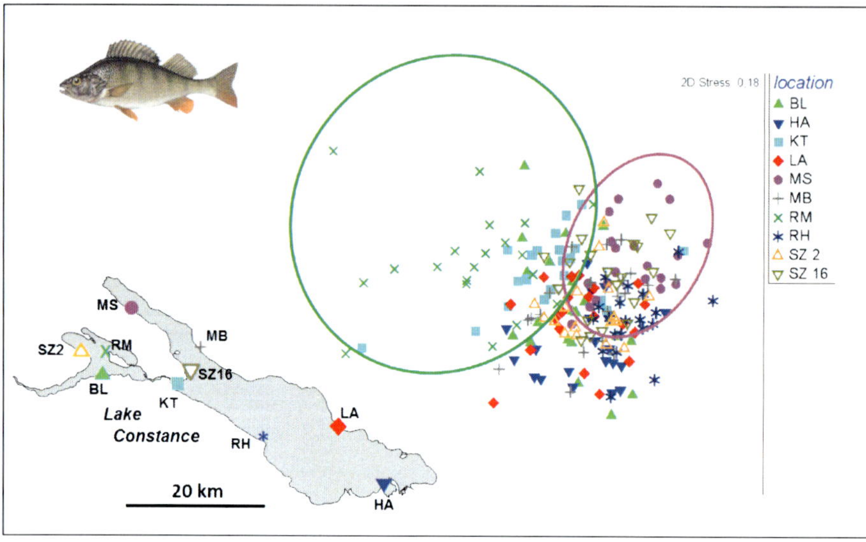

Fig. 3. Two-dimensional nonmetric multidimensional scaling (nMDS) plot calculated using Bray-Curtis ordination of parasite abundance data for adult European perch from 10 localities (different symbols) around Lake Constance sampled in 2008 (n = 199). Based on an analysis of similarity (ANOSIM) the parasite communities were significantly different ($R \geq 0.30$; $p < 0.05$). The difference is shown for two localities, RM and MS, encircling the 95% confidence area occupied by Bray-Curtis similarity data of individual fish.

8.3 Parasites of Perch

8.3.1 Protozoa

The protozoan parasites of fish include both ecto- and endoparasites. Among the endoparasitic species, the blood flagellates of the genera *Trypanosoma*, *Cryptobia* and *Trypanoplasma* spp. are of special importance. Most have a complex life cycle including transmission by a vector. In the case of protozoan fish parasites this is typically a leech, for example *Piscicola geometra* for European perch (Markevich 1963; Goater et al. 2014).

The ectoparasitic protozoa fall into three categories: First, there are opportunists with limited persistence as parasites. These are typically secondary infectors that mainly parasitize stressed or immune-depressed fish, for example ciliates of the genera *Ophryoglena*, *Tetrahymena*, *Hemiophrys* and *Glaucoma*. A second category includes ubiquitous parasites lacking host or site preference, for example the flagellate *Ichthyobodo necator* (formerly *Costia necatrix*), ciliates of the genus *Chilodonella* spp., *Ichthyophtirius multifilis* (commonly called "white spot", "ick" or "ich"), and several ubiquitous species of *Trichodina* and *Tripartiella* infecting the skin and dinoflagellates of the genus *Oodinium* (or *Piscinoodinium*) (Fig. 1). The third group are specialized parasites exhibiting restricted host specificity but with a pronounced

Table 2. Species, prevalence and mean infection intensity of parasites infesting European perch sampled from Lake Constance at different ages (wph = weeks post hatch). First three columns (2, 4 and 7 weeks post hatch (wph), grey box) represent the pelagic, the other two (8 and 12 wph) the littoral samples. *Proteocephalus* spp. includes *P. percae* and *P. longicollis*. Additional information is provided concerning invertebrate hosts (1.IH = first, 2.IH = second intermediate host), species names given in brackets where known, for Lake Constance from own observations (from Behrmann-Godel 2013).

	Age of perch wph (n analysed)					
	2(51)	4 (85)	7(52)	8(60)	12(60)	
Parasite	**Prevalence (mean intensities)**					**Invertebrate host**
Cestoda						
*Triaenophorus nodulosus**	0(0)	0(0)	56(1.3)	40(2.0)	50(2.0)	
*Triaenophorus nodulosus***	0(0)	0(0)	38(1.9)	10(1.0)	4(1.0)	1.IH: Copepods (*Cyclops* spp.)
Proteocephalus spp.	0(0)	12(1.1)	71(4.8)	40(2.5)	15(2.0)	1.IH: Copepods (*Cyclops* spp.)
Eubothrium crassum	0(0)	70(3.0)	63(3.5)	87(6.0)	45(3.0)	Copepods
Trematoda						
Bunodera luciopercae	0(0)	6(1.2)	63(2.7)	85(3.0)	95(12.5)	1.IH: Bivalves; 2.IH: Copepods, cladocerans, ostracods, amphipods, ephemeroptera
Diplostomum baeri	0(0)	0(0)	0(0)	15(4.0)	70(3.0)	1.IH: Snails
Tylodelphys clavata	0(0)	0(0)	0(0)	55(2.5)	100(70)	1.IH: Snails (*Radix auricularia, Radix labiata*)
Ichthyocotylurus variegatus	0(0)	0(0)	0(0)	0(0)	20(2.5)	1.IH: Snails (*Valvata piscinalis*)
Cotylurus pileatus	0(0)	0(0)	0(0)	8(1.0)	50(5.0)	1.IH: Snails
Hysteromorpha triloba	0(0)	0(0)	0(0)	26(2.5)	26(2.0)	1.IH: Snails
Bucephalus polymorphus	0(0)	0(0)	0(0)	5(2.0)	0(0)	1.IH: Snails
Nematoda						
Raphidascaris acus	0(0)	0(0)	0(0)	1(1.0)	0(0)	-
Maxillopoda						
Ergasilus sieboldi	0(0)	0(0)	0(0)	0(0)	1(1.0)	-
Argulus foliaceus	0(0)	0(0)	0(0)	5(2.0)	5(2.0)	-

* = plerocercoids; ** = procercoids

preference for a specific infection site, for example the highly specialized trichodines and several species of *Tripartiella*, all of which parasitize the gills of fish (Paperna 1991; Goater et al. 2014).

8.3.2 Myxozoa

The myxozoans are spore-forming parasites of freshwater and marine fishes. Previously they were classified as protozoans but recent molecular analyses, studies on specific functional specializations and the description of multicellular stages have led to their reclassification as a new phylum of metazoans with more than 2,100 species described to date (Yokoyama et al. 2012). Most myxozoans are not harmful to fish but certain members, especially ones that belong to the Class Myxosporea, can be highly fish-pathogenic.

A typical myxosporean life cycle includes two obligate hosts, an annelid (oligochaetes in freshwater and polychaetes in marine water) and a vertebrate, typically a fish (Fig. 4). In the fish host myxosporean spores develop by sporogony and are released into the water. If ingested by an annelid host, mature actinospores are formed via sporogony and released into the water where they can infect fish by skin, fin or gill contact. Subsequent invasion of the sporoplast allows the life cycle to be completed.

The best known fish pathogenic myxosporean species is *Myxobulus cerebralis*, which causes cranial lesions and spinal deformities symptomatic of "whirling disease" in rainbow trout (Halliday 1976). *Myxobulus sandrae* (Lom et al. 1991) and *Triangula percae* spp. nov. (Langdon 1987) have been identified as causative agents of rare localized lesions in the spinal cord, vertebral collapse and marked curvature of the

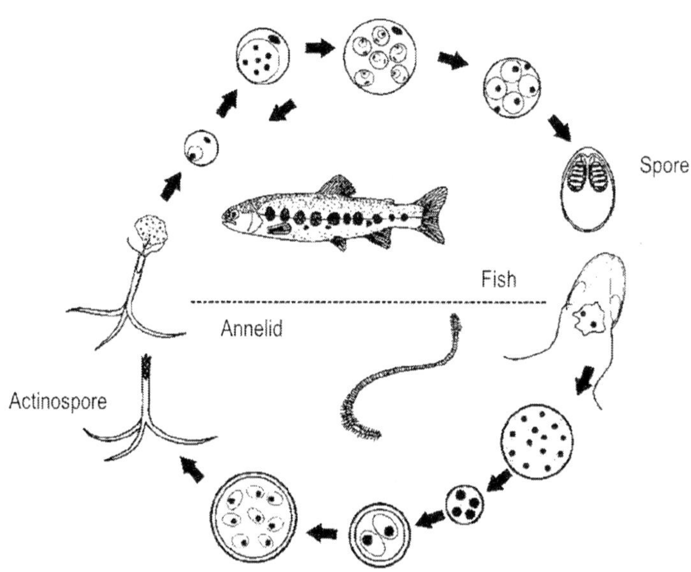

Fig. 4. Typical myxosporean life cycle (*Myxobulus cerebralis*) with alternating fish and annelid hosts, modified with permission after Yokoyama et al. (2012).

vertebral column in European perch. In yellow perch, *Myxobolus neurophilus* and *M. aureatus* have been found to infect the brain and may cause neurological symptoms in this species (Khoo et al. 2010).

More commonly, perch are found to be infected by a number of myxosporean taxa that form whitish, spore-containing nodes on the gills or skin, or in the muscle tissue. *Henneguya psorospermica* is one such species associated with European perch (Schäperclaus 1979) (Fig. 1) while *H. doori* infects yellow perch (Cone 1994).

8.3.3 Platyhelminthes

The Platyhelminthes are a species-rich group with three parasitic classes, the Trematoda, the Monogenea and the Cestoda.

8.3.3.1 Trematoda

Most digenean trematodes have a complex life cycle including several obligate hosts. The trematodes are also known as flukes, and categorized further as eye, blood, liver or lung flukes according to the infection site in the intermediate vertebrate host. In most species, the definitive host is a vertebrate such as a fish-eating bird like a gull or a cormorant. Within the definitive host, the trematodes usually inhabit the intestine. Here they reach maturity and reproduce sexually, laying eggs that are expelled into the environment along with the feces of the host. From the eggs hatch free-living motile miracidia that infect the first intermediate host by penetration of the body surface. For the majority of trematode species, the first intermediate host is an aquatic snail within which a redia or sporocyst will develop and reproduce asexually to yield large numbers of the next infective stage, the cercaria. Cercariae are released continually at a rate of several hundred to several thousand per day and shed from the snail via a special "birth pore" over an extended period of up to several months. The cercariae of most species are motile and actively search for the next host, which they again usually infect by penetration of the body surface. This next life cycle host might be a second intermediate host within which metacercariae are formed that are infective for the next host, or in some cases it might be the final host within which the adult worm develops. The number of obligate intermediate hosts within the life cycle of different trematode species varies from one to four (Goater et al. 2014). In Fig. 5 the life cycle of *Diplostomum spathaceum* is shown as an example.

Members of the *Diplostomum* species complex are eye flukes, infecting European perch and yellow perch as well as other fish species. They include several of the most common species of trematode infecting European and yellow perch, along with *Bunodera* spp. (Carney and Dick 2000a; Morozinska-Gogol 2008) (Table 1).

8.3.3.2 Monogenea

The Monogenea constitute a very diverse group almost exclusively parasitic on freshwater and marine fish species and highly host-specific. However under the dense

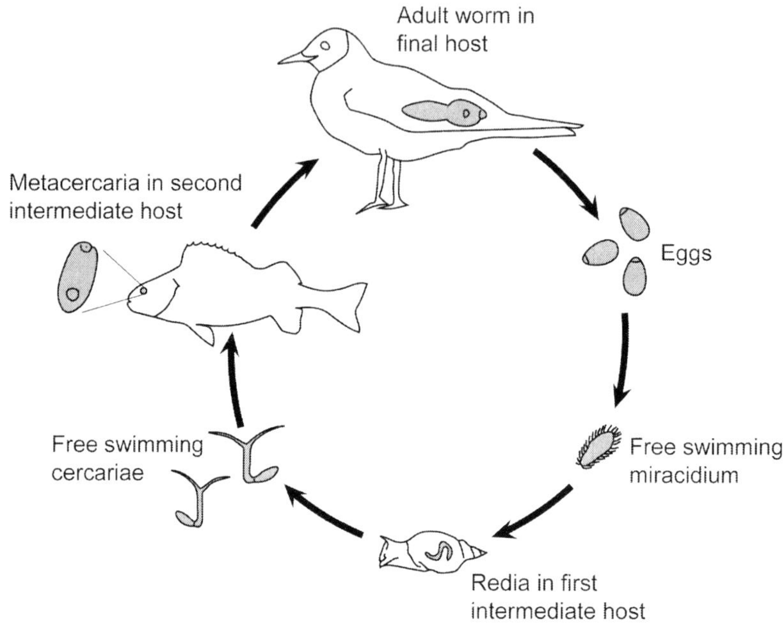

Fig. 5. Life cycle of *Diplostomum spathaceum*.

rearing conditions found in aquaculture or the aquaria used in the ornamental fish trade for example, members of the monogenean orders Dactylogyridea and Gyrodactylidea in particular can cause mass mortalities (Thoney and Hargis 1991). In contrast to the trematodes, most monogeneans have a direct life cycle, spending their entire life on a single host individual. They are either oviparous, like members of the Dactylogyridea, mainly infecting the gills of their hosts, or viviparous like the Gyrodactylidea, which infect the skin of their host fish (Chubb 1977). Several species of monogenea have been recorded parasitizing European perch and yellow perch, the most prominent being *Urocleidus adspectus* on yellow perch (Cone 1980; Cone and Burt 1985) and *Ancyrocephalus percae* and *Gyrodactylus gasterostei* for European perch (Morozinska-Gogol 2008) (Table 1). Figure 6 shows an exemplary life cycle for *Ancyrocephalus percae* infecting European perch.

8.3.3.3 Cestoda

The cestodes, or tapeworms, constitute a diverse group of approximately 3,400 parasitic species with vertebrate hosts (including 800 known species infecting teleosts). Adult cestodes may be the parasites that cause the most revulsion in non-parasitologists because of their large size (some species grow to several meters in

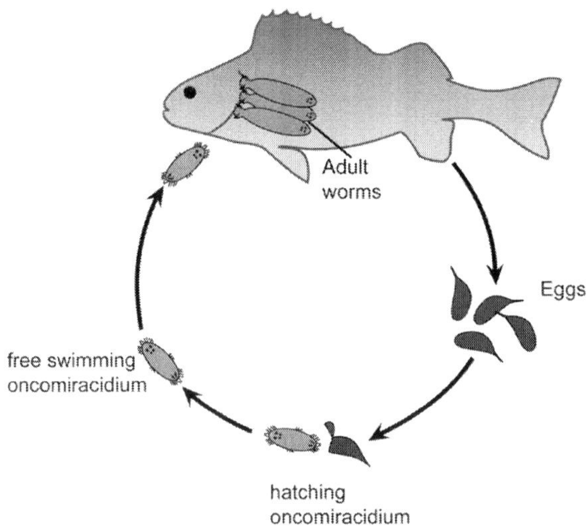

Fig. 6. Life cycle of *Ancyrocephalus percae*. Adult worms lay eggs that are shed into open water. Ciliated oncomiracidia hatch, seek a new host, infest the gills as postoncomiracidia and develop into adults.

length) and conspicuousness, and infestations are generally perceived by farmers, anglers and fishermen as detrimental to fish. In reality however, adult cestodes tend to be located exclusively within the intestine of their hosts, where they function like a "second gut", using suckers, hooks or other holdfast organs to anchor themselves and absorb nutrients directly from the gut content via their own skin. Thus besides a mostly marginal reduction in food or vitamin uptake for the host, low numbers of adult cestodes are of little inconvenience and maybe even be imperceptible to the host. However larval cestodes, which reside in the host's flesh, can be harmful, reducing the desirability, and hence profitability, of fishery produce, or even rendering stock unsuitable for human consumption.

Generally, cestodes have a complex life cycle including invertebrate intermediate hosts and vertebrate definitive hosts. The eggs are released from adult worms resident in the intestine of the definitive host. From these eggs, free-swimming oncomiracidia hatch and are consumed by the intermediate host, often a copepod. Depending on the species, the copepod can be ingested by a second intermediate host in which the parasite develops into a plerocercoid that is infective for the definitive host or it can be directly ingested by a definitive host where the adult worm develops in the intestine and the life cycle is completed. As an example, the life cycle of the pike tapeworm *Triaenophorus nodulosus,* which uses European and yellow perch as second intermediate host species (Kuperman 1973) is shown in Fig. 7.

The most commonly encountered cestodes of European and yellow perch are members of the *Proteocephalus* spp. species complex, *Eubothrium crassum*, *Bothriocephalus cuspidatus* and *Triaenophorus nodulosus* (Carney and Dick 2000a; Morozinska-Gogol 2008) (Table 1).

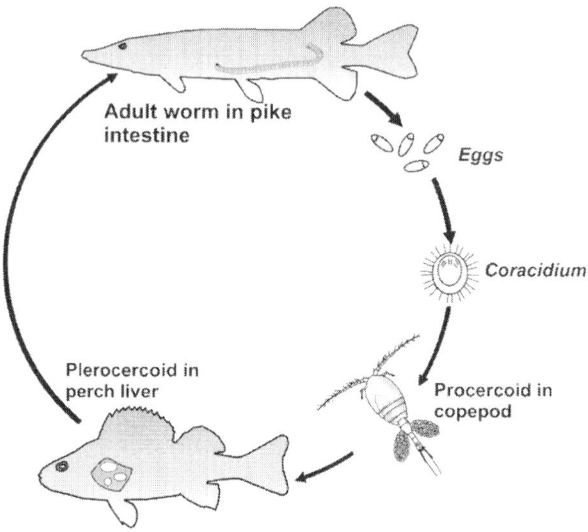

Fig. 7. Life cycle of *Triaenophorus nodulosus* (from Brinker and Hamers 2005).

8.3.3.4 Acanthocephala

The acanthocephalans (thorny-headed worms) are a relatively small group of parasites with about 1,100 species (Goater et al. 2014). They live as adults in the intestines of fish, amphibians, reptiles, birds and mammals and have a worldwide distribution. They employ a unique thorny structure, the proboscis (Fig. 8), to anchor themselves in the intestinal wall of their host, sometimes leading to an inflammatory reaction in heavily infected individuals (Schäperclaus 1979). Similar to the cestodes, the acanthocephalans absorb nutrients directly from the intestine of their host and lack both mouth and digestive tract. They vary in length from 1 mm to more than 60 cm. Acanthocephalans

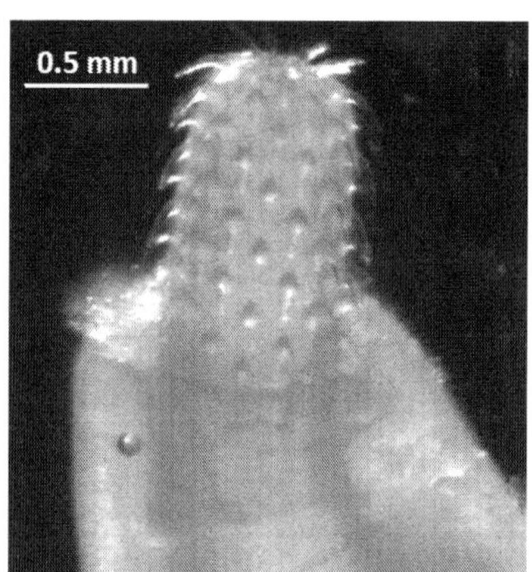

Fig. 8. Proboscis of *Acanthocephalus lucii* (adult worm).

are dioecious and often sexually dimorphic, with the females growing larger than the males. The life cycle is complex, and includes an arthropod intermediate host and a vertebrate definitive host. In some species, additional paratenic hosts may be included.

The most common acanthocephalans recorded in yellow and European perch are members of the *Pomphorhynchus* spp. and *Acanthocephalus* spp. complexes (Carney and Dick 2000a; Morozinska-Gogol 2008) (Table 1).

8.3.3.5 Nematoda

The nematodes (roundworms) comprise one of the most abundant phyla in the animal kingdom. Current estimates hover around 20,000 species, but the vast majority is as yet undescribed (Goater et al. 2014). Most nematode species are free-living, but there are a number of parasitic species, almost exclusively endoparasites of various animal and plant hosts. The nematodes are typically dioecious and often exhibit sexual dimorphism. They vary from 1 mm to more than 1 m in length, with the longest nematode described so far being *Placentonema gigantissima,* found in the placenta of sperm whales *Physeter catodon* where the females grow up to 8 m long and 2.5 cm in diameter. Goater et al. (2014) writes of the nematodes that "*perhaps no single group of related organisms on earth possess greater life history variability*." The broad range of possible nematode parasite life cycle strategies often involves free-living larval stages with behavioral adaptations to attract hosts and facilitate transmission. Common nematodes of European and yellow perch are members of *Camallanus* spp. and *Raphidascarus acus* (Moravec 1994; Goater et al. 2014). *Camallanus lacustris* typically resides in the intestine, specifically the pyloric cæca of European perch, but also in other hosts such as pike *Esox lucius*. Gravid female nematodes shed larvae that enter the water with the host's feces. These larvae are ingested by an invertebrate intermediate host, typically *Cyclops* spp. or *Asellus aquaticus*. When the intermediate host is predated by the definitive host, the nematode larvae can complete their development into adults. Planktivorous fish species such as sticklebacks can also be incorporated into the life cycle as paratenic hosts, which facilitate transmission to a piscivorous definitive host. *Camallanus lacustris* feeds on host blood, and has a very unique color, appearing red or pinkish with a conspicuous yellow oral capsule (Fig. 9).

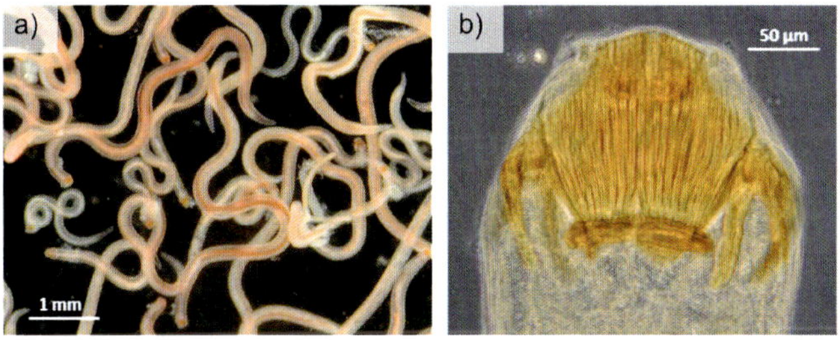

Fig. 9. *Camallanus lacustris* from European perch. The worms have a typical red or pinkish color (a) with a yellow oral capsule (b). Photos were kindly provided by Martin Kalbe, Max Planck Institute for Evolutionary Biology, Ploen, Germany.

8.3.4 Crustacea

The Crustacea are an extremely diverse group of arthropods comprising over 38,000 species. Crustaceans are primarily an aquatic group and a large number have a parasitic association with fish. The majority of fish parasitic crustaceans are ectoparasites that feed on host blood, mucus or skin cells. Many also parasitize the fish's gills, and mechanical disruption of the epidermis at the attachment and feeding sites may result in osmoregulatory or respiratory problems or create opportunities for secondary infectors including pathogenic fungi and bacteria to invade the host. One of the best-known crustacean parasites of freshwater fishes is *Argulus foliaceus* (Fig. 1), commonly known as the "carp louse." The pathogenicity of the carp louse lies not only in the mechanical damage to the fish's skin but also in irritation of the epidermis by digestive secretions. Additionally it has been shown that *A. foliaceus* can act as a vector for other pathogens, such as *Rhabdovirus carpio* which causes spring viremia of carp (SVC) (Ahne et al. 2002). *Argulus foliaceus* has a direct life cycle, in which adult females leave the host after copulation and deposit up to 500 eggs in a gelatinous string on any suitable submerged surface such as stones. They may then return to the same host individual or attach to another until ready to deposit the next clutch of eggs. Time to hatching is dependent on water temperature (within 8 days at 26°C), then newly hatched parasitic metanauplii set out to search for a host fish (Hoole et al. 2011). After attachment, nine larval moults follow in rapid succession until the adult stage is reached. In optimal conditions, the whole life cycle may be completed in less than 40 days, and thus as many as four generations can be realized in a single summer.

The most common crustacean parasites of yellow and European perch are *Argulus foliaceus* and members of the genus *Ergasilus*, which parasitize the gills (Carney and Dick 2000a; Morozinska-Gogol 2008) (Table 1).

8.4 Impacts of Parasites on Growth, Fitness and Survival of Perch

Perch show extensive plasticity in growth rates and age at maturity with variation in abiotic and biotic factors. In general, individuals grow more slowly in the northern hemisphere and at high latitudes, where temperature seems to be the most important abiotic factor affecting growth (reviewed in Craig 2000). Differences between the sexes are also apparent, with females growing faster and attaining much larger sizes than males. However males typically reach maturity one year earlier than females. In Lake Constance for example, most male European perch first spawn in their second year of life, while females mature in their third year (Eckmann and Schleuter 2013). Factors that influence perch growth have been studied intensively by many authors and include latitude (Heibo et al. 2005), prey type (Boisclair and Leggett 1989) and availability (Hjelm et al. 2000), lake productivity (Hayward and Margraf 1987; Abbey and Mackay 1991), duration and severity of winter (Johnson and Evans 1991), presence of competing species (Bergman and Greenberg 1994) and presence of predators (Magnhagen and Heibo 2004). Heibo and Magnhagen (2005) demonstrated that the same broad range of factors also affects age at maturity. The effect of parasites on perch growth, survival and reproductive potential in natural populations is less well studied, although it has been shown that distinct parasite species can impact negatively on perch growth (Johnson and Dick 2001; Cloutier et al. 2012) and can increase perch mortality (Deufel 1975), especially of young age classes (Szalai and Dick 1991; Szalai et al. 1992).

It is very difficult to link parasite infection with possible effects on viability, fertility and growth of fish populations in the field (Kennedy 1984; Lester 1984; Sindermann 1986). Predation and scavenging of moribund or dead animals by other fauna hinder attempts to assess the extent of an epidemic and observable non-lethal aspects of infection such as pathological alterations, behavioral changes, reduced reproductive success or increased contaminant load are very difficult to quantify (Lester 1984; Sindermann 1986). Causal links between parasite infection and target organ disorders can be difficult to separate from other factors, especially when the time elapsed since initial infection of the active pathogen is not known. However, parasites take nutrients from their hosts, and provoke reaction by their mere presence, by inflicting lesions or by actively destroying host tissue (Goater et al. 2014). In intermediate or paratenic hosts, parasites might even benefit from altering the host's behavior or weakening it such that the chances of transmission to the next host are improved (fish examples reviewed in Barber et al. 2000). Further possible effects include reduced fertility and fitness, in extreme cases resulting in castration of hosts (e.g., trematodes in water snails or *Schistocephalus* in sticklebacks) (Heins and Baker 2008; Baudoin 1975). In the laboratory, controlled experiments can be conducted in order to investigate single aspects of infection, though it usually remains unclear how well these trials reflect real life. However using the seminal work of Anderson and Gordon (1982) and Crofton (1971) increasingly sophisticated statistical tools, applied with appropriate data sets, might offer insights into even non-observable traits like parasite-induced mortality in the field.

8.4.1 Parasite Impacts on Adult Perch

As shown earlier in this chapter, a large number of macroparasite species target European and yellow perch as intermediate, paratenic or definitive hosts. For the majority of these parasites, little or no information is available regarding impacts on their host, but most probably, the effect of the majority is minimal as long as no epidemic outbreak occurs. The host is a resource that is exploited by the parasite to maximize its own fitness. The manner of exploitation (and thus the potential for harm to the host) depends on how the interests of the parasite are best served for maximizing fitness (Poulin 1998). Parasite species that use perch as definitive hosts usually do not benefit from a weakened or moribund host, because their own reproduction is optimal in a well fed, healthy, normally behaving individual. In contrast, parasite species that use perch as intermediate or transport host, may profit from weakening the host or by inducing behavior that renders the host vulnerable to predation by the parasite's next life cycle host. An excellent example of such manipulation of host behavior concerns the eye fluke *Diplostomum spathaceum* and its intermediate fish host, the rainbow trout *Oncorhynchus mykiss*. *Diplostomum spathaceum* has a complex life cycle (Fig. 5), including three obligate host species. Fish are used as second intermediate hosts, wherein *D. spathaceum* locates itself in the lens of the eye. Experiments with infected rainbow trout have shown that *D. spathaceum* metacercariae in the eye can induce cateract formation on the lens, which impairs the host's escape response and predisposes it to predation by birds, thereby improving the chances of transmission to the definitive host (Shariff et al. 1980; Seppälä et al. 2011). Although no comparable experiments have been done with perch, both European and yellow perch are among the wide range of fish taxa parasitized by members of the highly speciose *Diplostomum* species complex and a similar mechanism of behavioral manipulation can be assumed at least for the lens-infecting *Diplostomum* species. Between 1960 and 1971, regular lethal outbreaks of *Diplostomum* spp. where recorded in Lake Constance. Not only European perch but also pike, burbot *Lota lota* and cyprinids including bream *Abramis brama* were killed by massive infections in spring and early summer (Deufel 1975). In these fish kills, mortality was caused by penetration of the skin by excessive numbers of *Diplostomum* cercariae, a disease known as diplostomosis or cercariosis (Majoros 1999; Larsen et al. 2005). Diplostomosis is a recognised threat to fish in lake farms and in natural waters, and tends to be especially devastating to fish larvae and juveniles (Molnár 1974; Larsen et al. 2005).

Negative impacts on perch health have also been shown for the pike tapeworm *Triaenophorus nodulosus*, which infects both perch species as a second intermediate host. *T. nodulosus* has a complex life cycle (see Fig. 7) including three obligate hosts. Adult worms live in the intestine of pike, from where eggs are shed along with the host's feces and hatch into motile coracidia. These coracidia are ingested, mainly by copepods of the genus *Cyclops,* within which each coracidium develops into a fish-infective procercoid. Several fish species, but principally European and yellow perch, serve as the second intermediate host (Kuperman 1973) after ingesting infected copepods. The procercoids migrate to the liver of the intermediate fish host where they become encapsulated by a host tissue response and develop into plerocercoids (Fig. 10).

Plerocercoids are only rarely found in other organs (Kuperman 1973). Burrowing activity of the plerocercoid and the resulting lysis of host cell membranes causes pathological symptoms such as inflammation, atrophy, necrosis, hyperæmia, hæmorrhage and edema (Rosen 1918; Scheuring 1922; Kuperman 1973; Schäperclaus

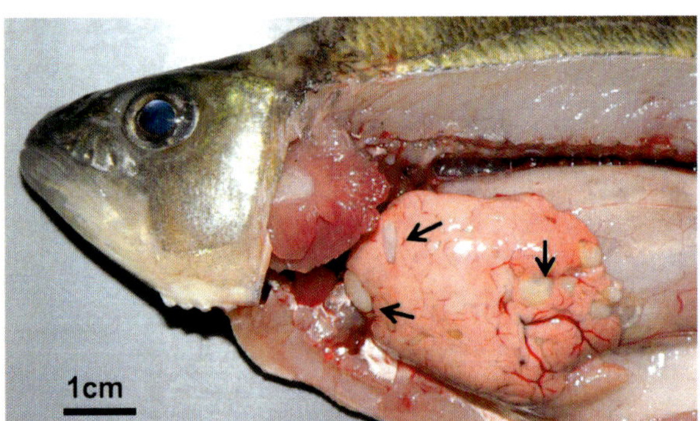

Fig. 10. Heavily infected adult European perch from Lake Constance. Several encapsulated pike tapeworms *Triaenophorus nodulosus* can be seen in the liver (arrows).

1979; Hoffmann et al. 1986). Further damage may result from pressure on surrounding tissues as large larvae are encapsulated (Scheuring 1922) and from osmotic and toxic stress (Read and Simmons 1963; Schäperclaus 1979). Predation on infected intermediate hosts by the pike completes the life cycle. Brinker and Hamers (2007) compared the growth of European perch with different intensities of *T. nodulosus* plerocercoid infection (varying from no infection, to normal infection with 1–3 plerocercoids and severe infection with >3 plerocercoids). Increasing infection intensity correlated with a reduction in perch growth (Fig. 11) amounting to a loss of ~10% potential mass for normal infection and ~16% for severe infection at legal gill net catch size—an effect large enough to be commercially significant. Furthermore, Brinker and Hamers (2007) and Molzen (2006) both reported a clear correlation between levels of pathological liver alteration and infection level, with knock-on negative effects on fertility, i.e., the gonadosomatic index was reduced by 20% or more (Molzen 2006). However, recent research in the field has shown that the prevalence of *T. nodulosus* in perch drops from ~90% to ~70% when co-infecting with the gill worm *Ancyrocephalus percae* (Roch unpublished results) indicating a subsequent increase in parasite-induced mortality.

Triaenophorus nodulosus shows a highly aggregated distribution in perch. Such distribution patterns are common in almost all parasite communities including passive parasite-intermediate host systems (Crofton 1971; Anderson and Gordon 1982; Lester 1984; Pacala and Dobson 1988) and are assumed to benefit the parasite in two ways: (1) a small proportion of the intermediate host population carries the bulk of parasites and suffers the negative impacts, but the host population as a whole is not endangered; (2) heavily infected second intermediate hosts are probably weakened and therefore become relatively easy prey for the definitive host (Kennedy 1984; Balling and

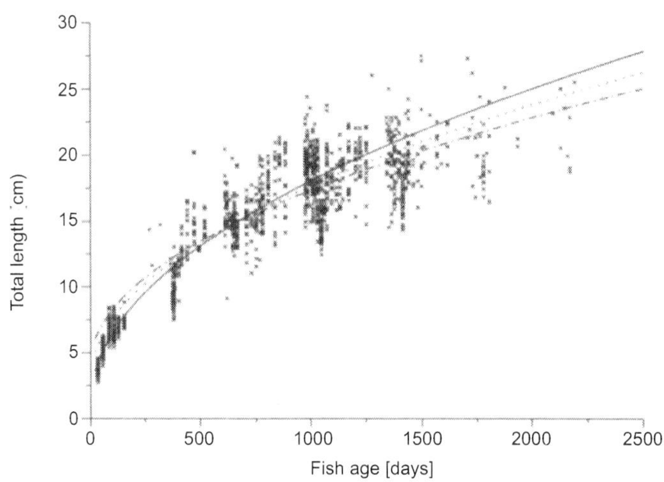

Fig. 11. Growth curves of European perch from Lake Constance. The three lines represent growth regressions for different infection categories: Solid line—no infection; dotted line—low infection; hatched line—heavy infection (for details see text). Modified with permission from Brinker et al. (2007).

Pfeiffer 1997). A consequence of aggregated distribution is thus that the probability of transmission increases (Sindermann 1986; Zander 1998).

Infection with *T. nodulosus* increases the relative risk of mortality in the host, as can be seen in spring in Lake Constance, when severely infected fish suffer from increased mortality following spawning stress (Brinker et al. 2007). This is further supported by recursive fitting of data for 1307 Lake Constance perch (Brinker unpublished data) to the negative binomial distribution, revealing parameter k larger than 1, which indicates parasite-induced mortality associated with high-intensity infection (Crofton 1971).

In the yellow perch, the nematode *Raphidascaris acus* also infects the liver and causes significant and quantifiable negative effects. The parasite has a complex life cycle, including aquatic invertebrates as paratenic hosts, fish as intermediate hosts and piscivorous fish as definitive hosts. In North America, *Perca flavescens* is the main intermediate host, and infection is mainly via ingestion of chironomid prey (Johnson and Dick 2001). The parasite causes extensive liver pathology and wider intensity-dependent symptoms. For example, if infestation exceeds 100 cysts per gram of liver tissue the yellow perch suffer from significant growth reduction (Johnson and Dick 2001). Especially striking however, are further age-related effects. Infection intensity tends to peak just before the host reaches sexual maturity and the combined pressure of parasite infection and the physiological demands of vitellogenesis and spermatogenesis lead to increased host mortality and reduced growth (Szalai 1991). It is even possible that the generally observed pattern of bimodal weight distribution in female yellow perch following vitellogenesis may be caused by the parasite. The significance of *R. acus* for yellow perch is convincingly demonstrated by observations from Dauphin Lake in Manitoba, where in the absence of the parasite natural mortality was reduced by about 50% (Szalai 1991). Thus it seems that tissue-invading parasites may play a significant role in the regulation of perch populations in both North America and Europe.

Another macroparasite that was shown to have a dramatic negative impact on the health of European perch from Lake Constance is the monogenean gill parasite *Ancyrocephalus percae*. These oviparous flatworms have a direct lifecycle (see also Fig. 6) and are hermaphroditic (Chubb 1977). Eggs are laid into the open water, followed by hatching of oncomiracidia that are able to directly infest a new host. *Ancyrocephalus percae* is distributed widely all over Europe, but infects only the gills of European perch (e.g., Andrews 1979; Bylund and Pugachev 1989; Morozinska-Gogol 2008). Little information about this parasite is available in the scientific literature, most likely because it appears not to cause severe problems for European perch outside Lake Constance.

At some point after the year 2008, *A. percae* was introduced into Lake Constance (Behrmann-Godel et al. 2014). The route into the lake is unknown, but having arrived, the parasite became highly invasive. It infects Lake Constance perch at the isthmus near the gills and causes oval-shaped wounds in the tissue (Fig. 12a), a behavior never described for this parasite before. All age classes of perch are affected, especially young fish. The wounds occurring on the isthmus of the hosts' gills can be dramatic. In extreme cases, high intensity infections can sever the isthmus, leading to partial decapitation (Fig. 12b).

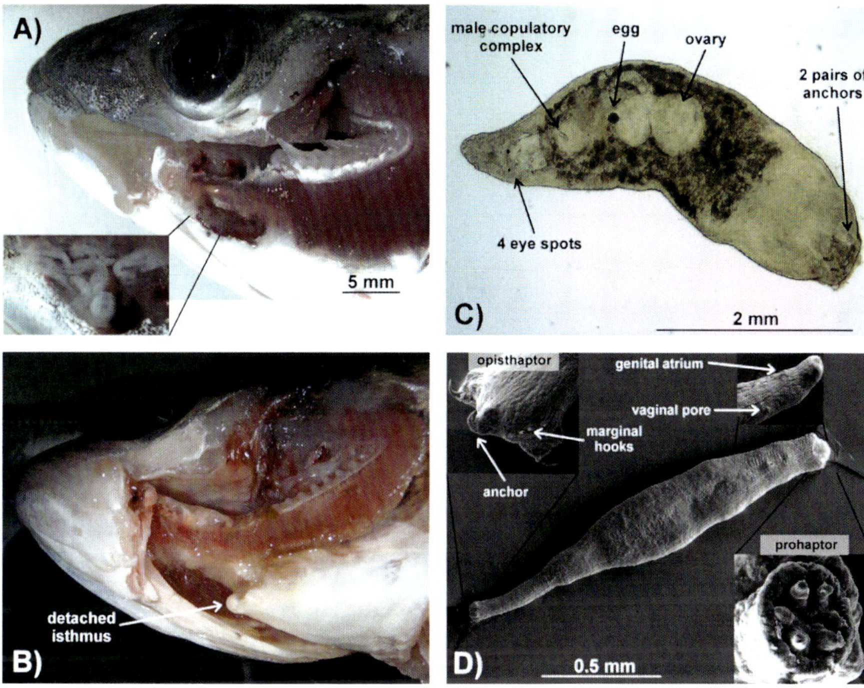

Fig. 12. European perch from Lake Constance infected with the gill worm *Ancyrocephalus percae*. (a) Aggregation of parasites in a wound at the isthmus. The operculum and distal part of the gills have been removed for a better view. (b) Heavily infected fish (isthmus is detached from lower jaw), operculum and parasites removed for better view. (c) Light microscopic view of an adult *A. percae* from perch showing morphological characters. (d) Scanning electron micrograph (Foto by V. Burkhardt-Gebauer) of ventral body of an adult *A. percae* showing morphological characters. Reprinted with permission from Behrmann-Godel et al. (2014).

Clear differences in the prevalence of *A. percae* infection can be found between two genotypes/morphotypes of European perch in Lake Constance distinguishable by fin color. While the majority of perch have yellow fins, there are increasingly regular catches of red-finned perch in certain areas of the lake. Red-finned perch occur alongside yellow-finned fish but are significantly less affected by *A. percae*, indicating a greater resilience to this parasite (Roch et al. 2015). Immunity plays a crucial role in parasite defence (Buchmann and Lindenstrom 2002), and the resilience of red-finned perch in Lake Constance is most likely due to a more effective immune response. Additionally there are occasional catches of adult yellow-finned perch with red areas on their fins. This unusual color pattern is not related to disease or bacterial infection, and it is likely that these mixed color morphotypes are hybrids between yellow- and red-finned types (Roch et al. 2015). Studies on prevalence of the gill worm *A. paradoxus* show that the mixed color types are significantly less affected by the parasite than yellow-finned relatives, but more so than the pure red-finned ones (Fig. 13). It seems that fin color is concomitant with some other trait that influences *A. percae* parasite burden. Early studies into the phenomenon have sought evidence

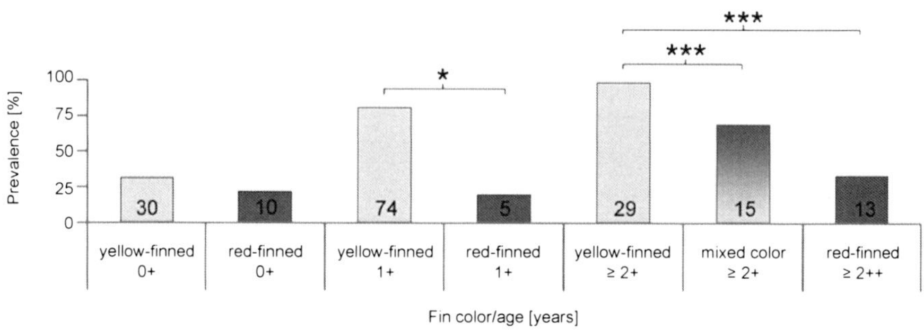

Fig. 13. Infection of three morphotypes of European perch from Lake Constance (yellow-finned, red-finned and mixed color) with the monogenean gill parasite *Ancyrocephalus percae*. Shown is the prevalence in three different age classes: young of the year (0+), one year old (1+) and older perch (>2+). Sample size is indicated in the bars, significant differences are indicated by asterisks (Fisher's exact test).

of a cellular immune response to the parasite. In chemotaxis assays, perch leukocytes showed a positive chemotactic reaction to *A. percae* exposure. Young of the year red-finned perch are also infested by *A. percae*, but seem to gain a resistance to the gill parasite within their first year. The yellow-finned morphotype does not develop a successful immune defence, resulting in infestation of all age classes (Fig. 13).

Measurements of *post mortem* muscle pH in yellow-finned perch showed that infection with *A. percae* has a negative effect on energy reserves in the muscle of infested individuals. Following death, energy-rich compounds like glucose are metabolized by bacteria in an anaerobic process producing lactate (Binke 2004), which subsequently lowers pH. This process has been shown to be faster when fish are stressed (Thomas et al. 1999) and the magnitude of pH loss is dependent on the stored amount of energy-rich compounds in muscle tissue (Unger et al. 2008). As uninfected perch show a comparable pH development during 24 h *post mortem* but higher pH drop than infected ones, this indicates energy depletion in muscle tissue of infected fish. Interestingly, one year old perch infected by *T. nodulosus* suffered a significantly lower incidence of *A. percae* infection, suggesting a possible immune priming effect whereby the presence of the pike tapeworm activates the perch immune system, which is then better able to react against *A. percae*. However the observation could also be explained by an increased death rate associated with infection by two harmful macroparasites. Follow up experiments will show whether red-finned perch in Lake Constance really do have an immune advantage that allows them to throw off infection better than yellow-finned perch.

8.4.2 Impact of Parasite Infection on Perch Larvae and Early Juveniles

The high levels of mortality suffered by larval and juvenile fish are due mainly to predation, but starvation, disease and parasitism also take a significant toll (Wootten 1974). It is easy to imagine that any debilitation or deviation from normal fish behavior will increase their risk of predation and hence increase larval and juvenile mortality.

An increasing number of studies show that the early developmental stages of fish are targeted by several species of endo- and ectoparasites (Balbuena et al. 2000; King and Cone 2009; Kuchta et al. 2009; Pracheil and Muzzall 2009; Skovgaard et al. 2009a; Behrmann-Godel 2013). Parasitic infections may be detrimental to fish fry by inducing malformations and increasing mortality (Johnson and Dick 2001; Skovgaard et al. 2009b; Grutter et al. 2010; Kelly et al. 2010; Nendick et al. 2011). However the costs of parasitism in terms of various life history parameters are far from well understood and the consequences of parasite infection of fish larvae and juveniles for population recruitment might be especially severe for threatened species (e.g., see Collyer and Stockwell 2004).

The timings of parasite succession relative to first parasite encounter are almost unstudied in perch (but see Kuchta et al. 2009). Behrmann-Godel (2013) recently observed that European perch fry were infected with a succession of 13 different parasite species during the first three months of development (Table 2). First infections have already occurred by the time perch larvae in the pelagic zone of Lake Constance start feeding on zooplankton including infected copepods, four weeks after hatching. Distinct changes in parasite community composition and abundance were found to be associated with perch fry age and with the ontogenetic habitat shift from pelagic to littoral nursery areas (Behrmann-Godel 2013).

Trophically transmitted parasites such as the pike tapeworm *T. nodulosus*, which is consumed along with infected prey, have the potential to reduce growth of adult perch, as described in the previous paragraph. In perch larvae (age: 42–152 days post hatch (dph)) however, experimental infections had no effect on growth, but did result in increased mortality of heavily infected larvae and juveniles. Infection intensity was higher in fish that died during the experiment than in the survivors (Fig. 14). For juvenile yellow perch, between 3–24 months of age (90–1550 dph) investigated from four Canadian Shield lakes, Johnson and Dick (2001) found negative effects of two parasite species on both, growth and survival of perch. The myxosporidian *Glugea* spp. and the trematode *Apophallus brevis* increased mortality and reduced the growth of YOY fish when occurring in high intensities (>100 parasites for *Glugea* spp.) and high densities (>50 and again with >100 cysts/g filet weight for *A. brevis*).

Interestingly, the youngest European perch larvae investigated in Lake Constance appeared not to be infected by skin penetrating trematode cercariae from snail intermediate hosts (2 weeks post hatch (wph), Table 2). Perch spawn early in spring (between April and June depending on latitude) in shallow littoral areas. The larvae hatch after 6–14 days, depending on water temperature and are soon transported to the pelagic zone (Wang and Eckmann 1994). As snail intermediate hosts in the littoral zone already shed cercaria by the time perch larvae developed and hatched (Deufel 1975; Behrmann-Godel 2013), perch larvae might be expected to be at risk of infection during their first days post hatching. The complete lack of skin-penetrating trematode cercariae in young sampled fish could indicate two possible scenarios. First, larval perch may not be recognized as potential hosts by the skin-penetrating trematode cercariae and thus not infected, or second, cercarial infection of fish larvae takes place but is rapidly fatal. Experimental infection of perch larvae with skin-penetrating cercariae of *Diplostomum spathaceum, Cyatocotylidea* sp. and *Tylodelphys clavata* led to death

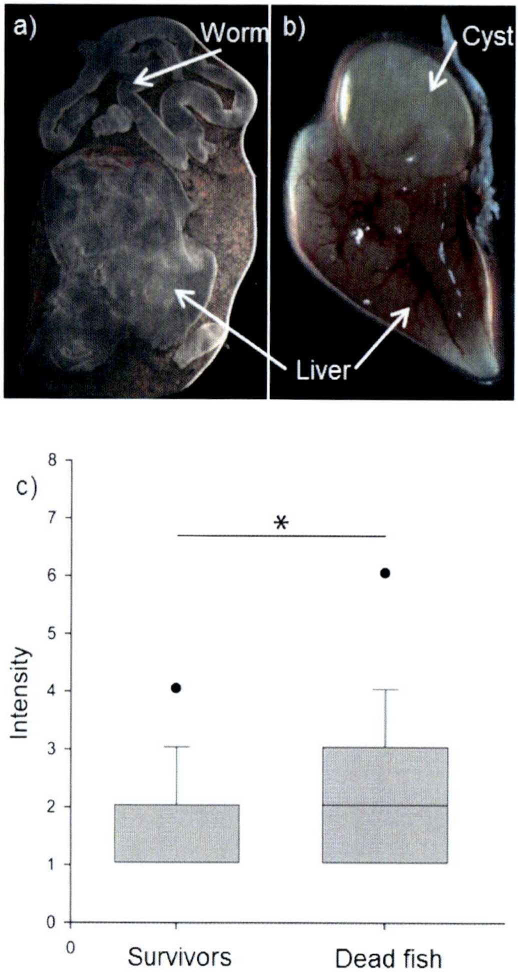

Fig. 14. Laboratory infection of larval and juvenile European perch with *Triaenophorus nodulosus*. (a) In some fish, the plerocercoids where not encysted (perch age: 48 dph (days post hatch)). Plerocercoid has been squeezed out of the liver to show burrows created by the worm. (b) In other host fish the plerocercoids were encysted by a host tissue reaction (perch age: 75 dph) (photos: Michael Donner). (c) Boxplots of *T. nodulosus* intensity during the infection experiment (Median, box = 50% percentiles, whiskers = 95% percentiles, dots = outliers). The density of *T. nodulosus* plerocercoids in host tissues was significantly lower in infected survivors than in infected fish that died during the experiment (n = 162 survivors, 77 dead fish, age: 42–125 dph; Welch ANOVA, $F = 4.38$, $p = 0.038$).

from cercariosis within 24 hours (Behrmann-Godel unpublished). Thus during the first days after hatching, even very low doses of cercariae infection by skin penetration may result in high mortality rates of perch larvae (Fig. 15). With increasing age of perch however, dose dependent mortality decreased and by 19 days old, mortality rates among perch infected with the mild doses used in the experiments (max. 10 cercariae) were very low (Fig. 15f).

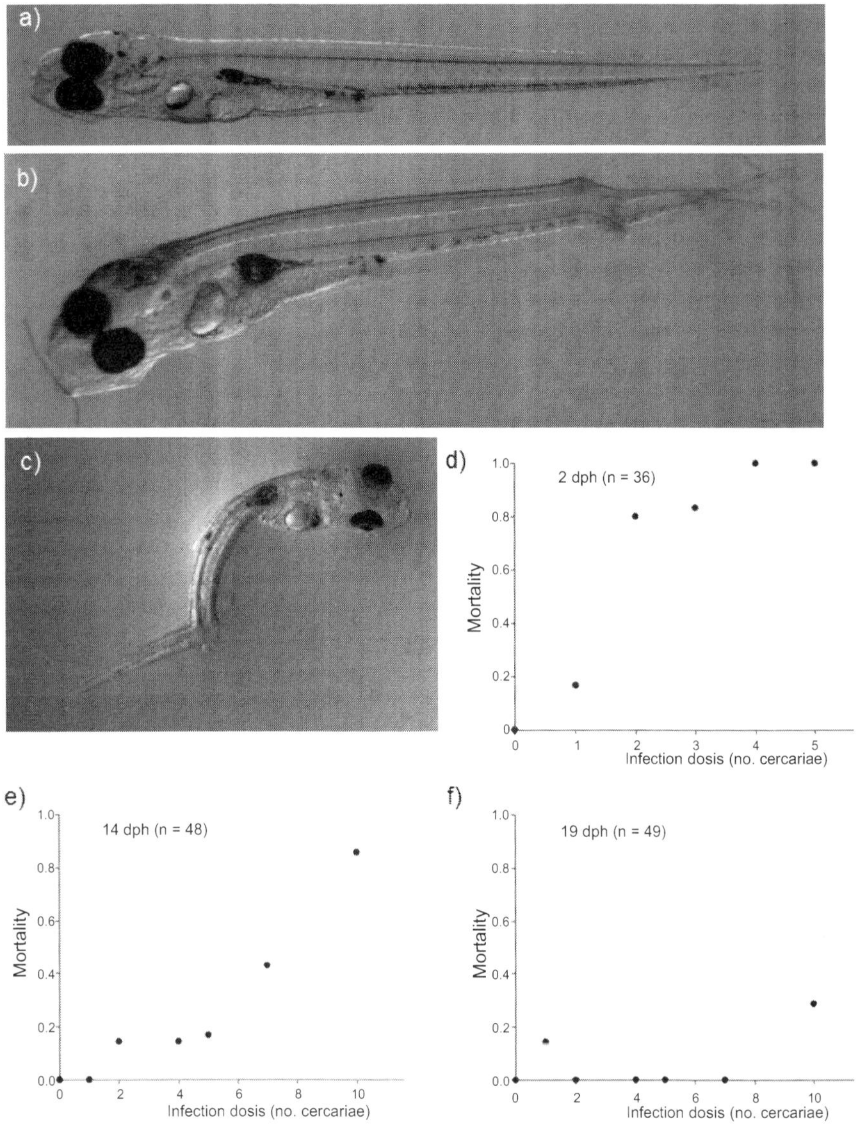

Fig. 15. Cercariosis caused by skin-penetrating trematode cercaria in European perch larvae and juveniles. *Diplostomum spathaceum* cercariae caused spinal cord malformations and death of infected three dph (days post hatch) perch larvae (b, c), while uninfected control larvae survived (a). In laboratory experiments perch larvae were individually infected with different doses of trematode cercariae (d–f, every dot represents perch mortality within 24 hours calculated from six (d) or seven (e, f) infection experiments with individual larvae). Skin penetration by *Tylodelphys clavata* (d) and *Cyatocotylidea* sp. (e, f) resulted in dose dependent mortality of perch larvae. Mortality rates decreased with perch age and were almost negligible at 19 dph (f).

Cercarial shedding from infected intermediate host snails is highly temperature dependent. It has been shown that with a slight increase of water temperature in spring, the number of cercariae shed by snails increases exponentially (Lyholt and Buchmann 1996 and personal observations). *Lymnaea stagnalis* snails infected with the trematode *Diplostomum spathaceum* started shedding cercariae at water temperatures of 4–6°C. At 10°C, cercariae were shed at a rate of 10,000 per snail per day, rising to 58,000 cercariae per snail per day at 20°C (Lyholt and Buchmann 1996). A slight increase in spring water temperatures as a result of global warming may therefore be expected to result in an increased shedding of trematode cercariae in spring. Rising spring water temperatures have already been shown for several lakes including Lake Constance (Stich and Brinker 2010; Straile et al. 2012), and it might be speculated that increasing numbers of cercariae in the littoral zone will negatively impact newly hatched perch larvae and result in high mortality rates.

8.5 Ecology and Evolution of Host-parasite Interactions

Parasite studies are increasingly incorporated into ecosystem research, and especially into investigations of aquatic ecosystems. Incorporating parasites into "classical" food webs from which they were previously excluded greatly increases the complexity of the models, but improves understanding by increasing apparent species richness, trophic chain lengths and connectivity (Marcogliese and Cone 1997; Lafferty et al. 2006, 2008).

Furthermore, parasites have long been overlooked as prey in their own right (Johnson et al. 2010). They may be consumed along with a host organism or via direct predation of free-living stages (Lafferty et al. 2006, 2008). Incorporating parasites into a food web for the Carpinteria salt marsh in California doubled trophic connectivity and quadrupled the number of links (Lafferty et al. 2006). Thus future ecosystem-level investigations of trophic interactions should consider parasites as a matter of course from the outset.

Community ecology is a central theme in parasitology but has yet to be extensively studied in fish (Bush et al. 1990; Kennedy 1990). Indeed, there is considerable uncertainty as to whether the freshwater fish parasite communities are stochastic in nature or follow ecological patterns (Kirk 1990). The topic in all its facets is beyond the scope of this chapter, but yellow perch are among the few fish in which this complex matter has been successfully addressed, we will briefly review the work and its important ecological implications. Carney and Dick (2000) and Johnson et al. (2004) provide evidence that certain aspects of parasite assemblages are predictable and follow decisive ecological patterns. For example, the coevolution of parasite and host provides some phylogenetic predictability in terms of host specificity (important parasite species being *Urocleidus adspectus*, *Bunodera sacculata* and *Proteocephalus pearsei*) (Johnson et al. 2004). However ecological predictability also occurs, as a result of stable infracommunity processes (Carney and Dick 2000). Parasites that are transmitted trophically via predation on intermediate host species can be used as natural indicators of trophic linkage within ecosystems (reviewed in Marcogliese and Cone 1997). Analysis of parasite communities associated with yellow perch in several

Canadian Lakes of differing trophic status, size and invertebrate abundance showed that the parasite infracommunities were non-random, highly nested associations of species that were predictive in several ecologically important ways, providing an extensive insight into host feeding habits and dietary preferences and indicative of the richness of the invertebrate community of the habitat (Carney and Dick 2000). Similar results from Jonson et al. (2004) suggested parasite fauna to be good indicators of the trophic status of yellow perch in the Canadian Shield lakes. Interesting in this context is the observation that parasite assemblages are generally richer in waters with complex invertebrate communities, whereas the complexity of fish communities seems to have no influence on parasite richness (Johnson and Dick 2000). This observation has yet to be fully explained, but as rule of thumb, schooling host and exploiting exclusive trophic categories tend to homogenize parasite infracommunities, while overlap with other species tends to increase parasite richness. Thus the data for yellow perch, though at odds with Kennedy's (1990) assessment that local compound communities of freshwater fish are stochastic, suggest that stable infracommunity processes do act in the complex environment of fish communities and therefore could be valuable tools in ecological research.

In the light of anthropogenic environmental change, several parasite species have been shown to accumulate trace metals, making them interesting candidates for environmental pollution indicators (Sures 2003; Vidal-Martinínez 2010). Acanthocephalans, for example, have been shown to accumulate toxic metals such as arsenic, cadmium and lead at concentrations hundreds to a thousand times greater than in the tissue of their host fish (Sures et al. 1999; Sures 2001). The trace metal load of parasites may be therefore useful in identifying anthropogenic pollution in the host environment at levels that are otherwise undetectable (Nachev 2013). Interestingly however, fish infected by some parasites, including acanthocephalans, have lower trace metal concentrations in their tissue than unparasitized fish from the same polluted area (Sures 2008; Vidal-Martínez 2010). This raises the possibility that some endoparasites may act as "metal sinks," effectively cleansing the host tissues of trace metals and eventually benefiting the overall health status of the host fish.

All these studies strongly suggest that environmental change and anthropogenic environmental pollution in particular, can impact considerably on parasite-host interactions. Any detrimental effect of parasites can be exacerbated when the host is stressed by pollutants in the environment. Marcogliese et al. (2005, 2010) studied the combined effects of parasites and pollution on yellow perch sampled at sites in the St. Lawrence River, Quebec, Canada that differed in the degree of pollution (mainly by trace metals). The results of that study show that perch in more heavily polluted sites generally expressed higher levels of oxidative stress biomarkers (lipid peroxidation) than perch from non-polluted waters (Marcogliese et al. 2005). Within the same polluted site, parasitized perch were in a worse state of health than non-parasitized fish, with higher levels of oxidative stress biomarkers. Further study (Marcogliese et al. 2010) indicated that, despite comparable numbers of parasites in yellow perch from contaminated and uncontaminated sites, the pathogenicity of single parasite species was enhanced under polluted conditions. Thus while environmental pollution may not affect resistance to specific parasite species, tolerance is likely to be reduced under polluted conditions.

A comparison of both the phylogenetic and biogeographic data of perch and their associated parasites was used to understand historic ecological parasite-host interactions and host switches, and to test different hypotheses of the phylogeographic origin of the yellow perch in North America. Carney and Dick (2000b) have used phylogenetic and biogeographic information of both yellow perch and three of their associated parasites *Crepidostomum cooperi*, *Proteocephalus pearsei* and *Urocleidus adspectus* to investigate whether the parasite-host associations are solely based on ecology or on co-speciation between the partners. They could show for all three investigated parasite-host interactions that they are not based on parasite-host co-speciation but have all arisen by a host switch and must have had a North American ancestor. *Crepidostomum cooperi* forms a monophyletic clade with *C. ictaluri* and *C. cornutum*, two parasites of endemic North American Centrarchidae and Ictaluridae. *Urocleidus adspectus* was found in a clade with uncertain relationships that is a sister clade to the *Ligictalurids*, a group of parasites of the North American ictalurid catfishes. Finally *P. pearsei* form yellow perch and *P. percae* from European and yellow perch were not found to be sister taxa based on the Proteocephalid phylogeny, suggesting no co-speciation to have occurred between these two parasite species but also a host switch of *P. pearsei* to yellow perch from another North American endemic host species. These findings of Carney and Dick (2000b) together with a study on *Bunodera* spp. biogeography from yellow perch and North American sticklebacks (Choudhurs and Règagnon 2005) strongly support the hypothesis of a Laurasian origin for the Percidae as first proposed by Wiley (1992) (see also Chapter 2 of this book for molecular-level corroboration). This hypothesis predicts the existence of a North American-European ancestor for both the yellow and the European perch that dispersed into both continents. Divergence between the two species happened later during vicariance due to the separation of both continents after the opening of the North Atlantic during the Miocene (20 Myr BP).

8.6 Conclusions and Future Perspectives

The parasite fauna of perch, including European and yellow perch, is highly diverse, with numerous representative species of a variety of genera. The diversity of the parasite community of perch reflects the variable autecology of the host fish species, their widespread distribution, broad habitat preferences (including streams, rivers, lakes, estuaries and low salinity marine areas such as the Baltic Sea) and concomitant nonselective feeding behavior. At the perch host individual level, the variability of the parasite infracommunity is also high because individual fish tend to migrate between habitats and consume food from different sources and trophic levels during ontogenetic development and in later life. Meanwhile, many associations between perch and their parasites are quite well studied and a diverse array of analytical tools is now available for eco-parasitological investigations. Immunological techniques recently developed for measuring physiological reactions to parasite infection in perch include the analysis of oxidative stress biomarkers (Marcogliese et al. 2005, 2010), measurements of *post mortem* muscle pH and chemotaxis assays with perch leukocytes (see 3.1 above). All these new techniques and methods are yielding intriguing new insights and answers to eco-parasitological questions. Perch and their parasites are thus an increasingly valuable model system for research areas including facilitation phenomena in parasite infrapopulations, cost-benefit analyses and the study of parasite-host co-evolutionary processes.

The scope of parasitological study, once an almost exclusively descriptive biological discipline, has widened considerably in recent decades into a variety of new and intriguing areas, not least evolutionary biology, where parasite-host interactions allow unique insights and opportunities for study. Fish and their parasites are particularly promising in this respect, given the relative ease of culture, fecundity and short generation time relative to other vertebrate hosts and the extensive reservoir of biological background knowledge.

One of the most prominent fish model organisms studied in this context is the three-spined stickleback *Gasterosteus aculeatus,* whose natural parasites include the tapeworm *Schistocephalus solidus* and the monogenean *Gyrodactylus gasterostei* (Barber 2013). Several aspects of parasite-host co-evolution have been investigated using these model systems, including infection dynamics (Kalbe and Kurtz 2006; Raeymaekers et al. 2011) and parasite-host co-adaptation, mainly including the study of genes of the major histocompatibility complex (MHC) (Eizaguirre et al. 2009, 2011; Lenz et al. 2013). The expanding range of model parasite-host systems will further benefit our understanding of co-evolutionary processes. Recently a similar methodology was developed to study genes of the MHC in European perch (Michel et al. 2009; Oppelt and Behrmann-Godel 2012). Major histocompatibility complex receptors (especially class II receptor genes) present antigens derived from extracellular pathogens such as parasites to helper T cells of the immune system, and induce an adaptive immune response to parasitic invaders (Janeway et al. 2005). This means that the specific MHC setting of an individual directly determines its ability to resist or defend against parasitic infections and as such is expected to be a major target trait of antagonistic parasite-host co-evolution. Björklund et al. (2015) were able to compare MHC variability in two perch populations exposed to different temperature conditions over 35 years. One population, living in the vicinity of a nuclear power station, was isolated and exposed to artificially warm water used as a coolant, while the other nearby population served as an unmanipulated control. They concluded that isolation and heating has led to a change in the selection regime imposed by parasites. It resulted in observable changes in MHC allele variability and cycling patterns. These observations were supported by the finding that the current parasite communities of the two perch populations now differ significantly from each other (Marian Schmid, M.Sc. Thesis, University of Konstanz).

Despite recent advances, major deficits remain in some important scientific aspects of perch knowledge, including: (1) the lack of comparative studies for European and yellow perch concerning parasite-host interactions and the impact of parasites on host fitness, (2) understanding of perch immunity, in particular specific immune reactions against parasitic invaders, (3) the physiological consequences of parasitic infections for perch, and especially of multiple infections with a diverse range of parasitic species, or of invasive parasite species new to the ecosystem, and (4) the variable susceptibility to infection of different perch genotypes. This last field in particular will benefit greatly from the development and widening availability of new genetic techniques, such as whole genome sequencing. The rapid development of new molecular techniques has revolutionized many fields of biology, including parasitology, where it promises to facilitate and support urgently needed taxonomic approaches, such as the identification of "new" parasites and pathogens or the description of cryptic species. Such methods may also help to clarify the origins of diseases and infer transmission routes of pathogens. Additionally, it may elucidate host ecology, as recently demonstrated by Criscione et al. (2006), who showed that the source of a population of steelhead trout could be evaluated more precisely using the mitochondrial genotypes of its trematode parasites (*Plagioporus shawi*) than those of the trout itself.

It is fitting to end this chapter by highlighting the importance of multidisciplinary approaches. The development of synergistic scientific projects combining interesting new and "classic" fields of biology including ecology, parasitology, immunology and molecular genetics, is likely to be key in driving forward understanding of parasite-host interactions in complex ecosystems.

8.7 Acknowledgements

We thank Patrice Couture and Gregory Pyle for inviting us to contribute to this volume. We thank students Daniela Harrer (PhD), Michael Donner (PhD), Samuel Roch (MS), Martina Knaur (MS), Marian Schmid (MS), Melody Reithmann (BS) and Dennis Rosskothen (BS) for supplying results from their graduate and undergraduate degree work in the Fish Ecology Group at the University of Konstanz and the Fisheries Research Station at Langenargen. Warm thanks to Amy-Jane Beer for reviewing the manuscript and for English correction. We thank three unknown reviewers for their fruitful contributions. JBG is thankful to Reiner Eckmann for his kind support. Funding came from the German science foundation (DFG) within the CRC 454 "littoral of Lake Constance" as well as from the Stiftung für Umwelt und Wohnen and the University of Konstanz.

8.8 References

Abbey, D.H. and W.C. Mackay. 1991. Predicting the Growth of Age-0 Yellow Perch Populations from Measures of Whole-Lake Productivity. Freshwater Biol. 26: 519–525.

Ahne, W., H.V. Bjorklund, S. Essbauer, N. Fijan, G. Kurath and J.R. Winton. 2002. Spring viremia of carp (SVC). Dis. Aquat. Org. 52: 261–272.

Alvarez-Pellitero, P. 2008. Fish immunity and parasite infections: from innate immunity to immunoprophylactic prospects. Vet. Immunol. Immunopathol. 126: 171–198.

Anderson, R.M. and D.M. Gordon. 1982. Processes influencing the distribution of parasite numbers within host populations with special emphasis on parasite-induced host mortalities. Parasitol. 85: 373–398.

Andrews, C. 1979. Host specificity of the parasite fauna of perch *Perca fluviatilis* L. from the British Isles, with special reference to a study at Llyn Tegid (Wales). J. Fish Biol. 15: 195–209.

Balbuena, J.A., E. Karlsbakk, A.M. Kvenseth, M. Saksvik and A. Nylund. 2000. Growth and emigration of the third-stage larvae of *Hysterothylacium aduncum* (nemtoda: anisakidae) in larval herring *Clupea harengus*. J. Parasitol. 86: 1271–1275.

Balling, T.E. and W. Pfeiffer. 1997. Frequency distribution of fish parasites in the perch (*Perca fluviatilis* L.) from Lake Constance. Parasitol. Res. 83: 370–373.

Barber, I., D. Hoare and J. Krause. 2000. Effects of parasites on fish behaviour: a review and evolutionary perspective. Rev. Fish Biol. Fish. 10: 131–165.

Barber, I. 2013. Sticklebacks as model hosts in ecological and evolutionary parasitology. Trends Parasitol. 29: 556–566.

Baudoin, M. 1975. Host castration as a parasitic stragtegy. Evolution 29: 335–352.

Behrmann-Godel, J. 2013. Parasite identification, succession and infection pathways in perch fry (*Perca fluviatilis*): new insights through a combined morphological and genetic approach. Parasitol. 140: 509–520.

Behrmann-Godel, J., S. Roch and A. Brinker. 2014. Gill worm *Ancyrocephalus percae* (Ergens 1966) outbreak negatively impacts the Eurasian perch *Perca fluviatilis* L. stock of Lake Constance, Germany. J. Fish Dis. 37: 925–930.

Bergman, E. and L.A. Greenberg. 1994. Competition between a planktivore, a benthivore, and a species with no ontogenetic diet schifts. Ecology 75: 1233–1245.

Binke, R. 2004. From muscle to meat. Fleischwirtschaft 84: 224–227.

Björklund, M., T. Aho and J. Behrmann-Godel. 2015. Isolation over 35 years in a heated biotest basin causes selection on MHC class IIß genes in the European perch (*Perca fluviatilis* L.). 5(7): 1440–1455.

Boisclair, D. and W.C. Leggett. 1989. Among-Population Variability of Fish Growth: II. Influence of Prey Type. Can. J. Fish. Aquat. Sci. 46: 468–482.

Brinker, A. and R. Hamers. 2007. Evidence for negative impact of plerocercoid infection of *Triaenophorus nodulosus* on *Perca fluviatilis* L. stock in Upper Lake Constance, a water body undergoing rapid reoligotrophication J. Fish Biol. 71: 129–147.

Britton, J.R., J. Pegg and C.F. Williams. 2011. Pathological and Ecological Host Consequences of Infection by an Introduced Fish Parasite. PLoS ONE 6(10): e26365.

Buchmann, K. and T. Lindenstrom. 2002. Interactions between monogenean parasites and their fish hosts. Int. J. Parasitol. 32: 309–319.

Buchmann, K. 2012. Fish immune responses against endoparasitic nematodes—experimental models. J. Fish Dis. 35: 623–635.

Bush, A.O., K.D. Lafferty, J.M. Lotz and A.W. Shostak. 1997. Parasitology meets ecology on its own terms: Margolis et al. revisited. J. Parasitol. 83: 575-583.

Bylund, G. and O.N. Pugachev. 1989. Monogenea of fish in Finland (Dactylogyridae, Ancyrocephalidae, Tetraonchidae). *In:* O.N. Bauer (ed.). Parasites of Freshwater Fishes of North-West Europe: Materials of the International Symposium within the Program of the Soviet Finnish Cooperation 10–14 January 1988: Institute of Biology & Zoological Institute, USSR Academy of Sciences.

Carney, J.P. and T.A. Dick. 1999. Enteric helminths of perch (*Perca fluviatilis* L.) and yellow perch (*Perca flavescens* Mitchill): stochastic or predictable assemblages? J. Parasitol. 85: 785–795.

Carney, J.P. and T.A. Dick. 2000a. Helminth communities of yellow perch (*Perca flavescens* (Mitchill)): determinants of pattern. Can. J. Zool. 78: 538–555.

Carney, J.P. and T.A. Dick. 2000b. The historical ecology of yellow perch (*Perca flavescens* [Mitchill]) and their parasites. J. Biogeogr. 27: 1337–1347.

Chubb, J.C. 1977. Seasonal occurrence of helminths in freshwater fishes. Part 1. Monogenea. Adv. Parasitol. 15: 133–199.

Cloutier, V.B., H. Glémet, B. Ferland-Raymond, A.D. Gendron and D.J. Marcogliese. 2012. Correlation of Parasites with Growth of Yellow Perch. J. Aquat. Anim. Health 24: 100–104.

Coles, T. 1981. The distribution of perch, *Perca fluviatilis* L. throughout their first year of life in Llyn Tegid, North Wales. Arch. Fisch. Wiss. 15: 193–204.

Collyer, M.L. and C.A. Stockwell. 2004. Experimental evidence for costs of parasitism for a threatened species, White Sands pupfish (*Cyprinodon tularosa*). J. Anim. Ecol. 73: 821–830.

Combes, C. 2001. Parasitism: The Ecology and Evolution of Intimate Interactions. University of Chicago Press, Chicago.

Cone, D.K. 1980. The Biology of *Urocleidus adspectus* (Monogenea) Parasitizing *Perca flavescens* PhD Thesis, The University of Brunswick, Canada.

Cone, D.K. and M.D.B. Burt. 1985. Population biology of *Urocleidus adspectus* Mueller, 1936 (Monogenea) on *Perca flavescens* in New Brunswick. Can. J. Zool. 63: 272–277.

Craig, J.F. 2000. Percid Fishes Systematics, Ecology and Exploitation. Blackwell Science, Osney Mead, Oxford.

Criscione, C.D., B. Cooper and M.S. Blouin. 2006. Parasite genotypes identify source populations of migratory fish more accurately than fish genotypes. Ecology 87(4): 823–828.

Crofton, H.D. 1971. A quantitative approach to parasitism. Parasitol. 62: 179–193.

Deufel, J. 1975. Der Wurmstar (Diplostomum-Krankheit) und die Schwarzfleckenkrankheit der Fische. Fisch und Umwelt, Schriftenreihe fuer Fischpathologie und Fischoekologie 1: 97–104.

Eckmann, R. and D. Schleuter. 2013. Der Flussbarsch. Die neue Brehmbücherei, Band 677, Westarp Wissenschaften-Verlagsgesellschaft mbH Hohenwarsleben, Hohenwarsleben.

Eizaguirre, C., T.L. Lenz, R.D. Sommerfeld, C. Harrod, M. Kalbe and M. Milinski. 2011. Parasite diversity, patterns of MHC II variation and olfactory based mate choice in diverging three-spined stickleback ecotypes. Evol. Ecol. 25: 605–622.

Eizaguirre, C., T.L. Lenz, A. Traulsen and M. Milinski. 2009. Speciation accelerated and stabilized by pleiotropic major histocompatibility complex immunogenes. Ecol. Lett. 12: 5–12.

Ewald, P. 1995. The evolution of virulence: a unifying link between parasitology and ecology. J. Parasitol. 8: 659–669.

Goater, T.M., C.P. Goater and E.W. Esch. 2014. Prasitism the Diversity and Ecology of Animal Parasites, 2nd Edition. Cambridge University Press.

Grutter, A., T. Cribb, H. McCallum, J. Pickering and M. McCormick. 2010. Effects of parasites on larval and juvenile stages of the coral reef fish *Pomacentrus moluccensis*. Coral Reefs 29: 31–40.

Halliday, M.M. 1976. The biology of *Myxosoma cerebralis*: the causative organism of whirling disease of salmonids. J. Fish Biol. 9: 339–357.

Hayward, R.S. and F.J. Margraf. 1987. Eutrophication effects on prey size and food available to Yellow Perch in Lake Erie. Trans. Am. Fish. Soc. 116: 210–223.

Heibo, E. and C. Magnhagen. 2005. Variation in age and size at maturity in perch (*Perca fluviatilis* L.), compared across lakes with different predation risk. Ecol. Fresh. Fish 14: 344–351.

Heibo, E., C. Magnhagen and L.A. Vollestad. 2005. Latitudinal variation in life-history traits in Eurasian perch. Ecology 86: 3377–3386.

Heins, D.C. and J.A. Baker. 2008. The stickleback–*Schistocephalus* host–parasite system as a model for understanding the effect of a macroparasite on host reproduction. Behaviour 145: 625–645.

Hjelm, J., L. Persson and B. Christensen. 2000. Growth, morphological variation and ontogenetic niche shifts in perch (*Perca fluviatilis*) in relation to resource availability. Oecologia 122: 190–199.

Hoffmann, R.W., J. Meder, M. Klein, K. Osterkornj and R.D. Negele. 1986. Studies on lesions caused by plerocercoids of *Triaenophorus nodulosus* in some fish of an alpine lake, the Königssee. J. Fish Biol. 28: 701–712.

Hoole, D., P. Bucke, P. Burgess and I. Wellby. 2011. Diseases of carp and other cyprinid fishes. Fishing News Books, Blackwell Science, Oxford.

Janeway, C.A., P. Travers, M. Walport and M.J. Shlomchik. 2005. Immunobiology: The Immune System in Health & Disease. Garland Science, London.

Johnson, M.W., P.A. Nelson and T.A. Dick. 2004. Structuring mechanisms of yellow perch (*Perca flavescens*) parasite communities: host age, diet, and local factors. Can. J. Zool. 82: 1291–1301.

Johnson, P.T.J., A. Dobson, K.D. Lafferty, D.J. Marcogliese, J. Memmott, S.A. Orlofske, R. Poulin and D.W. Thieltges. 2010. When parasites become prey: ecological and epidemiological significance of eating parasites. Trends Ecol. Evol. 25: 362–371.

Johnson, M.W. and T.A. Dick. 2001. Parasite effects on the survival, growth, and reproductive potential of yellow perch (*Perca flavescens* Mitchill) in Canadian Shield lakes. Can. J. Zool. 79: 1980–1992.

Johnson, T.B. and D.O. Evans. 1991. Behaviour, energetics, and associated mortality of Young-of-the-Year White Perch (*Morone americana*) and Yellow Perch (*Perca flavescens*) under Simulated Winter Conditions. Can. J. Fish. Aquat. Sci. 48: 672–680.

Jones, S.R.M. 2001. The occurrence and mechanisms of innate immunity against parasites in fish. Dev. Comp. Immunol. 25: 841–852.

Kalbe, M. and J. Kurtz. 2006. Local differences in immunocompetence reflect resistance of sticklebacks against the eye fluke *Diplostomum pseudospathaceum*. Parasitology 132: 105–16.

Kelly, D.W., H. Thomas, D.W. Thieltges, R. Poulin and D.M. Tompkins. 2010. Trematode infection causes malformations and population effects in a declining New Zealand fish. J. Anim. Ecol. 79: 445–452.

Kennedy, C.R. 1978. The biology, specificity and habitat of the species of *Eubothrium* (Cestoda: Pseudophyllidea), with reference to their use as biological tags: a review. J. Fish Biol. 12: 393–410.

Kennedy, C.R. 1984. The use of frequency distributions in an attempt to detect host mortality induced by infections of diplostomatid metacercariae. Parasitol. 89: 209–220.

Kennedy, C.R. 1990. Helminth communities in freshwater fish: structured communities or stochastic assemblages? pp. 131–156. *In:* G. Esch, A. Bush and J. Aho (eds.). Parasite Communities: Patterns and Processes. Springer, Netherlands.

King, S.D. and D.K. Cone. 2009. Infections of *Dactylogyrus pectenatus* (Monogenea: Dactylogyridae) on Larvae of *Pimephales promelas* (Teleostei: Cyprinidae) in Scott Lake, Ontario, Canada. Comp. Parasitol. 76: 110–112.

Khoo, L., F.A. Rommel, S.A. Smith, M.J. Griffin and L.M. Pote. 2010. *Myxobolus neurophilus*: morphologic, histopathologic and molecular characterization. Dis. Aquat. Organ. 89: 51–61.

Kuchta, R., M. Cech, T. Scholz, M. Soldánová, C. Levron and B. Skoríková. 2009. Endoparasites of European perch *Perca fluviatilis* fry: role of spatial segregation. Dis. Aquat. Organ. 86: 87–91.

Kuperman, B.I. 1973. Tapeworms of the genus *Triaenophorus,* Parasites of Fishes. Amerind Publishing Co. Pvt. Ltd., New Delhi.

Langdon, J.S. 1987. Spinal curvatures and an encephalotropic myxosporean, *Triangula percae* sp. nov. (Myxozoa: Ortholineidae), enzootic in redfin perch, *Perca fluviatilis* L., in Australia. J. Fish Dis. 10: 425–434.

Lafferty, K.D., A.P. Dobson and A.M. Kuris. 2006. Parasites dominate food web links. PNAS 103: 11211–11216.

Lafferty, K.D., S. Allesina, M. Arim, C.J. Briggs, G. De Leo, A.P. Dobson, J.A. Dunne, P.T.J. Johnson, A.M. Kuris, D.J. Marcogliese, N.D. Martinez, J. Memmott, P.A. Marquet, J.P. McLaughlin, E.A. Mordecai, M. Pascual, R. Poulin and D.W. Thieltges. 2008. Parasites in food webs: the ultimate missing links. Ecol. Lett. 11: 533–546.

Larsen, A.H., J. Bresciani and K. Buchmann. 2005. Pathogenicity of *Diplostomum* cercariae in rainbow trout, and alternative measures to prevent diplostomosis in fish farms. Bull. Eur. Ass. Fish Pathol. 25: 20–27.

Lenz, T.L., C. Eizaguirre, B. Rotter, M. Kalbe and M. Milinski. 2013. Exploring local immunological adaptation of two stickleback ecotypes by experimental infection and transcriptome-wide digital gene expression analysis. Mol. Ecol. 22: 774–786.

Lester, R.J.G. 1984. A review of methods for estimating mortality due to parasites in wild fish populations. Helgoländer Meeresuntersuchungen 37: 53–64.

Locke, S.A., J.D. McLaughlin and D.J. Marcogliese. 2010a. DNA barcodes show cryptic diversity and a potential physiological basis for host specificity among Diplostomoidea (Platyhelminthes: Digenea) parasitizing freshwater fishes in the St. Lawrence River, Canada. Mol. Ecol. 19: 2813–2827.

Locke, S.A., J.D. McLaughlin, S. Dayanandan and D.J. Marcogliese. 2010b. Diversity and specificity in *Diplostomum* spp. metacercariae in freshwater fishes revealed by cytochrome c oxidase I and internal transcribed spacer sequences. Int. J. Parasitol. 40: 333–343.

Lom, J., A.W. Pike and I. Dyková. 1991. *Myxobolus sandrae* Reuss, 1006, the agent of vertebral column deformities of perch *Perca fluviatilis* in northeast Scottland. Dis. Aquat. Org. 12: 49–53.

Lyholt, H.C.K. and K. Buchmann. 1996. *Diplostomum spathaceum*: Effects of temperature and light on cercarial shedding and infection of rainbow trout. Dis. Aquat. Org. 25: 169–173.

Magnhagen, C. and E. Heibo. 2004. Growth in length and in body depth in young-of-the-year perch with different predation risk. J. Fish Biol. 64: 612–624.

Majoros, G. 1999. Mortality of fish fry as a result of specific and aspecific cercarial invasion under experimental conditions. Acta Vet. Hung. 47: 433–450.

Marcogliese, D.J. 2001. Implications of climate change for parasitism of animals in the aquatic environment. Can. J. Zool. 79: 1331–1351.

Marcogliese, D.J., L.G. Brambilla, F. Gagnè and A.D. Gendron. 2005. Joint effects of parasitism and pollution on oxidative stress biomarkers in yellow perch *Perca flavescens*. Dis. Aquat. Org. 63: 77–84.

Marcogliese, D.J., C. Dautremepuits, A.D. Gendron and M. Fournier. 2010. Interactions between parasites and pollutants in yellow perch (*Perca flavescens*) in the St. Lawrence River, Canada: implications for resistance and tolerance to parasites. Can. J. Zool. 88: 247–258.

Marcogliese, D.J. and M. Pietrock. 2011. Combined effects of parasites and contaminants on animal health: parasites do matter. Trends Parasitol. 27: 123–130.

Marcogliese, D.J. and D.K. Cone. 1997. Food webs: A plea for parasites. 1997. Trends Ecol. Evol. 12(10): 394–394.

Markevich, A.P. 1963. Parasite fauna of freshwater fish of the Ukrainian S.S.R. Israel Program for Scientific Translations.

Molnár, K. 1974. On diplostomosis of grasscarp fry. Acta Vet. Adac. Sci. Hung. 24: 63–71.

Molzen, B.U. 2006. Die Auswirkung des Befalls mit Plerocercoiden des Hechtbandwurms (*Triaenophorus nodulosus* (P.)) auf den Flussbarsch (*Perca fluviatilis* L.) im Bodensee-Obersee, PhD Thesis, Ludwig-Mximilians-Universität München, Germany.

Moore, J. 2002. Parasites and the Behavior of Animals. Oxford University Press, New York.

Moravec, F. 1994. Parasitic Nematodes of Freshwater Fishes of Europe. Kluwer Academic Publishers, Prag.

Morozinska-Gogol, J. 2008. A check-list of parasites of percid fishes (Actinopterygii: Percidae) from the estuaries of the Polish coastal zone. Helminthologia 45: 196–203.

Michel, C., L. Bernatchez and J. Behrmann-Godel. 2009. Diversity and evolution of MHII beta genes in a non-model percid species—the Eurasian perch (*Perca fluviatilis* L.). Mol. Immunol. 46: 3399–3410.

Nachev, M., G. Schertzinger and B. Sures. 2013. Comparison of the metal accumulation capacity between the acanthocephalan *Pomphorhynchus laevis* and larval nematodes of the genus *Eustrongylides* sp. infecting barbel (*Barbus barbus*). Parasite. Vector. 6: 21.

Nendick, L., M. Sackville, S. Tang, C.J. Brauner and A.P. Farrell. 2011. Sea lice infection of juvenile pink salmon (*Oncorhynchus gorbuscha*): effects on swimming performance and postexercise ion balance. Can. J. Fish. Aquat. Sci. 68: 241–249.

Oppelt, C. and J. Behrmann-Godel. 2012. Genotyping MHC classIIB in non-model species by reference strand-mediated conformational analysis (RSCA). Con. Gen. Res. 4: 841–844.

Pacala, S.W. and A.P. Dobson. 1988. The relation between the number of parasites/host and host age: population dynamic causes and maximum likelihood estimation. Parasitol. 96: 197–210.

Paperna, I. 1991. Diseases caused by parasites in the aquaculture of warm water fish. Ann. Rev. Fish Dis. 1: 155–194.

Paull, S.H. and P.T.J. Johnson. 2011. High temperature enhances host pathology in a snail–trematode system: possible consequences of climate change for the emergence of disease. Freshwater Biol. 56: 767–778.

Poulin, R. 1998. Evolutionary Ecology of Parasites. Chapman & Hall, London.

Poulin, R. 2006. Global warming and temperature-mediated increases in cercarial emergence in trematode parasites. Parasitol. 132: 143–151.

Poulin, R. and S. Morand. 2004. Parasite Biodiversity. Smithsonian Inst. Press, USA.

Poulin, R. and E.T. Valtonen. 2002. The predictability of helminth community structure in space: a comparison of fish populations from adjacent lakes. Int. J. Parasitol. 32: 1235.

Pracheil, B.M. and P.M. Muzzall. 2009. Chronology and development of juvenile bluegill parasite communities. J. Parasitol. 95: 838–845.

Raeymaekers, J.A.M., K.M. Wegner, T. Huyse and F.A.M. Volckaert. 2011. Infection dynamics of the monogenean parasite *Gyrodactylus gasterostei* on sympatric and allopatric populations of the three-spined stickleback *Gasterosteus aculeatus*. Folia Parasit. 58(1): 27–34.

Read, C.P. and J.E. Simmons. 1963. Biochemistry and physiology of tapeworms. Physiol. Rev. 43: 263–305.

Roch, S., J. Behrmann-Godel and A. Brinker. 2015. Genetically distinct colour morphs of European perch *Perca fluviatilis* in Lake Constance differ in susceptibility to macroparasites. J. Fish Biol. 86: 854–863.

Rosen, F. 1918. Recerves sur le dévelopment des Cestodes: 1. Le cycle évolutif es *Bothricéphales*. Bulletin de la Societe Neuchateloise des Sciences Naturelles 1–47.

Schäperclaus, W. 1979. Fischkrankheiten. Akademie-Verlag, Berlin.

Scheuring, L. 1922. Studien an Fischparasiten; *Triaenophorus nodulosus* (Pallas) Rud. und die durch ihn im Fischkörper hervorgerufenen pathologischen Veränderungen. Zeitschrift für Fischerei und deren Hilfswissenschaften XXI: 93–204.

Schmid, M. 2014. Impacts of an increased water temperature on the parasite infestation of the Eurasian perch (*Perca fluviatilis* L.) M.S. Thesis, University of Konstanz, Germany.

Scholz, T., V. Hanzelova, A. Skerikova, T. Shimazu and L. Rolbiecki. 2007. An annotated list of species of the *Proteocephalus* Weinland, 1858 aggregate sensu de Chambrier et al. (2004) (Cestoda: Proteocephalidea), parasites of fishes in the Palaearctic Region, their phylogenetic relationships and a key to their identification. Syst. Parasitol. 67: 139–156.

Seppälä, O., A. Karvonen and E.T. Valtonen. 2011. Eye fluke-induced cataracts in natural fish populations: is there potential for host manipulation? Parasitol. 138: 209–214.

Shariff, M., R.H. Richards and C. Sommerville. 1980. The histopathology of acute and chronic infections of rainbow-trout *Salmo gairdneri* Richardson with eye flukes, *Diplostomum* spp. J. Fish Dis. 3: 455–465.

Sindermann, C.J. 1986. Effects of parasites on fish populations: practical considerations. Int. J. Parasitol. 17: 371–382.

Skovgaard, A., Q.Z.M. Bahlool, P. Munk, T. Berge and K. Buchmann. 2009a. Infection of North Sea cod, *Gadus morhua* L., larvae with the parasitic nematode *Hysterothylacium aduncum* Rudolphi. J. Plankton Res. 33: 1311–1316.

Skovgaard, A., I. Meneses and M.M. Angélico. 2009b. Identifying the lethal fish egg parasite *Ichthyodinium chabelardi* as a member of Marine Alveolate Group I. Environ. Microbiol. 11: 2030–2041.

Sonnenberg, R., A.W. Nolte and D. Tautz. 2007. An evaluation of LSU rDNA D1-D2 sequences for their use in species identification. Front. Zool. 4: 6.

Stich, H.B. and A. Brinker. 2010. Oligotrophication outweighs effects of global warming in a large, deep, stratified lake ecosystem. Global Change Biol. 16: 877–888.

Straile, D., R. Adrian and D.E. Schindler. 2012. Uniform Temperature Dependency in the Phenology of a Keystone Herbivore in Lakes of the Northern Hemisphere. PLoS ONE 7: e45497.

Sures, B. 2008. Host-parasite interactions in polluted environments. J. Fish Biol. 73: 2133–2142.

Sures, B. 2003. Accumulation of heavy metals by intestinal helminths in fish: an overview and perspective. Parasitol. 126(07): S53–S60.

Sures, B. 2001. The use of fish parasites as bioindicators of heavy metals in aquatic ecosystems: a review. Aquat. Ecol. 35(2): 245–255.

Sures, B., W. Steiner, M. Rydlo and H. Taraschewski. 1999. Concentrations of 17 elements in the zebra mussel (*Dreissena polymorpha*), in different tissues of perch (*Perca fluviatilis*), and in perch intestinal parasites (*Acanthocephalus lucii*) from the subalpine lake Mondsee, Austria. Environ. Toxicol. Chem. 18: 2574–2579.

Szalai, A.J. and T.A. Dick. 1991. Role of predation and parasitism in growth and mortality of Yellow Perch in Dauphin Lake, Manitoba. Trans. Am. Fish. Soc. 120: 739–751.

Szalai, A.J., W. Lysack and T.A. Dick. 1992. Use of confidence ellipses to detect effects of parasites on the growth of Yellow Perch, *Perca flavescens*. J. Parasitol. 78: 64–69.

Thomas, P.M., N.W. Pankhurst and H.A. Bremner. 1999. The effect of stress and exercise on post-mortem biochemistry of Atlantic salmon and rainbow trout. J. Fish Biol. 54: 1177–1196.

Thoney, D.A. and W.J. Hargis, Jr. 1991. Monogenea (Platyhelminthes) as hazards for fish in confinement. Ann. Rev. Fish Dis. 1: 133–153.

Unger, J., A. Brinker and H.B. Stich. 2008. Transmission of neozoic *Anguillicoloides crassus* and established *Camallanus lacustris* in ruffe *Gymnocephalus cernuus*. J. Fish Biol. 73: 2261–2273.

Vidal-Martínez, V.M., D. Pech, B. Sures, S.T. Purucker and R. Poulin. 2010. Can parasites really reveal environmental impact? Trends Parasitol. 26: 44.

Wang, N. and A. Appenzeller. 1998. Abundance, depth distribution, diet composition and growth of perch (*Perca fluviatilis*) and burbot (*Lota lota*) larvae and juveniles in the pelagic zone of Lake Constance. Ecol. Freshwater Fish 7: 176–183.

Wang, N. and R. Eckmann. 1994. Distribution of perch (*Perca fluviatilis* L.) during their first year of life in Lake Constance. Hydrobiologia 277: 135–143.

Woo, P.T.K. 1992. Immunological responses of fish to parasitic organisms. Annu. Rev. Fish Dis. 2: 339–366.

Wootten, R. 1974. Studies on the life history and development of *Proteocephalus percae* (Müller) (Cestoda: Proteocephalidea). J. Helmithol. 48: 269–281.

Yokoyama, H., D. Grabner and S. Shirakashi. 2012. Transmission biology of the Myxozoa. *In*: E.D. Carvalho, G.S. David and R.J. Silva (eds.). Health and Environment in Aquaculture. InTech, Rijeka, Croatia. DOI:10.5772/29571.

Zander, C.D. 1998. Parasit-Wirt-Beziehungen. Springer Verlag, Berlin, Heidelberg, New York.

Zehnder, M.P. amd J. Mariaux. 1999. Molecular systematic analysis of the order Proteocephalidea (Eucestoda) based on mitochondrial and nuclear rDNA sequences. Int. J. Parasitol. 29: 1841–1852.

9

Behaviour of Perch

C.A.D. Semeniuk,[1,*] C. Magnhagen[2] and G. Pyle[3]

ABSTRACT

Perca spp. are an ecologically significant component of many North American and European freshwater food webs. Their success can be attributed to the array of diverse behaviours they exhibit that make them well adapted to the lakes, ponds, creeks and rivers—both native and non-native—they inhabit. The differences in behaviour observed among conspecific populations reflect interacting drivers, both biotic and abiotic, that shape the adaptive behaviours of perch: genetic differentiation due to selection, the competitive environment, predation pressure, life history change, and habitat complexity. Resultantly, perch habitat-selection movements, competitive abilities, antipredator strategies, predation impacts, social behaviours and behavioural phenotypes can drive, in turn, ecosystem-level dynamics. This chapter explores the behaviour of perch and their impacts, and urges the continuation of research into the behavioural responses of perch to human-induced, rapid environmental change since such discoveries will advance both fundamental research, in addition to enhancing its applied value.

Keywords: Habitat selection, movement ecology, intracohort competition, interspecific competition, predator avoidance, predation-sensitive foraging, predator-mediated trophic cascades, individual specialization, personality

[1] Great Lakes Institute for Environmental Research, University of Windsor, Ontario, Canada N9B 3P4.
E-mail: semeniuk@uwindsor.ca

[2] Department of Wildlife, Fish, and Environmental Studies, Swedish University of Agricultural Sciences, Umeå, Sweden 901 83.
E-mail: Carin.Magnhagen@slu.se

[3] Dept. of Biological Sciences, 4401 University Drive, University of Lethbridge, Lethbridge, Alberta, Canada T1K 3M4.
E-mail: gregory.pyle@uleth.ca

* Corresponding author

9.1 Introduction

Fish are by far the vertebrate group with the most strikingly diverse repertoire of behaviours and behavioural adaptations. As such, their study provides a fascinating insight into the complexities of animal behaviour. Percids are no exception, seemingly rife with contradictions in their behaviours. From active dispersers to spawning-site homers, cannibals to prey, strong competitors as consumers to weaker ones as top predators, facultative shoalers to solitary predators, and from risk-averse to risk-prone phenotypes, perch have a varied collection of behaviours that enable them to facilitate their occupancy and adaptation to a myriad of habitats across a wide geographical range. Their behaviours furthermore influence the population dynamics of their predators, prey, and competitors; and equally, the ecological context in which perch are found shapes their behavioural repertoire. In terms of species differentiation, European (*Perca fluviatilis*) and yellow (*P. flavescens*) perch are overwhelmingly similar. The same abiotic and biotic effects limit distributions and growth capacities of both species, and they occupy comparable trophic positions. As a result, they display a similar suite of parallel behaviours, and can be considered behaviourally equivalent. The following chapter reviews both European and yellow perch behaviour in terms of their habitat selection movements, competitive abilities, antipredator strategies and conversely, their impacts as predators, and lastly, their social behaviours. Although presented individually in sub-sections within the chapter, each type of behaviour cannot be considered without acknowledging the interacting drivers, both biotic and abiotic, that shape the adaptive behaviours of perch in their attempts to maximize lifetime fitness. The unifying theme throughout this chapter, therefore, is that the adaptiveness of each element of perch behaviour must be considered in the context of the genetic predisposition, and how individuals make tradeoffs within and between the competitive environment, predation pressure, life history changes, and habitat suitability. Each section of this chapter therefore discusses specific behaviours of perch within this framework. The chapter ends by demonstrating how knowledge of perch behaviour can contribute to how individuals will fare in their responses to cumulative effects of multiple stressors in today's changing environments.

9.2 Habitat Selection and Movement

The act of occupying a habitat involves a process of behavioural responses that can result in the disproportionate use of habitats that influence survival and fitness of individuals. Selected habitats are characterized by many variables such as food availability, predator occurrence, ease of defense, likelihood of offspring survival, and microclimate changes. The types of habitats animals use will also vary depending on life-history stage and/or age. In recent years, several studies have gone beyond describing selection in terms of chosen features, and have focused on behavioural processes involved in habitat selection, such as the role of familiarity with a location and/or individuals, and the organism's movement ecology.

9.2.1 Spawning-group Fidelity—Influence of Site and Kin

In general, perch aggregate to spawn on shallow reef complexes or in slow-moving tributaries (Craig 2000). Females drape egg masses on submerged macrophytes or rock, which are then fertilized by 2–5 males that have moved into the nest area prior to the arrival of females (Robillard and Marsden 2001). While perch neither display alternative reproductive strategies nor provide parental care, there is evidence suggesting that they display spawning group fidelity that results in fine-scale population substructures. For instance, individual perch from different parts of a lake or tributary do not consistently form a single panmictic population, but rather group into genetically distinct local subpopulations (Gerlach et al. 2001; see Chapter 2, this volume).

On an even finer scale, many yellow perch spawning groups within a given body of water are genetically divergent despite the apparent potential for dispersal and gene flow among them (Sepulveda-Villet and Stepien 2012). One possible cause of genetic isolation of local populations is philopatric behaviour—i.e., the tendency of animals to remain near a particular area. Interestingly, perch display distinct seasonal patterns in terms of their spatial distribution and movement rates between littoral and pelagic habitats from summer to winter; consequently, simple proximity to spawning sites is not an explanatory factor in spawning habitat selection. Aalto and Newsome (1990) found that yellow perch egg mass removals from spawning sites led to fewer fish returning to their natal spawning grounds in subsequent years than at control sites. Both MacGregor and Witzel (1987) and Glover et al. (2008) reported that yellow perch captured and tagged during spawning season frequently returned to the same site—in the former study after being released many kilometres away, and in the latter study, with 35%–80% of recaptured individuals returning to their specific marking site. These studies suggest that yellow perch are capable of homing behaviours and actively return to the same spawning populations year after year.

Population structuring is not solely a consequence of conserved homing behaviours to spawning sites; both yellow (Leung et al. 2011) and European perch (Gerlach et al. 2001; Table 1) are known to form stable kin-related groups. In certain populations, genetically similar individuals tend to aggregate with one another and in close proximity to other aggregations of genetically similar individuals (Bergek et al. 2010). These related perch stay together throughout their early development and later in life. Notably, reproductive success is significantly lower in non-kin groups (i.e., hybrids from two cross-bred populations), reducing pre-zygotic and post-zygotic fitness (Kocovsky et al. 2013), whereas kin-groups exhibit higher fertilization rates and higher hatching success. Causes for this outbreeding depression include incompatible genotypes in hybrid offspring, contributing to the disruptive divergence of populations (Behrmann-Godel and Gerlach 2008). It is this assortative mating that underlies the observed genetic relatedness among individuals in mating groups. In all likelihood, the affinity for spawning *groups* found in perch (i.e., the genetic isolation of local populations) is an outcome of either homing to natal sites to spawn, remaining in genetically and morphologically differentiated groups throughout their lives, or both (Kocovsky et al. 2013). An outcome of this population structuring is divergent yet sympatric groups of perch that can be detected in relatively small lakes both in Europe and in North America (Behrmann-Godel et al. 2006; Sullivan 2013).

Table 1. Kin structure for each subpopulation of *P. fluviatilis* in Lake Constance, Switzerland. To evaluate the number of shared alleles of full-sibs, 1000 pairs of individuals were randomly chosen, and two offspring each were generated and the number of matches of alleles between them was calculated, i.e., the average number of matches per locus between individuals *x* and *y*: M*xy*. Using a logistic regression analysis the probabilities for each M*xy* to be related or not was statistically analyzed. Data from Gerlach et al. (2001).

Local populations	1 OG	2 LA	3 RO	4 KR	5 GN	6 US
No. of full-sibs	65	88	40	32	84	49
M*xy* *P*-value	<0.001	0.036	0.19	0.44	< 0.001	0.043
	***	*	NS	NS	***	*
No. full- + half-sibs	97	126	76	57	129	92
M*xy* *P*-value	0.02	0.001	NS	NS	< 0.001	0.05
	*	**			***	*

*P < 0.05; **P < 0.01; ***P < 0.001; NS = not significant.
1 OG = Obere Güll; 2 LA = Langenargen; 3 RO = Romanshorn; 4 KR = Kreuzlingen; 5 GN = Gnadensee; 6 US = Untersee.

In addition to homing to natal sites and fidelity to kin groups, there may exist cryptic barriers to population isolation as well (e.g., distinct physico-chemical water masses, Leclerc et al. 2008). While genetic differentiation between local populations of fish typically occurs through mechanisms of geographic reproductive barriers and/or of the existence of kin groups, sympatric speciation in multiple subpopulations of European perch has been found in either small (24 km^2) homogeneous lakes in central Sweden (Bergek and Björklund 2007), and along the coast of the Baltic Sea (Bergek et al. 2010). These authors suggest that the small-scale genetic differentiation amongst subpopulations can be caused by open water that is situated between littoral zones of the lake in the former case; and in the latter study, by environmental conditions such as water temperature differences among habitats during spring spawning, with each constituting important barriers to the dispersal of perch. Nevertheless, all three mechanisms (homing, kin groups, and cryptic population isolation) provide an intriguing challenge to the management of subpopulations (i.e., stocks) of perch (Sepulveda-Villet and Stepien 2011; Kocovsky et al. 2013; Fig. 1).

9.2.2 Habitat Selection and Movement of Larvae

In aquatic organisms such as fish, larval dispersal is crucial to determine, among other things, the chances of individuals to settle in optimal habitats, the level of connectivity among populations, and/or the degree of stability of local communities (Sale 2004). Dispersal is an active or passive transport process between two sites and includes three distinct phases (departure–transport–settlement; Bennetts et al. 2001), and has both benefits (e.g., first access to prime habitat) and costs (e.g., high mortality rate) to the population. Fish larvae therefore depart their natal sites to find rich, relatively safe nursery habitats or environments, and resultantly settle in areas with optimal temperature, current speeds and food availability that minimize competition and cannibalism (Humphries 2005).

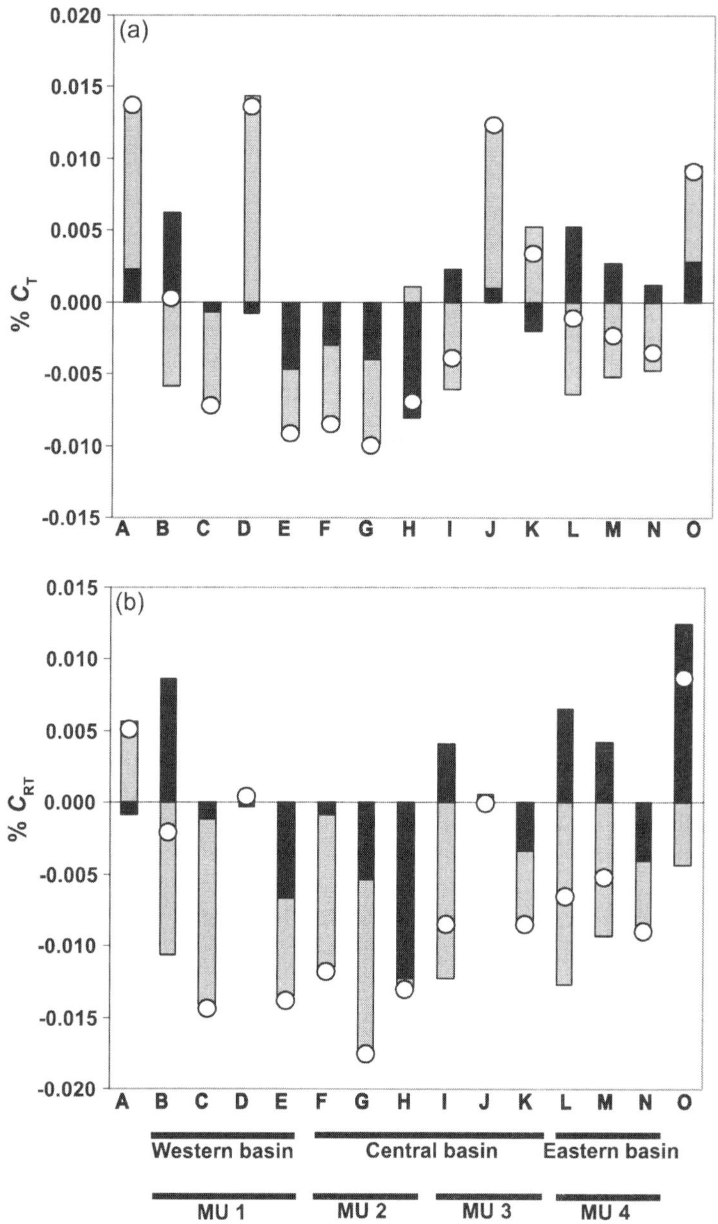

Fig. 1. Contribution of Lake Erie *P. flavescens* samples to total genetic diversity (C_T) and allelic richness (C_{RT}). Each sample's contribution to total diversity or allelic richness (circles) is divided into a diversity component (C_S; black bars) and a differentiation component (C_D; grey bars). Bars with positive values indicate that the spawning sample significantly contributed to overall lake-wide diversity or significantly diverged from other groups. Results show some spawning groups contribute more to overall lake-wide genetic diversity than do others. Partitioning of the yellow perch's genetic structure shows little congruence to lake basins or to current management units (MU) (Sepulveda-Villet and Stepien 2011).

Once fry emerge in late spring, both yellow and European perch show a clear larval pelagic phase with a high potential to drift offshore over long distances (Dettmers et al. 2005). In larger lakes driven by marine-like hydrodynamics (like the Laurentian Great Lakes), current patterns that vary spatially and temporally have a large influence on the timing of larvae settlement. Indeed, yellow perch larvae have been found more than 50 km offshore in Lake Michigan (Miehls and Dettmers 2011). In smaller lakes with an absence of strong currents, non-directional settlement may occur, and this limited dispersal might additionally contribute to the genetic similarity found within localized areas.

While passive dispersal may seem a plausible mechanism in the habitat-selection process, perch larvae may play a more active role during the pelagic phase of their development. In both marine and freshwater habitats, the transport of larvae can in fact combine passive elements related to abiotic factors such as currents, discharge, windforcing or temperature, with a suite of active behavioural reactions to environmental cues (i.e., olfaction, vision, vertical and horizontal movements, phototaxis, habitat choice and orientation; Lechner et al. 2013a).

In particular, active drift exhibited by fish is linked to behavioural 'choices' made to satisfy particular physiological or life-history needs (Lechner et al. 2013b). For example, in a small lake on the outer archipelago of the Baltic Sea, European perch actively and immediately departed for the pelagic area after hatching without waiting to fill their swim bladder first, presumably to reduce predation by invertebrates (Urho 1996). This behaviour is in contrast to the more typical phenomenon that can be found in a sympatric species, the white roach (*Rutilus rutilus*), that hatches at the same time as perch, shares similar development rates, but remains in the littoral zone until the swim bladder fills to reduce the energetic costs of swimming (Urho 1996). In a recent study by Bertolo et al. (2012), the authors explicitly modelled the spatial patterns of yellow perch larvae with directional (e.g., due to water currents) and non-directional (e.g., due to active aggregation) habitat characteristics in a large shallow fluvial lake of the St. Lawrence River. Their analyses revealed that yellow perch showed a strong habitat association with aquatic vegetation at unexpectedly small larval sizes (12 mm), independent of spatial structure. These results suggest that habitat selection is a major factor determining yellow perch larvae distributions at least two weeks after hatching. The motivations behind this active habitat-selection behavior are still unresolved; percids, however, appear to be adapted to variable environments and, therefore, the habitat shift between littoral to pelagic zones can be driven by various drivers such as genetic predisposition (Post and McQueen 1988), predation, energetic-driven ontogenetic diet change, or a combination (Urho 1996).

9.2.3 Movement Patterns of Young and Adult Perch

The movement ecology of animals results from the dynamic interplay of four basic components, of which three are attributes of the animal itself, the internal state of the organism, its motion capacity (i.e., the ability and choice to move on its own accord), and its navigation capacity (the ability to orient and navigate, including the implied use of memory or inherited capacity). The fourth component is the external environment. The external environment can modify animal movements because of landscape attributes, the distribution of resources, differing environmental conditions, and the presence of other organisms (Holyoak et al. 2008; Revilla and Wiegand 2008).

As adults, perch movement patterns are affected by their prey and by physicochemical factors such as water temperature, light intensity, and dissolved oxygen (Radabaugh et al. 2010). However, as juveniles, perch movement behaviours are equally as complex and more diverse. After a short staging period in the offshore pelagic zone of lakes, young-of-the-year perch return to the littoral zone and settle in habitats that optimize the tradeoff between the risky activity of food searching and the requirement for protection against predation (Brosse and Lek 2002; Miehls and Dettmers 2011). As perch change habitats during their first year, they undergo an ontogenetic niche shift from planktonic to benthic prey (e.g., Chironomidae larvae; Roswell et al. 2013; Fig. 2). By the time they reach age two, perch will have switched predominantly to piscivory, and shoals of perch will converge to exploit the limnetic zone (open water) habitats if available (Parker et al. 2009), exhibiting routine homing behaviours to and from the safer littoral zone at night (Zamora and Moreno-Amich 2002).

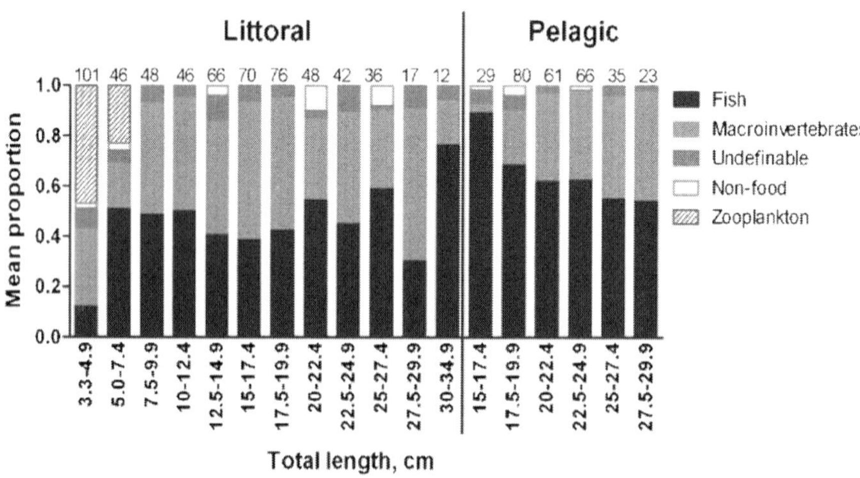

Fig. 2. Mean proportions of the different food categories in the stomach contents of *P. fluviatilis* for different length categories in littoral and pelagic zones. The numbers above the bars indicate the number of fish in the respective length category. Data from Mustamäki et al. (2014).

Simple versus complex habitat structures also influence young perch movement rates, with fish moving greater distances in simple vs. vegetated habitats to find preferred habitats such as refuge areas, or the more uniform distribution of prey items. Consequently, habitat structure can impose a potential cost to perch with increasing exposure to predation and a decrease in foraging efficiency as vegetation density diminishes (Diehl 1993; Bauer et al. 2009). Perch movement ecology is further affected by prey resource availability in the absence of habitat features. In an aquarium experiment with European perch, Olsson and colleagues (2007) studied the plastic activity responses of perch exposed to different resource levels and simulated habitat types. They found that perch exposed to low-resource treatments had higher activity compared to fish exposed to higher resources irrespective of habitat type; and that these highly active fish consequently had lower growth rates. These behavioural changes suggest that perch are trading off high metabolic costs of swimming for increased prey-encounter opportunities. In sum, perch display a varied range of habitat/resource selection behaviours—from spawning-group affinity to larval habitat-selection, to subsequent movement ecology that all reflect, in turn, the complex interactions between the environment and resource acquisition.

9.3 Competition

Competition is one of the most important mechanisms responsible for density-dependent population dynamics and population regulation (Begon et al. 1996). The ways in which individual *Perca* spp. compete with members of their own cohort (intracohort competition), their own species (intercohort competition), and with co-occurring species (interspecific competition) both as prey and predators have been intensely studied, with impacts on the wider ecological community just now beginning to be realized. These studies on perch competition have been carried out both under field and laboratory settings, using both observational and manipulative techniques, and have been instrumental in helping to elucidate the unique life-history traits of perch. The following section deals exclusively with the mechanisms and consequences of competition over food resources.

9.3.1 Intracohort Competition

In many lakes perch, a food generalist, undergoes ontogenetic diet niche shifts. Diets and diet shifts of young percids are variable across systems and among years (Roswell et al. 2013); and environmental conditions and prey availability can strongly influence perch diets and ontogenetic transitions, since perch need to attain a suitable size at which they can effectively use fish as prey (Lott et al. 1996). Perch feed on zooplankton after hatching, and by the end of their first year of life they have shifted to macroinvertebrate prey. They then generally switch to a piscivorous diet by two years of age or at a total length of approximately 150 mm. These functional niches that develop during ontogeny set constraints on the morphology and growth rates of perch (Urbatzka et al. 2008). However, this feeding and growth pattern is not strictly maintained, and in recent years researchers have begun to note a significant shift towards bimodality in size distributions taking place within the 0+ (young-of-year, YOY) age cohort. This contrasting morphology of large and small fish sizes is typically accompanied by contrasting trophic position (and hence diet), limnology, and behaviour. Rapid diet shifts during early ontogeny have been traced either using stomach-contents or with stable-isotope analyses and reveal intraindividual variation in within-cohort diets with splits shown between zooplankton and benthos (Roswell et al. 2013) or planktivory and piscivory (Borcherding 2006). In the latter situation, the differential growth cohorts develop a morphological polymorphism with a deep-bodied piscivorous dietary strategy and a slender planktivorous morph to accompany their diet shift (Beeck 2003); and piscivorous YOY exhibit faster growth rates as well (Borcherding et al. 2010). There are a multitude of suggested mechanistic hypotheses directly and indirectly related to competition to explain the causes of intra-cohort variation in perch size distribution:

1. Interference competition
2. Niche partitioning
3. Individual foraging behaviour and specialization
4. Phenology and cannibalism
5. Patchy prey distribution
6. Habitat complexity

Competition between individuals can manifest itself in two ways: interference, in which competition occurs directly between individuals via aggression when an individual interferes with the foraging attempts of others; or via exploitation, where resource depletion by one individual depletes the amount available to others. Post et al. (1997) found cohort splitting in a wild population of YOY yellow perch, and consequently assessed the effects of ontogeny and density-dependency on spatial distribution, growth, and diet. They noted that the high-density cohort appeared to split into a faster growing littoral component and a slower growing pelagic component that persisted into the following spring (Fig. 3). The differences in spatial behaviour between the two cohorts could not be explained by spatial patterns in prey availability, and high overwinter mortality rate was recorded for the high-density

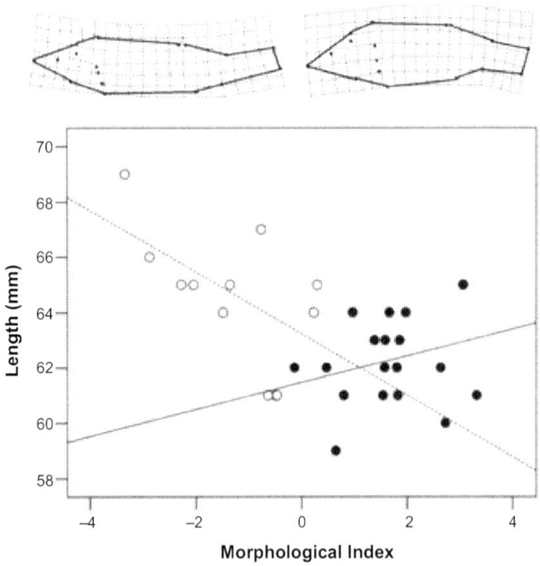

Fig. 3. The length of the juvenile perch at year 1 in relation to the morphological index from the discriminant function analysis. Filled symbols and solid lines represent perch from the littoral zone whereas open symbols and dashed lines represent perch from the pelagic zone. Visualization of the morphological scores above the graph represents artificial individuals of scores –4 (left) and 4 (right) to make visualization easier. Data from Svanbäck and Eklöv (2013).

cohort as well. The authors ascertained that the distribution and growth data support the occurrence of density-dependent interference competition in which competitively superior individuals can monopolize optimal habitats, relegating other individuals to suboptimal ones. Subsequent differences in growth rates were thereby increased by intense competition for food resources, resulting in growth delays among the weaker competitors forced into in the pelagic zone, while the fitter individuals remaining in the littoral zone were only slightly influenced by their competitors.

In contrast, Staffan and colleagues (2002) experimentally manipulated both the resource amount and ranking of competing YOY European perch and noted that the variation of individual food intake rate was large in small groups and unrelated to body size, suggesting asymmetrical foraging success, but this disparity was reduced when fish assigned the same feeding rank were placed together. Because no aggressive interactions were detected during feeding or defense of the resource, the authors postulated that variation in food intake rates might be explained by increased foraging efficiency of specific individuals via differences in morphology, metabolism and/or non-aggressive behaviours. Further, a long-term study showed that individual growth was correlated over several months, indicating a consistent difference in food intake among the marked perch in the tanks (Staffan et al. 2005).

In the previous study of Staffan et al. (2005), YOY size divergence increased from a coefficient of variation (CV) of 24% in January to 52% in August. In contrast to the laboratory set-up, perch may be able to avoid direct competition in the wild by diverging in their foraging strategies through niche partitioning. In this instance, alternative use of food resources can be caused by inherent individual variation and not through interference competition between members of the same cohort. In two similar stocking studies, Urbatzka et al. (2008) and Heermann et al. (2013) both observed intra-cohort variation in the growth of European YOY perch, with a broad, bimodal size distribution that ranged 136 mm and 110 mm between the smallest and largest individuals in these two studies, respectively. This size divergence was caused by the alternative use of food resources: some individuals switched to piscivory and cannibalism resulting in increased growth rates; and the small-sized YOY perch remained planktivorous or benthivorous (Fig. 4). No overt mechanism was evident, such as habitat partitioning, or variation in prey distribution. However, intrapopulation niche partitioning can come about as the result of a complex interaction between resource traits, resource abundance, and the individual's phenotype (variable morphological, behavioural, or physiological capacity). Indeed, this functional tradeoff of increased efficiency in one environment against a reduction in another promotes divergence and enables distinct morphotypes to coexist (Ward et al. 2006). These traits then interact with prey availability, their escape rates, environmental heterogeneity, and social interactions to mold the individual's actual resource use; and consequently, drive individuals into differential patterns within the same environment (Bolnick et al. 2003).

The suggestion of inherent food preferences through distinct behavioural phenotypes has also been posited by Borcherding (2006) and Westerberg et al. (2004). For instance, boldness (i.e., actively exploratory with novel objects or in novel situations) in perch has been positively correlated to body mass change and additionally, bolder fish learn to feed on new food items earlier than shyer fish (Magnhagen and Staffan 2003). It is not then presumptuous to assume that bolder individuals will gain

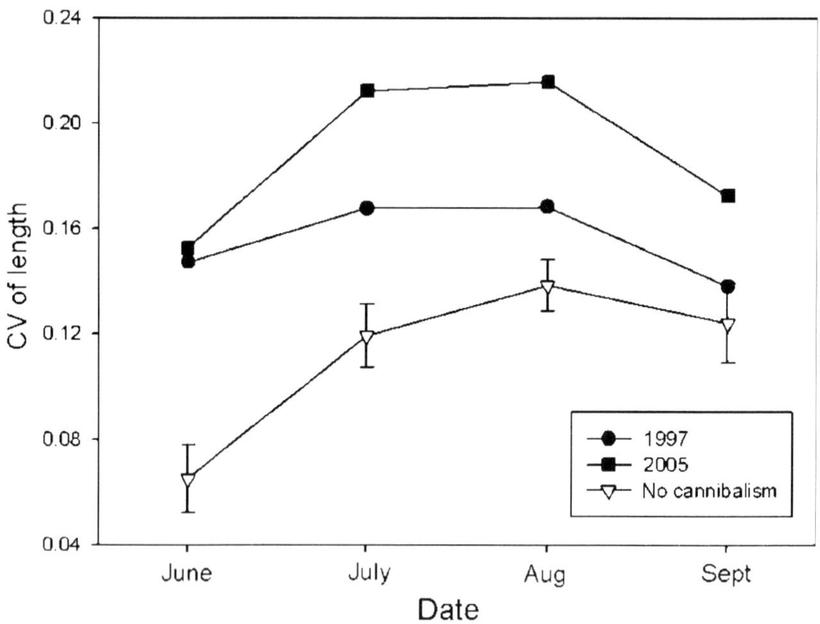

Fig. 4. Seasonal development for coefficient of variation (CV) for average length of YOY *P. fluviatilis* in years with (1997, 2005) and years without intra-cohort cannibalism (mean ± 1 SE). Data from Byström et al. (2012).

greater access to food resources (and display a competitive advantage/dominance over shyer, less aggressive individuals), and subsequently shift earlier to a more benthic or piscivorous diet. These innate personalities—defined as a broad domain of behavioural individuality involving the widest range of consistent and enduring behavioural traits (Budaev and Brown 2011)—can shape food preferences and foraging strategies and thus promote the dichotomy in sizes and trophic levels found within perch cohorts. Similarly, individual diet specialization and inter-individual niche variation can occur when there are efficiency trade-offs in using alternative resources (Heermann et al. 2013). When competitive pressure increases, previous sub-optimal resources may now impart a benefit, and natural selection will favour individuals who switch to new resources (Bolnick et al. 2003; Svanbäck and Persson 2004). This tradeoff, however, is dependent on densities of foragers in relation to food resources, with individual specialization increasing only with increasing intraspecific competition (Heermann et al. 2013).

Within-cohort splitting has also been attributed to phenological differences in the timing of hatching combined with intra year-class cannibalism. For instance, asynchrony in larval hatching promotes initial size variation that could enable cannibalism by perch larvae on their smaller siblings resulting in a bimodal size distribution (Urbatzka et al. 2009). A study by Brabrand (1995) observed that cohorts of perch larvae, hatched within 24 h, developed into a bimodal body size distribution as early as 6 days after commencement of external food uptake. At this development stage, intra-cohort cannibalism occurred among larval perch individuals of a body size as small as 10.5 mm on even smaller siblings (Fig. 5a,b). Phenological differences in hatching between predators and prey can also drive within-cohort polymorphism. In a modeling exercise followed by field validation, Borcherding and colleagues (2011) found that no size divergence occurred when perch either hatched earlier than their prey (bream, *Abramis brama*), or when the related size differences were so large that no gape-size limitations existed when both predator and prey met in the littoral area. Equally, the same null effect on size differentiation was found when bream hatched much earlier than perch, preventing predation from occurring (since the bream reached too large a size for consumption), which, in turn, increased the competitive interaction of perch with its prey. In between the two phenological extremes, however, only the largest individuals of the age cohort were able to prey on the alternative food resource, while the smaller siblings ended up competing with one another and also with bream, resulting in even further differences in growth and hence size classes.

The last set of hypotheses to explain the phenomenon of size/foraging divergence as an outcome of intracohort competition is the influence of heterogeneity in biotic and abiotic features. Beeck (2003) suggests that patchy prey distribution (in terms of abundance or catchability) may cause random effects that enable some juveniles to forage on better prey from a bioenergetic point of view. This variation in resource availability can be modulated by habitat complexity. For instance, Roswell and

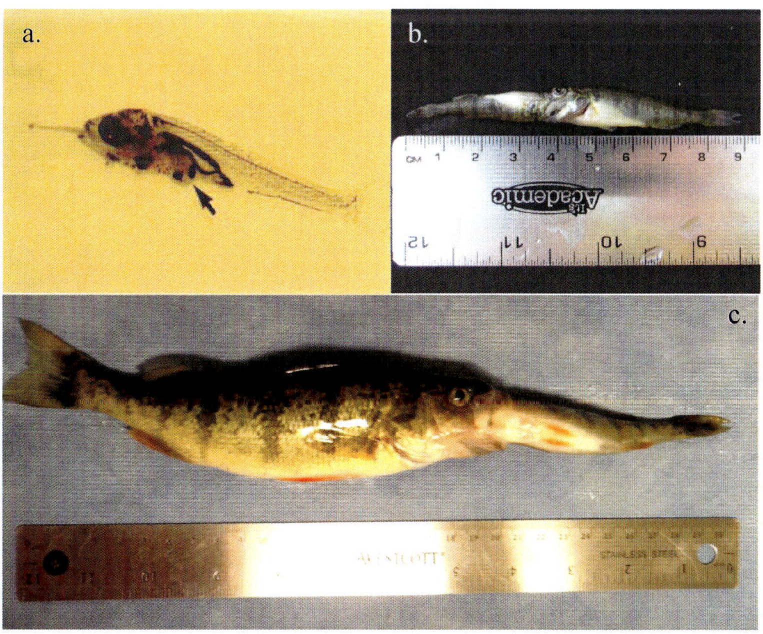

Fig. 5. (a) Intracohort cannibalism of *P. fluviatilis*: 10.5 mm (cannibal) and 5.3 mm (prey). Arrow indicates eyes of prey observed in stomach. Data from Braband 1995. (b) Intracohort cannibalism of *P. flavescens*: 55 mm (cannibal) and 40 mm (prey). Photo courtesy of M. McCloskey. (c) Intercohort cannibalism of *P. flavescens*: 22.0 cm (cannibal) and 12.5 cm (prey). Photo courtesy of C. Semeniuk.

colleagues (2013) studied interindividual variation in the diets of YOY yellow perch and attributed the observed foraging-mode split between zooplankton and larger benthic prey to variation in spatial patterns and specialization within given sites. The authors did not find any strong explanatory power provided by differences in prey abundance or ontogenetic diet shifts. Instead, fine-grained physical site characteristics, such as substrate and depth that differed within and amongst sites, were proposed to explain the within-site differences in YOY perch diets since these attributes could promote a contrasting profile of suitable prey. Relatedly, environments providing limited number of refuges (i.e., simple, structured habitats) could promote competition amongst fish for shelters; and subsequent aggressiveness can trigger the establishment of social hierarchy and resultant size divergence among young perch. Mikheev and colleagues (2005) found that interference competition over limited refugia (but not for food) was the mechanism responsible for the density-dependent cohort splitting they observed in their study of European perch. Inferior competitors were then presumably driven to lower quality habitats with lower growth potential (and higher predation risk), and size discrepancies within a cohort then emerged.

As evidenced by the array of proposed hypotheses that are not all mutually exclusive, patterns of diet shifts between members of a cohort can be variable between sites and years, therefore signifying that the flexibility in perch foraging behaviours is extremely labile, and is driven by both abiotic and biotic factors.

9.3.2 Intercohort (Intraspecific) Competition

Intraspecific food competition exerts powerful selective forces on all individuals, as has been shown by the body-size and foraging-behaviour polymorphisms observed *within* an age cohort in perch. Between individuals of different ages, the nature and the intensity of competition similarly vary according to: resource characteristics and distributions in time and space, the ecological context, and the relative competitive abilities of the foragers (Ward et al. 2006). From an individual standpoint, organisms are predicted to become opportunistic when resources are scarce and intraspecific competition is high (i.e., optimal foraging theory; Stephens and Krebs 1986). As a consequence, the individuals' diets shift and broaden, and the niche breadth of the entire population is also predicted to increase as well (Araujo et al. 2008). Similarly, from a population dynamics perspective, the strength and direction of competitive interactions *between* fish age classes can result in Ricker-type responses, where negative pressure on one age class results in the overcompensation of another, with ultimate impacts being observed in overall population size, population growth rate, and recruitment (Nislow et al. 2011). To encapsulate this process, 'trophic polymorphism' has been coined and represents a response to competitive pressures within a population that results in an expansion of the niche via character displacement or release, either morphological, behavioural, or both, as 'morphs' or cohorts separate into distinct ecological niches (Ward et al. 2006). This ability to respond adaptively to ecological conditions confers significant (fitness) benefits, and can affect community structure (Swanson et al. 2003).

To illustrate this concept, in a combination of field and laboratory experiments Persson (1987a) investigated the effects of resource availability and distribution on size class interactions in European perch. Persson observed that when alone, 1- and 2-year-old perch utilized the same food resources but when they coexisted, their diets differed. In high-resource years, intercohort cannibalism and competition were intense, since foraging success led to increased energetic gains and hence increased fecundity, which amplified competition over resources (Fig. 5c). In low-resource years, young perch (1 year old) gained the advantage since they were able to capture specifically sized prey as a result of their greater swimming capacity and gape size. The efficiency of larger perch (2 years old) in catching smaller prey decreased as a result of decreased ability to see small prey, in addition to reduced maneuverability and ability to retain small prey once caught. These energetic-, search- and handling-time tradeoffs in age/size resulted in the larger, older perch having a foraging advantage in the littoral habitat instead. This habitat was dominated by macroinvertebrates, whereas smaller perch preferred the pelagic area.

Population niche width cannot only expand when intraspecific competition increases, but when resources become scarce, as well. In a follow-up study in another Swedish lake, Svanbäck and Persson (2004) documented a population of European perch over 9 years that switched between a phase that was dominated by adult perch (low-food years) and a phase that was dominated by juvenile perch driven by intercohort competition and cannibalism. In fact, individual foraging patterns and the niche breadth of the population were highest when the adult population density was high. In other words, diminished resources tended to promote individual specialization in diet niche.

In a similar longitudinal study of yellow perch year-class representation in a structurally simple habitat, Sanderson et al. (1999) discovered that older juvenile year-classes suppressed younger cohorts through both prey competition and cannibalism (Fig. 6). Equally, zooplankton biomass tended to be lower in years with large YOY cohorts, and a reduction in the limited and shared zooplankton resources by abundant YOY negatively impacted adult survival. These repeated oscillations in which one or two out of three age classes would dominate the population over time (YOY, juveniles, and adults) were driven predominantly by pulses of abundant, reproductive, adult perch. As young perch grew to juveniles, they excluded the possibility of survival of successive cohorts through competitive and cannibalistic interactions. This exclusion occurred until the juveniles became reproductively mature, when they were outcompeted by YOY over limited resources, and the cycle would then repeat.

Perch are renowned for this variable recruitment leading to strong year-classes or cohort dominance; and the macroinvertebrate-feeding phase is thought to be the competitive bottleneck promulgating these population dynamics since prey may limit the extent to which predatory individuals are recruited to larger predatory stages (Byström et al. 1998; Fig. 7a,b). The density and size of zooplankton is influenced by the availability of different habitat types that in turn, along with season and temperature, influence the accessibility of different prey types and sizes. For instance, eutrophic lakes (low summer food availability) or ones of simple habitat complexity are purported to exhibit stunted perch populations due to high intraspecific competition caused by restricted littoral zones or limited larger-prey diversity, since these traits and conditions favour smaller perch. In complex lakes with submersed vegetation covering large areas, larger-sized zooplankton persist, and the subsequent higher foraging efficiency of larger

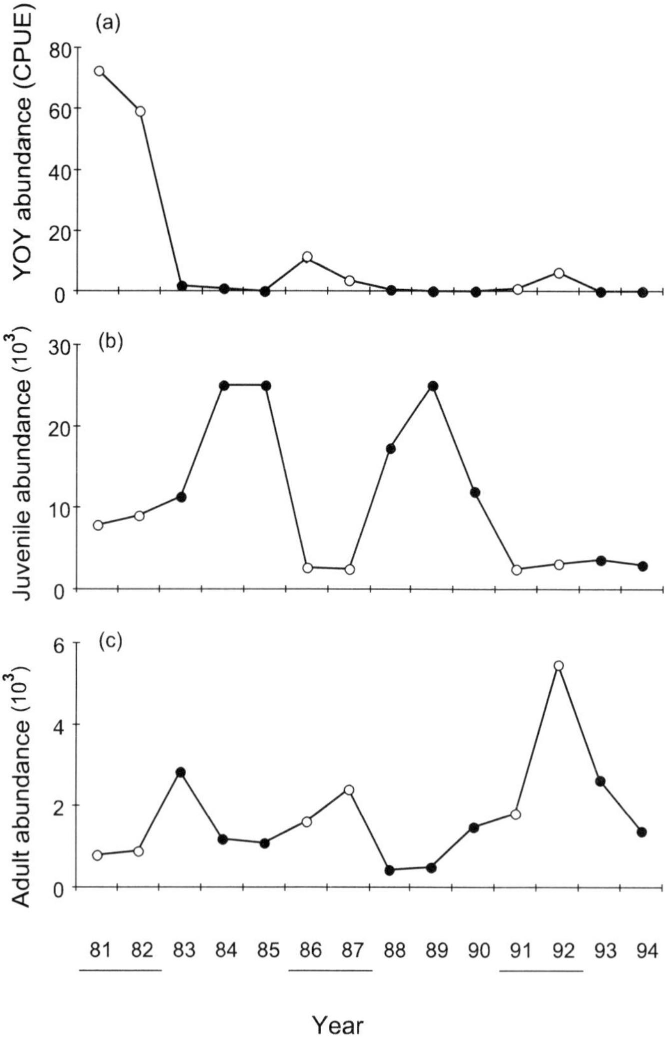

Fig. 6. Abundance of (a) YOY, (b) juvenile, and (c) adult *P. flavescens*. YOY data are CPUE estimates, and juvenile and adult abundances were estimated using acoustic data. Years in which YOY were abundant and continued to dominate the population over the following 3–4 years are underlined and represented by open circles. Data from Sanderson et al. (1999).

perch on macroinvertebrates changes the relative competitive abilities of different size classes in favour of larger perch (Persson 1987b). Accordingly, different age (size) classes can more easily segregate by habitat, reach the required size class necessary for piscivory, and engage in intraspecific predation, thereby potentially reducing juvenile density, increasing higher-energy consumption, and alleviating intraspecific competition. As a consequence of these exploitative competitive interactions, size class structure and abundance of entire populations are affected, and niche widths expand and contract accordingly.

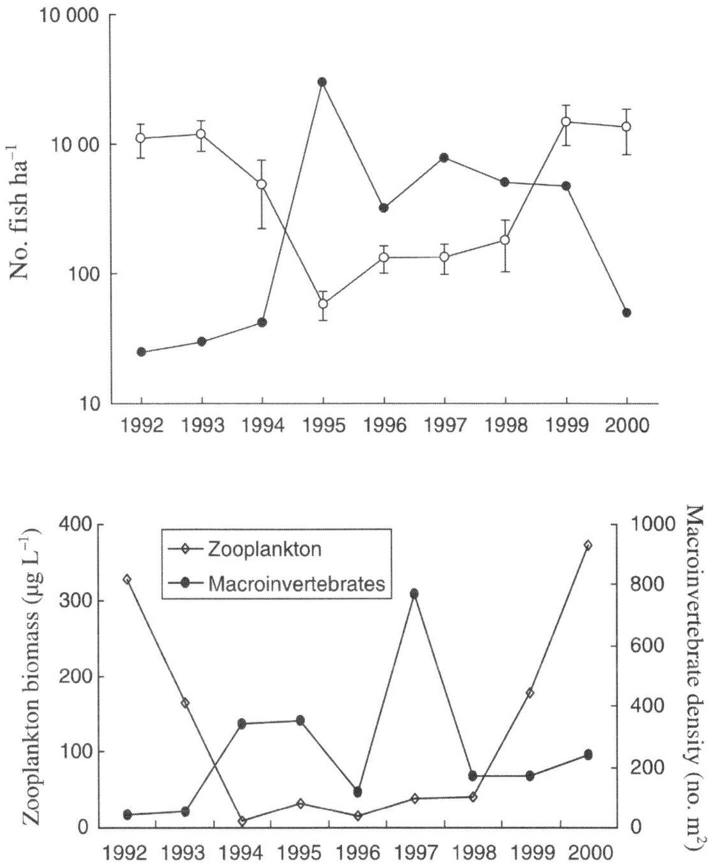

Fig. 7. (a) Changes in population density for 1-year-old *P. fluviatilis* (means, filled circles) and perch ≥ 2 years old (means 95% CL, open circles) during the study period. Note log scale on *y* axis. (b) Changes in average resource biomasses (pelagic zooplankton and macroinvertebrates) during the study period. Zooplankton biomasses are averages over the growing season. Data from Svanbäck and Persson (2004).

9.3.3 Interspecific Competition

Whereas intraspecific competition is expected to generate selection for increased total niche width as individuals or cohorts specialize, interspecific competition has the opposite effect. In the former case, adding new resources reduces the overall severity of resource limitation; and in the latter, limiting a population from certain resources because species get competitively excluded decreases population niche width and associated growth and/or abundance (Araújo et al. 2011). The magnitude of these effects is modulated by phenology (the order and magnitude of emergence), seasonal effects (environmental patterns of resource availability), and habitat attributes.

Perca spp. show morphological and behaviour-related feeding efficiency adaptations to three major niche shifts during their life history, starting from planktivory, switching to a macroinvertabrate diet, and ending up as a piscivorous predator. During the early life stage, the order of arrival and abundance of perch and their competing species can influence interspecific competition for available resources, affecting growth or survival of either or both populations. Earlier-emerging YOY yellow perch have been known to slow the growth rate of bluegill sunfish (*Lepomis macrochirus*) due to competition for high-energy prey, *Daphnia* spp. (Kaemingk et al. 2012a). In contrast, yellow perch lose their competitive advantage when emergence overlaps, since a follow up study by Kaemingk and colleagues (2012b) demonstrated that perch shifted feeding patterns to a diet higher in zooplankton compared to macroinvertebrates when in sympatry with bluegills, resulting in slower growth. Additionally, Schoenebeck and Brown (2010) inferred that competition continues to exist between bluegill and yellow perch at the adult stage for prey resources.

Young European perch find themselves in a similar interaction situation competing for available food resources with roach (*Rutilus rutilus*) during times of low resource availability. Roach tend to be favoured in systems with low food availability due to their ability to prey on a broad range of diet items at a lower vegetative planktonic trophic level and smaller zooplankton compared to European perch that feed mainly on invertebrates and small fishes (Persson 1983). The distributions of the species in the lake typically correspond to their niche width and resource availability. The degree of resource limitation varies in turn with season and is typically highest in summer, caused by the increased metabolic demands at higher temperatures. From a community perspective, Bergman (1990) observed an indirect interaction effect on perch by roach caused by a third competitor, ruffe (*Gymnocephalus cernuus*). As older perch begin to shift habitats to the benthic zone, they can be excluded by already established benthic competitors such as ruffe that competitively delay perch from moving into these newer habitats. Perch therefore involuntarily remain longer in their previous, pelagic niche, and while no longer directly competing with roach in the pelagic zone, perch will nonetheless experience slower growth rates feeding on less profitable zooplankton instead of fish prey.

Perca spp. are visually oriented foragers, and are known, regardless of whether at the zooplankton-, macrozoobenthos-, or fish-feeding stage, to forage mainly from dawn to dusk. Consequently, the ability to efficiently locate and consume food will not only necessitate how perch interact with competitors, but how they trade off habitat attributes with the presence of heterospecifics. In studying the effects of habitat complexity on foraging success of European perch in competition with ruffe, Dietericht et al. (2004) found that substrates covered with zebra mussels (*Dreissena polymorpha*) affected the foraging success of these competitors. Perch, as visually oriented predators, had few problems detecting caddisfly larvae (*Tinodes waeneri*), regardless of whether they were attached to bare stones or hidden in dense mussel beds, and consequently out-competed ruffe, a more mechano-sensory oriented forager. In an experiment where limited food was supplied only during the night, ruffe outcompeted perch, resulting in a reduced specific growth rate, because of the limited visual acuity of foraging of perch in low light conditions (Schleuter and Eckmann 2006). While European perch may be the superior competitor over complex substrates (e.g., dense littoral vegetation; Persson 1993), another experiment—this time involving roach—demonstrated that in macrophyte-poor, highly coloured water, perch prey-selectivity was affected, with perch displaying difficulty in targeting larger prey sizes. They were additionally outcompeted by roach, and subsequent reductions in growth and delayed second niche-shift from benthivory to piscivory were observed (Estlander et al. 2010).

While perch foraging success can be hampered by the synergistic effects of interspecific competitors and humic, low-light water conditions, niche differentiation can still occur if the opportunity presents itself, in the form of habitat configuration. In a field study conducted in a large and deep reservoir by Kahl and Radke (2006), European perch and roach were able to avoid interspecific competition by segregating into distinct spatial niches first, followed by specialization on differing food resources. These separations resulted in lower overall density, which manifested in low inter- and intraspecific competition of both species; consequently, this dual partitioning resulted in a widening of the juvenile competitive bottleneck to such an extent that perch were not forced to compete with their older conspecifics during their first year of life. Perch were thus able to become large and accordingly shift without delay to piscivory.

Percids not only compete with other fish species, but also influence whole community structure (terrestrial and aquatic) through their direct impacts on the invertebrate population. Specifically, this phenomenon is known to occur in boreal lakes. In these lakes, which are typically characterized as small, oligotrophic, acidic, and poorly vegetated, perch can nonetheless persist. They function as a keystone predator of aquatic invertebrates in the absence of (i) vegetation (which serves as cover for invertebrate prey) and (ii) perch predators (e.g., pike, *Esox* spp.; see Elmberg et al. 2010). In boreal lake ecosystems, the lack of predation on perch not only affects invertebrate prey but also vertebrate species that use the same resources, such as ducks, via exploitative competition. For example, a field experiment by Nummi et al. (2012) demonstrated that the use of the three experimental lakes by goldeneye broods (*Bucephala clangula*) decreased after European perch were introduced. This reduction was attributed to the inability of ducklings to acquire enough energetic resources (i.e., invertebrate prey) to survive because of interspecific competition with perch. These species interactions are identical with goldeneye ducks in North America interacting with yellow perch. Eadie and Keast (1982) observed that goldeneye and perch co-occurred on all three of their study lakes but there was an inverse relationship between their abundances on the majority of study plots within those lakes with higher perch densities. This reciprocal density trend was not due merely to differences in habitat use by perch and goldeneye, but instead to food-resource limitation during the brood-rearing season. Indeed, *Perca* spp. are such strong competitors in these systems that when considering the impacts of three fish species independently on duckling recruitment (two potential competitors, roach and European perch; and one predator, pike) in Southern Finland, Väänänen and colleagues (2012) reported a negative association between perch and two species of waterfowl. The authors suggest that food competition with perch is a more important factor than pike predation in affecting lake use by ducks in oligotrophic boreal environments.

9.3.4 Perch Competition with Invasive Species

Intentionally and unintentionally introduced aquatic species comprise an increasing proportion of the fauna of freshwater lakes, and many of these are having negative consequences for native fish. Coexistence between indigenous and non-native species is possible if: (i) habitat use by fishes differs substantially; (ii) feeding primarily occurs in different habitats, at different times, or under different conditions even with partial to full overlap; or (iii) neither habitat space nor food resources are limiting. Eurasian ruffe are invading habitats in the North American Great Lakes, and this species, sympatric with European perch, has also been shown to have dietary overlap with yellow perch (Fullerton et al. 2006). Consequently, the exotic ruffe has been the subject of many competition/diet studies. While increased fish densities have been shown to lead to declines in both ruffe and yellow perch growth in laboratory settings (Fullerton et al. 2000), the full extent of impacts is still unknown. Fullerton and colleagues (2006) suggest that if ruffe and yellow perch share a habitat (e.g., during invasion or because of predation risk), competition for space will be weak or absent because they exhibit diel differences in foraging behaviours (night vs. day, respectively); but under limiting food resources, competition for food may occur because neither species has a clear advantage in its ability to consume invertebrates in any habitat.

Indigenous European perch must also contend with introduced, warm-adapted species into its native range, and research has been conducted on the competitive effects of one of the most successful introduced species, the pumpkinseed (*Lepomis gibbosus*) from North America. In a study to investigate the competitive interactions between the two species under ambient and increased temperatures predicted by current climate change models, Fobert et al. (2011) found promising results. When perch and pumpkinseed were reared in allopatry they had very similar diets and hence similar prey preferences. However, when reared in sympatry, they had less similar diets, and even fuller stomachs, indicating a clear partitioning of food resources. Furthermore, elevated temperatures accentuated these dietary shifts, because warmer temperatures increase the energy demand of the fish, and in sympatry it would also increase the competitive pressure, forcing a greater segregation of prey resources (if unlimited). Invasions can nonetheless have cascading effects on lake community structure. As an example in a North American lake, rainbow smelt (*Osmerus mordax*) invaded Crystal Lake in northern Wisconsin, USA, in the 1980s. This lake had a native planktivore community dominated by yellow perch that were quickly replaced by rainbow smelt as of 1994 (Beisner et al. 2003). A large overlap in diets of these fish species occurred during the invasion, and detailed studies showed that exploitative competition with smelt led to perch extirpation (Hrabik et al. 1998).

As a final note, perch themselves can be considered invasive species, affecting community structure and dominance in freshwater lakes. European perch were introduced to New Zealand in 1868 as an angling species (McDowall 1990) and have recently been studied for their potential to significantly alter community dynamics. In an experimental setup, Attayde and Hansson (2001) showed that perch, through nutrient recycling (i.e., by influencing the N:P ratio released by zooplankton) and direct grazing of zooplankton, was an important driver affecting phytoplankton community structure and favouring cyanobacteria dominance in lakes, and hence eutrophication. In North America, yellow perch have been introduced to Pacific Northwest watersheds and are now considered a well established and expanding nonindigenous species. Populations of yellow perch, along with those of other non-native species are being considered for mitigation since they have the potential to pose serious risks to the recovery of pacific salmonids via predation of juveniles in freshwater tributaries (Sanderson et al. 2009; see also Chapter 4, this book).

9.4 Antipredator Behaviour of Perch

Every aspect of an animal's life exposes it to the risk of predation, from foraging to finding mates. Therefore, individuals must avoid or defend themselves against predators in order to survive and successfully reproduce. Animals can use behavioural strategies to alleviate predation pressure such as reduced activity, cover seeking, increased vigilance, group formation, and direct predator avoidance. Because these anticipatory antipredator tactics can incur fitness costs in the form of energetic expenses, time unavailable for searching for food or mates, and the associated costs of grouping, an animal's antipredator behaviour is designed to minimize its costs and maximize its effectiveness for surviving predation (Rosier and Langkilde 2011).

9.4.1 Predator Detection and Response

Part of an individual's repertoire of antipredator behaviours is being able to detect predators before an actual encounter, and extract reliable information about current predation risk in order to balance its current activities against future lost opportunities due to predator avoidance. Young fish are particularly susceptible to predation while foraging, and therefore must trade off foraging against safety requirements (Wanzenbock et al. 2006). Foraging behaviour that is too cautious would deprive fish of necessary food, and alternatively, searching for food and handling prey would reduce vigilance. Prey can therefore use general cues of the presence of a predator threat, or species-specific cues to identify a predator's identity and determine its motivation and therefore the most appropriate behavioural response (Mikeev et al. 2006). These responses can be additionally modulated by innate behavioural phenotypes (see section, "Social Behaviour of Perch" in this chapter). Notwithstanding, it has been demonstrated that the level of chemosensory assessment used by juvenile perch is quite sophisticated, in that juvenile perch will increase antipredator behaviours such as shelter use when exposed to chemical cues of adult perch (cannibalistic predators) fed juvenile perch or heterospecific shoaling members, over predators fed on a diet of non-interacting species (Mizra and Chivers 2001). Perch are also capable of responding to chemical alarm cues from injured conspecifics. Groups of juvenile perch will increase shoal cohesion and movement towards the substrate, increase shelter use, and freeze their movements after detecting conspecific alarm cues (Mizra et al. 2003). Foraging YOY perch further respond to indirect nonspecific habitat cues of predation risk by reduction of intake of large prey, which are costly in terms of handling time (Mikeev et al. 2006). This allows fish to be more vigilant without ceasing their foraging activity even in potentially dangerous situations (Wazenbock et al. 2006). For naïve fish, however, multiple cues may be necessary to impart reliable information. Odour has been shown to act as a modulatory stimulus enhancing the effects of visual cues, which trigger an innate response in perch, and it is this combination of cues that cause a significant shift in prey-size selection for young perch without previous exposure to predators (Mikeev et al. 2006).

Given the importance of olfactory cues to the behavioural ecology of yellow perch, anything that could potentially disrupt their ability to perceive and respond to chemosensory information could have deleterious consequences to perch populations. Yellow perch from metal contaminated lakes have impaired chemosensory function resulting in impaired behavioural responses to antipredator cues and food cues relative to fish from pristine lakes (Mirza et al. 2009; Azizishirazi et al. 2013, 2014). This impaired ability to respond to chemosensory cues is not related to neuron density at the olfactory bulb (Mirza et al. 2009) or gene expression (Azizishirazi et al. 2014). Rather, it is likely due to contaminant-specific targeting of olfactory sensory neuron classes in the olfactory epithelium (Dew et al. 2014). For example, copper targets ciliated olfactory sensory neurons, which are used to perceive antipredator cues, while microvillus olfactory sensory neurons perceive food cues. Therefore, yellow perch inhabiting copper contaminated habitats may be more vulnerable to predation than those from clean sites, whereas perch from nickel contaminated sites may have an impaired ability to locate food compared to those from clean sites. Understanding the specific way that some environmental contaminants can affect the perception of distinct chemosensory cues improves our ability to evaluate risk to yellow perch populations inhabiting polluted environments.

9.4.2 Habitat-selection Under Predation Risk

Refuge use as a means to procure protection or shelter once a predator has been detected can alter the dynamics between predator and prey and between conspecific and heterospecific interactions as well. The outcome of these interactions depends on the structure of the environment. Predator preferences for prey are dependent on search and handling time as well as energetic profitability. When assessing the capture success by predators in habitats without prey refuge and in vegetation vs. bottom-crevice refuges, Christensen and Persson (1993) demonstrated that predators were more efficient at capturing juvenile perch in the absence of structure and with vegetation refuges. For prey to compensate, it was shown in a field enclosure experiment (Eklöv and Persson 1995) that juvenile perch used different parts of the prey refuge in a flexible way depending both on presence of predators and refuge type (structure forming a partial refuge, and structure forming a complete refuge), with juvenile perch survival increasing with refuge efficiency. Perch increased their time spent in the refuge in the presence of predators and reduced the number of switches between the open-water habitat and the prey refuge. This increase in survival may potentially be traded off in terms of growth rates, since access to food is possibly diminished. However, in a follow-up experiment, Persson and Eklöv (1995) demonstrated that juvenile perch compensated for lost foraging opportunity in the open water via increased exploitation of structure-associated prey in refuges. The decision of perch to exploit cover, however, additionally depends on other abiotic conditions: turbid waters can modulate perch antipredator behaviours and subsequent cover use. Snickars et al. (2004) in a habitat-utilization study under threat of predation, found that juvenile European perch avail themselves to less vegetated cover as turbidity increases since they presumably perceive that visual detection by predators is reduced in turbid conditions, allowing perch to continue their foraging behaviours instead.

With regards to the effect of predation on *intraspecific* competition, Eklöv and Svanbäck (2006) exposed YOY European perch to a predatory adult perch constrained to one of two habitats (pelagic or littoral). Without predators, YOY perch used both habitats and showed strong individual specialization. With predators, YOY perch remained in the safer habitat, suppressing diet variation, which suggests that the presence of predators may alter the habitat and food resource use of young fish and thereby reduce individual specialization. As a result, this predator-induced habitat and diet shift by juvenile perch may alter *interspecific* competitive interactions found within the safer habitat.

Indeed, in a laboratory experiment by Persson (1991) it was shown that piscivorous predators reversed the outcome of competitive interactions between juveniles of roach (*Rutilus rutilus*) and European perch, by behaviourally affecting their use of two available habitats, an open water habitat and a structurally complex refuge. Typically, roach are the superior competitors in unstructured or open habitat; however, by being forced into a structurally complex prey refuge along with their perch competitor, perch gained the competitive advantage. The reversal in competitive relationship was demonstrated both with respect to foraging rate and growth rate (roach decreased in both) and resulted from the prey refuge interfering with the roach's swimming performance. In all, the presence of structure can have substantial effects on the structure of fish communities, by affecting relative and absolute predation pressures from piscivorous perch on prey species (Christensen and Persson 1993), as well as the competitive output between competitors whose niche use are driven to overlap.

9.4.3 Predation-sensitive Foraging

The selection of safer habitats in the presence of a predator is not a universal response amongst perch prey, but can be modulated by energetic state, experienced risk-taking, individual specialization, and innate boldness. The level of nourishment is an important factor that quickly alters risk-taking behaviour. Borcherding and Magnhagen (2008) kept 1+ perch on two different food rations for an extended period of time and then assessed their willingness to forage vs. avoid predators. Undernourished individuals were much more risk-prone than their well-fed conspecifics. In addition, the morphological shape of perch differed significantly between feeding treatments. At low food levels perch developed a more slender body, while at high food levels they became deeper-bodied with a relatively smaller head. This morphological variation is purported to be a consequence in slender, starved fish of (i) all energy available used for metabolic maintenance with no surplus energy available for morphological modulation, and (ii) to aid in food capture, as this body shape also displays higher swimming speeds (Olsson et al. 2007). Similarly, in a physiological study of activity levels of yellow perch as prey, Kaufman et al. (2006) revealed that perch show clear metabolic adaptations to increased predation pressure. When perch are the main prey species, they invest most energy into mechanisms aiding in escape behaviours such as burst speeds and sustained swimming. When predation pressure is diluted by a larger, more profitable prey (e.g., cisco—*Coregonus artedi*), perch employ different strategies to balance the risk of predation while still foraging themselves, from the timing of ontogenetic diet shifts, to increasing time spent not moving, to increasing their use of cover.

It was traditionally believed that prior experience with predators is likely to have little influence on the phenotypic response to predation, since failing to react correctly to a predator may mean death to the prey and no second chance to learn and correct the behaviour (Hellström and Magnhagen 2011). However, this assertion negates widespread adaptive behavioural responses of individuals in face of biotic perturbations (see Dall et al. 2005 for a review). A longitudinal study by Olson et al. (2001) investigated changes in yellow perch growth associated with the establishment of a walleye (*Sander vitreus*) predator population. Prior to walleye stocking, yellow perch growth rate was low and independent of body size. Within a year of stocking, yellow perch growth rates increased concomitantly with walleye abundance. Growth became size-dependent as well since decreased growth was observed to occur initially in small perch as walleye became established. This rapid switch to size-dependent growth indicates a rapid behavioural response (not directional selection/numerical response) to predators. Next, in a common-garden experiment, in which perch from two different populations experiencing different predation regimes were raised under identical and dissimilar predation risks, Hellström and Magnhagen (2011) investigated the influence of inheritance and experience on risk-taking behaviours. The authors found that risk-taking behaviour, as opposed to staying in safer refugia, appeared to a large extent to be caused by fish adjusting to the current environmental conditions, such as predation pressure. Habituation to their environment also occurred, as risk taking increased and predator inspection decreased between runs, indicating a habituation effect to the constant presence of a predator over time. To sum, individuals are plastic in their behavioural repertoire to predation risk, and can indeed modulate their responses even though 'programmed' from their joint co-evolution with predators.

Resource polymorphism has been shown to promote differences in individual competitive abilities, habitat selection, and even length-frequency distributions. It is not presumptuous to then also expect divergent resources to induce complex behavioural consequences that lead to phenotypic differences in behaviour during predator encounters. In a laboratory experiment in which juvenile perch were raised on either zooplankton or fish, Heynen and colleagues (2011) recorded that planktivorous and piscivorous YOY perch exhibited an overall physiological difference that affected their antipredator behaviours. In the absence of a predator, planktivores displayed an overall high activity (presumably to acquire sufficient energetic gains for metabolic maintenance), while piscivores stayed primarily in the vegetation. In the presence of a predator, planktivorous perch decreased their activity while piscivorous perch increased theirs, but only when presented with fish larvae as prey. This individual specialization within perch thus has downstream effects with larger piscivore prey more willing (and potentially more able) to tradeoff predation risk with food rewards. These behavioural modifications can subsequently affect prey growth rates, which in turn can further affect other prey life-history traits such as investment in and scheduling of reproduction (Abrams and Rowe 1996), that can then feedback on intracohort competition and intercohort specializations, etc.

While resource availability affects prey antipredator behaviours in a bottom-up fashion, so too, can predators modify behaviours and life-history tradeoffs of prey in a top-down, non-consumptive way. Under a predation regime by a gape-limited predator, prey can adopt two contrasting strategies: one is to grow rapidly through the vulnerable size range and achieve a size greater than the gape of their predators (Biro et al. 2005). Potential costs include increased exposure to predators as increased foraging rates of prey will necessitate more conspicuous movements, and lack of use of refuges. The second tactic is for prey to modify their behaviour to reduce detection by predators, and involves a decrease in prey foraging rates. Each strategy can influence the amount of surplus energy available for growth and reproduction (Rennie et al. 2005). To evaluate the importance of non-consumptive effects of predators on perch life histories under natural conditions, Rennie et al. (2010) compared life-history variables, and rates of prey energy acquisition and allocation to predator abundance. By examining maximum size at maturity in both male and female yellow perch, life span, and reproductive investment as estimated from bioenergetics models, the authors observed lower growth rates and growth efficiencies in populations with fewer predators, despite increased consumption of resources. Intriguingly, their results suggest the possibility that reduced activity in the face of high predation risk increased prey growth rates. In specific, the energetic savings incurred by reduced activity allowed for a greater amount of excess energy per unit food consumed. Consequently, lower prey activity and foraging coupled with high-density predators can in turn influence resource availability and quality. This study provides a nice illustration of how behavioural modifications induced via predation risk can shape life histories of prey such as perch.

An underlying mechanism for the willingness of perch to take risks in the face of predation in the absence of undernourishment/resource availability is their innate personality (i.e., individual risk-taking tactics). Indeed, individuals that choose to spend more time in open habitat, or individuals that are more risk prone when faced with foraging decisions (Utne et al. 1997) are termed the bolder fish. These personalities can be consistent through time or context (forming a syndrome), but can also be modified by predation pressure (Fig. 8). Borcherding (2006), in an experiment of predation risk on YOY perch found larger sized perch to be more 'courageous' than smaller fish in using the open water side arm of the experimental setup when food was offered there. In a pond experiment, Heynen and colleagues (2014) studied the effects of differential predation on age, body shape and boldness. They found changes in morphology across two perch age classes, where individual YOY and 1-year-old perch developed a distinctly deeper and more downward bent body under high predation risk (to evade gape-limited predators). However, boldness differed between age classes, with YOY perch being significantly bolder and more consistent in their behaviours than their year-old conspecifics, with the latter age class behaving shyer with increasing predation risk intensity. Together, these results reveal that perceived predation (even in the absence of consumptive effects) can have fast and strong direct effects on the amount and distribution of phenotypes within a prey population. Further, the use of antipredator strategies by 'bold' and 'shy' individuals and how they are traded off against food acquisition can vary over age-classes in perch (Magnhagen and Borcherding 2008).

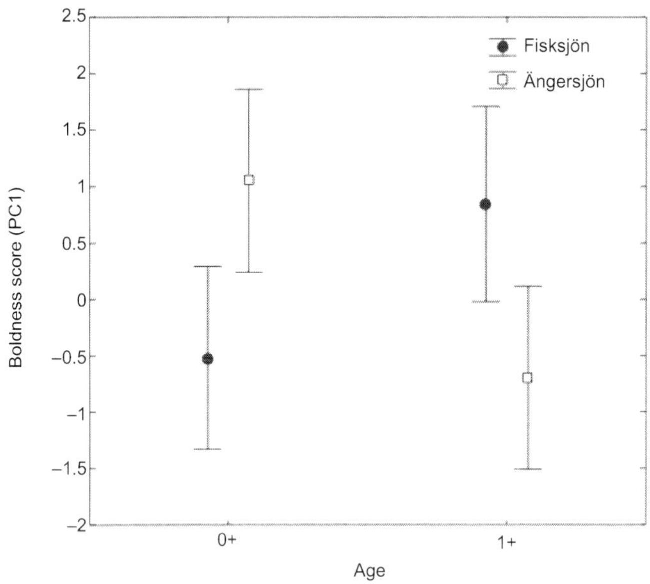

Fig. 8. Boldness score PC1 (mean ± 95% CI) estimated by PCA for *P. fluviatilis* of two age classes (0+, 1+) and two lakes (Fiskjön—high predation pressure at age 0+ only, and Ängersjön—high predation pressure at age 1+ only). High scores indicate long time in the open, short latency to start feeding and long duration of first feeding bout. Number of replicates was eight for each category except for Fiskjön 1+ perch, where $N = 7$. Data from Magnhagen and Borcherding (2008).

We have already seen that survival and growth patterns through juvenile developmental stages as mediated through habitat-selection and competitive behaviour, have profound effects on the population/community dynamics of size-structured populations of perch. The effects of predation—both consumptive and non-consumptive—also have the potential to significantly shape the structure of communities as predation risk can act as a selective force by indirectly affecting characteristics of the resource base, and in determining perch cohort size, morphology, and the outcome of intra- and inter-specific competition.

9.5 Perch Predator Behaviour

The ontogenetic shift in diet to piscivory in many fish species has been shown to induce both density- and trait-mediated direct and indirect effects on whole ecosystems. Indeed, the consumptive and non-consumptive effects of predation as an evolutionary selective force have wide-ranging implications for individuals, populations and communities. *Perca* species develop into visually hunting piscivore generalists, and play an important role in structuring communities by influencing the abundance of prey fish available to other predators, by reducing intraspecific competition and increasing the growth of prey populations (Graeb et al. 2005), and by influencing the recruitment of cohorts within a population.

9.5.1 Density- and Trait-mediated Direct Effects on Prey

Perch hunting behaviours are flexible, and are capable of compensating for antipredator behaviours exhibited in their prey by altering their foraging modes or by exploiting alternative prey in natural and experimental systems. For instance, when refuges in the form of submerged vegetation are present, large perch can switch from an active group foraging mode to an inactive solitary mode or stay in pairs to take advantage of any escapees from vegetative refuges. Alternatively, perch also adopt solitary active swimming modes to more efficiently track spatial changes in resource levels (Eklöv 1997). Perch as predators consequently induce non-consumptive, trait-mediated responses in their prey, as evidenced by organisms from cladocerans (*Daphnia* spp.) and roach to younger perch that alter their life histories (size- and age at maturity, Hülsmann et al. 2004), habitat selection strategies (Martin et al. 2010), and vigilance and avoidance behaviours (Diehl and Eklov 1995; Wisenden et al. 2004) in response to the olfactory and visual cues of perch (Fig. 9).

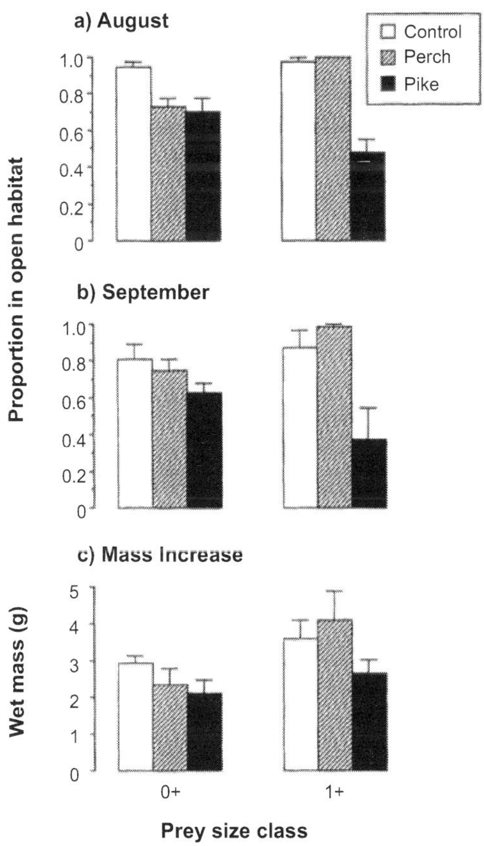

Fig. 9. Proportion of prey fish (0+ and 1+ *P. fluviatilis*) in the open habitat in August (a) and September (b), and increase in individual wet mass of prey fish over the experimental period (c) in the three piscivore treatments. Means ±1 SE, $N = 4$ enclosures in all cases. Data from Diehl and Eklöv (1995).

9.5.2 Multiple-predator Effects

The interactions between YOY percids and their predators can shift from predation to competition when perch become adults. Consequently, two predator species occupying the same trophic level often compete for resources by consuming the same prey, but they may also interact by inducing trait shifts in their targeted prey that make food more or less available to each other (Sih et al. 1998). Perch in North America and in Europe often occupy the same trophic level with other piscivores such as walleye, bass (largemouth, *Micropterus salmoides*) and northern pike. Prey responses to these multiple predators, and therefore predator successes, depend on predator foraging mode (ambush vs. active swimming), the divergence between perceiving a predation threat versus ambiguity in knowing where attacks will come from, and the availability of habitat within which to retreat. Indeed, structural complexity can both qualitatively and quantitatively change the interaction between piscivorous predators as well as with their prey. For example, in an attempt to avoid vegetative habitats where ambush predators (e.g., pike) lie in wait, prey such as roach (*Rutilis rutilis*) may move into more open habitats and be vulnerable to active perch predators (Martin et al. 2010). Conversely, actively chased prey may escape to 'refuges' only to be preyed on by sit-and-wait predators. Accordingly, the combined predator mortality on prey can be higher than predicted based on the additive effects of each predator alone. In most documented cases, the alternative predator is the one to benefit from the multiple-predator pairing while perch benefits remain unchanged or reduced. Eklöv and VanKooten (2001) demonstrated that growth rates of perch were similar when experimental enclosures contained only perch with their prey, roach, or when they contained perch combined with pike. In contrast, the growth rate of pike was higher when they were together with perch compared to when alone with roach. This one-sided facilitation between predators was caused by conflicting antipredator behaviour of roach, with pike representing a greater threat to roach than adult perch (Jacobsen and Perrow 1998). In a detailed whole-lake study to examine the effects of a new predator introduction on the residential piscivorous fish community, Schulze et al. (2006) investigated the spatial distribution, diet composition, growth, and consumption rates of piscivores before and after pikeperch (*Sander lucioperca*) introduction. They specifically tested how both density-dependent and trait-mediated responses affected interactions among three predators: the pikeperch, northern pike, and the European perch. The authors observed increases in piscivore biomass and annual consumption after the introduction that were attributable to the stocked pikeperch and increased northern pike abundances. In response to competition with pikeperch, piscivorous perch instead shifted their habitat use from open habitats towards the littoral lake areas. Furthermore, all piscivores increasingly fed on small perch; and in combination with the forced habitat shift of adult perch, these dynamics led to an eventual decreased abundance of large perch, attributable to the compensatory effects of intraguild predation and cannibalism.

9.5.3 Predator-mediated Trophic Cascades

Perhaps the most pervasive, ecosystem-wide effects of perch as predators come from their role as cannibals. Cannibalism in percids is a common, universal occurrence and is of substantial importance in year-class fluctuations, going so far as to drive dynamics of whole-lake perch populations (Persson and de Roos 2013). The population-level effects of cannibalism in freshwater fish populations are multifold: first, cannibalistic individuals in a population may persist under food limitation provided the costs of cannibalism (additional mortality) are smaller than the benefits, which mainly include extraction of energy sufficient for reproduction. Second, should increased fecundity occur because of increased energy uptake, population destabilization could happen by inducing cycles, with boom and bust demographics arising at alternate life stages. Alternatively, mortality imposed on victims may dampen population oscillations should cannibalism weaken other density-dependent factors responsible for this cycling (Mitchell and Walls 2008).

The time window during which cannibalistic behaviours in perch occur most likely coincides with (i) an initial broad size distribution (augmented through hatching asynchrony), (ii) a crash in the zooplankton resource following a strong perch recruitment year, and (iii) very slow growth of potential prey (Byström et al. 2012). Resultantly, the size differences observed within the year class can affect community-level predator-prey cycling if cannibalism pressure persists on younger stages as the cohort ages (Sanderson et al. 1999). Because cannibalism occurs both within a perch cohort and between, and perch cannibals often share a common resource with their cohort/prey (Persson et al. 2004), the cannibal benefits both from feeding on its victims and from reduced intraspecific competition (Polis 1988). An upshot of this phenomenon is that they often grow faster than the average individual (Byström 2006); nevertheless, Claessen et al. (2000) showed that perch deriving energetic profit from cannibalism do not permanently control their prey in terms of abundance. Consequently, there is still a high competition for the shared resource, and only a few individuals can grow to very large sizes.

The abundance and proportion of cannibals within a single cohort can have important implications for organisms that need to reach a certain size in time to successfully recruit to later life stages (Urbatzka et al. 2008; Byström et al. 2012). As an illustrative example, a juvenile-driven cycle with one cannibalistic cohort dominating the population has been empirically documented in yellow perch (Sanderson et al. 1999). In this case, the single year class totally dominated the population up to its maturation; and the competitive effect of the superior individuals was reflected in a massive die-off of adults following the birth of this dominating year class.

The intensity of the interactions of more than one cannibalistic cohort can also cascade throughout the community and affect lower trophic levels. In a study of the dynamics of a cannibalistic fish population of European perch in Lake Abborrtjärn, northern Sweden, both experimental and modelling analyses have shown that the cyclic dynamics of populations with more than one cannibalistic cohort can be separated into two phases. One ('*stunted*') phase is characterized by high densities of cannibals imposing a high mortality on cannibalistic perch prey at the lower trophic level. This leads to low survival and densities of 1-year-old perch the following year, as well as a small asymptotic size of cannibals (Persson et al. 2003; Fig. 10). The changes in perch population size structure, in turn, cascade down to phytoplankton through zooplankton, with high biomasses of zooplankton and a lower phytoplankton biomass due to the gape-limited smaller perch in the population. The other phase ('*giant*') is

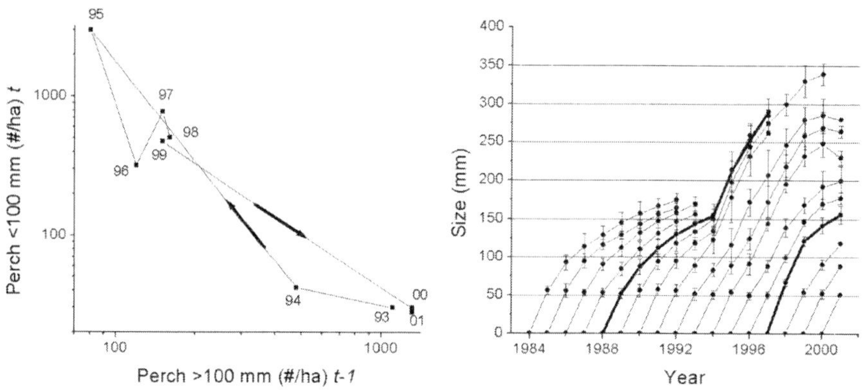

Fig. 10. (Left) Phase plot of the observed density of perch <100 mm vs. the density of perch >100 mm the previous year. Years refer to perch <100 mm. (Right) Observed growth trajectories (means ± 1 SD) of perch during 1984–2001. Thick growth curves show dominating year classes born in 1988 and 1997. High survival in YOY perch cohorts and giant growth among cannibals was observed during 1994–1998. The negative growth rates observed for year classes >200 mm during 2000–2001 may be caused by either actual negative growth or size-dependent mortality of larger perch or both. Data from Persson et al. (2003).

characterized by low cannibal density of the older cohort, leading to higher survival of cannibal perch YOY, a subsequent large number of 1-year-old perch the following year, and a large size divergence of cannibalistic individuals driven by intense competition. These dynamics result in low biomasses of zooplankton instead (Persson et al. 2014).

When cannibalistic perch share their environment with other lake predators, additional complex food web interactions ensue. Wahlström et al. (2000) investigated trophic-level effects of European perch cannibals in the presence and absence of a top predator, the northern pike. The abundance of small perch (young-of-the-year and 1-year old) was lower in lakes with only perch suggesting that intense cannibalism reduced these size classes to low levels in lakes lacking pike. With pike present, older perch abundance was relatively lower due to intraguild predation. Furthermore, the zooplankton communities in lakes with only perch were dominated by relatively small species and had higher total zooplankton biomass. These patterns are related to (1) the low foraging efficiency of large perch on very small zooplankton and (2) the low abundance of small zooplanktivorous perch (due to cannibalism) in lakes with only perch. Wahlström and colleagues surmised cannibalism introduces a vertical heterogeneity to food webs that causes consumer-resource dynamics that are not predictable from linear food chain models.

The effects of cannibalism on the population dynamics of perch are complex and challenging to predict. Impacts depend on whether cannibalism occurs within a cohort only, between cohorts, whether each cohort itself contains cannibals, the presence of other predators, and the availability of resources. Results include population cycling with boom and bust alternative year-classes, alternating individual specialization vs. decreased niche breadths, and stunted vs. giant cannibals.

9.6 Social Behaviour of Perch

9.6.1 Shoaling

Despite displaying competitive and/or predatory behaviours against conspecifics, fish belonging to the perch family (*Perca* spp.) nevertheless have a social lifestyle (Craig 2000), generally occurring in shoals or small groups (Fig. 11), but without the cohesiveness and synchronised movements seen in species like herring (e.g., *Clupea* spp.) or mackerel (e.g., *Scomber* spp.). Shoal size is usually large in early life, but decreases with age, and large individuals are sometimes found in small groups or even alone (Bruylants et al. 1986; LeCren 1992). Shoaling of young-of-the-year perch varies over the daily cycle. European perch aggregate in dense shoals in the epilimnion during daytime and disperse evenly below the surface at night (Probst et al. 2009). Yellow perch show a similar breakup of groups at twilight (Helfman 1979). Many fish are dependent on vision for maintaining cohesive groups, and this may be one reason for the reduced shoaling in darkness (Ryer and Olla 1998). Since both European and yellow perch are diurnal foragers (e.g., Helfman 1979; Jamet and Lair 1991; Zamora and Moreno-Amich 2002), the foraging benefits of shoaling may be greater in the daytime. Also predation risk from visual piscivores is reduced during the night (Post et al. 1998; Linehan et al. 2001).

The advantages of living in a group include protection from predation, due to dilution and confusion, but also the increased chance of detecting a predator early with many vigilant individuals (e.g., Pitcher and Parrish 1993; Krause and Ruxton 2002). Indeed, small perch are very vulnerable to predators since they have to move continuously while feeding on benthic organisms, and therefore face a high risk of encountering predators (Schulze et al. 2006). Consequently, another advantage to shoaling is that fish in a group may locate food more rapidly than when alone (the 'many-eyes' hypothesis; Pitcher et al. 1982), and social information transfer may lead to efficient decision-making in shoals (Ward et al. 2011). There are, of course, also costs of living in a group, as competition for food and other resources will increase with shoal size. Still, in pond experiments, piscivorous European perch had higher growth rates when they occurred in groups of five compared to when they were alone (Eklöv 1992). Also, in the laboratory, juvenile perch showed higher growth rates in groups than when they were kept individually, even though food intake was higher when they occurred alone (Strand et al. 2007). Solitary fish may experience stress due to a perception of risk, thus increasing energy costs. This was actually found in

Fig. 11. Shoaling YOY *P. flavescens*. Photo courtesy of K. Janisse.

another study, in which solitary perch had higher respiration rates compared to those in groups (Schleuter et al. 2007), which was explained with a calming effect of the presence of conspecifics.

Another mechanism for shoaling is via kin-selection. In Lake Constance two genetically different populations of European perch occur sympatrically (Behrmann-Godel et al. 2006). Genetic studies on the pelagic perch larvae showed that shoals during early ontogeny were kin-structured. Despite females spawning in close proximity to each other, siblings stay together (Gerlach et al. 2001). This observation might suggest that European perch demonstrate kin recognition and prefer kin to non-kin. Choice tests in a fluvarium showed preference for odours of unfamiliar kin vs. unfamiliar non-kin, and the authors suggested that assortative mating (i.e., individuals with similar genotypes and/or phenotypes mate with one another) may be causing the divergence in lake Constance perch (Behrmann-Godel et al. 2006).

9.6.2 Agonistic Interactions

Aggressive interactions are rarely seen among juvenile perch in a shoal (Staffan et al. 2002) although they do occur occasionally (Westerberg et al. 2004). However, competition for limited resources would be a cause for aggression if the benefits exceed the costs (Grant 1993). In aquarium studies on small groups of YOY European perch, no aggression was found during feeding, but the fish competed aggressively for shelter (Fig. 12), creating social hierarchies within the groups (Mikheev et al. 2005). In pond experiments looking at growth of YOY perch at different densities, a size divergence was found even at low densities, which was explained by dominance structures and social interactions (Huss et al. 2008). Similarly, at high densities, YOY yellow perch split into groups with faster growing littoral fish and slower growing pelagic fish, respectively, probably due to individual differences in competitive ability (Post et al.

Fig. 12. YOY *P. flavescens* seeking cover, with certain individuals relegated to more exposed positions (outside of submerged refugia). Photo courtesy of K. Janisse.

1997). In the laboratory, when young European perch were fed, some individuals took more of the food than others, and their share of the total food provided was consistent over an experimental period of ten days (Westerberg et al. 2004), and even over a time-span of eight months (Staffan et al. 2005). Thus, even in the absence of overt aggression, social interactions may lead to differences in food intake and growth.

9.6.3 Individual Personality in Groups

Upon review of perch behaviour, it has become obvious that not all individuals within a species react in the same way when exposed to the same type of stimulus, even if they are of the same cohort age, sex, genetic background and initial size (Magnhagen 2012). Individual differences in behaviour is currently a much studied phenomenon in many taxa (e.g., Koolhaas et al. 1999; Sih et al. 2004; Réale et al. 2007). An early study on yellow perch showed that schooling tendency differed among tagged fish that spent between 20 and 100% of their time in shoals of more than two individuals (Helfman 1984). In young European perch, consistent differences along the bold/shy gradient have been found (e.g., Magnhagen 2007). However, European perch is a social species, and individual behaviour within a group is often modified by social influences (Sumpter et al. 2008; Conrad et al. 2011). In aquarium studies looking at trade-offs between foraging near a predator and hiding in an area without food, the behaviour of young European perch was affected by the group (Magnhagen 2007; Magnhagen and Bunnefeld 2009; Fig. 13). The risk-taking behaviour was more similar within groups than between groups (Fig. 14), and individuals also behaved bolder in the presence of conspecifics than when alone (Magnhagen and Bunnefeld 2009). The degree to which perch adjust their individual behaviours is also a function of their phenotype. For example, shy European perch adjust their behaviour to the behaviour of other shoal mates more so than bold ones since they change their behaviour more with habituation and predation risk (Westerberg et al. 2004; Magnhagen and Staffan 2005).

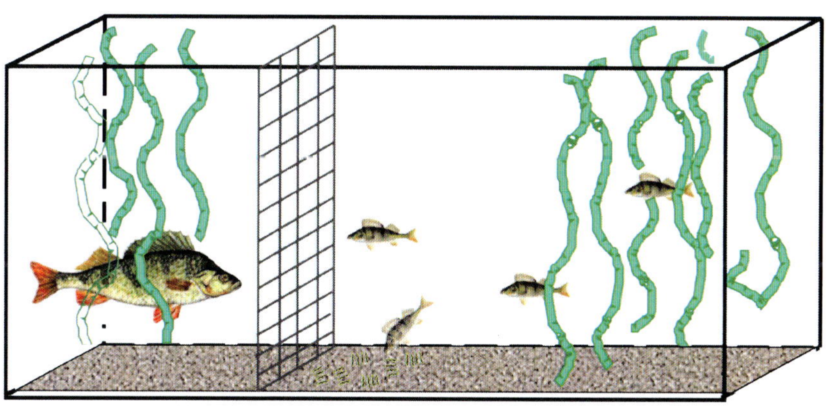

Fig. 13. The aquarium set-up used in the study by Hellström and Magnhagen (2011) to measure boldness in young of the year *P. fluviatilis*. The young perch could choose to stay in the vegetated area, far from the predator or enter the open area to feed. A plastic net separates the predator from the prey.

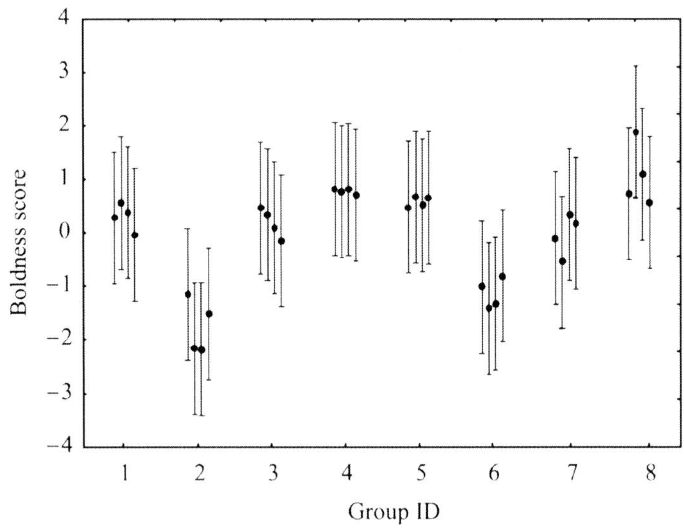

Fig. 14. Boldness scores (PC1) estimated from principal component analyses (PCA), including time in the open, latency to start feeding and duration of first feeding bout. Scores of individual one-year-old *P. fluviatilis* (mean ± 95% CI), tested in groups of four, during three repeated runs, are shown. Data from Magnhagen and Borcherding (2008).

9.7 Conclusion

Perca spp. have proven to be the ideal test subject when assaying for ecotoxicological effects of pollutants and toxic chemicals (see Chapter 10, this book); they are an ideal test species for bioenergetics research; and are an incredibly important recreational and commercial fish species whose management has garnered much attention (Chapters 4, 5, and 6). Today, rapid environmental change is affecting species population dynamics at an unprecedented rate as changed habitat characteristics, altered species interactions, disruptors of physiological processes, and altered sensory environments are affecting individuals' reproduction and survival. Behaviour is oftentimes a key first response as individuals attempt to mediate these changes through managing novel predators, competitors, diseases, and abiotic stressors, adopting new habitats and forage, and timing key life-history events such as dispersal and reproduction, and/or space use and movement patterns, to better fit changed spatiotemporal conditions. An in-depth study of the behavioural repertoire of perch can help us to anticipate the effects multiple stressors can have on their survival and population persistence. Perch are facing such stressors as hypoxic environments (Robb and Abrahams 2002), increased turbidity (Jacobsen et al. 2014), changes in community composition from introduced and restored populations of competitors, predators and prey (Fobert et al. 2010), new endocrine-disrupting chemicals (Brown et al. 2005; Blair et al. 2013), and of course, continued harvesting (van Kooten et al. 2010). Whether perch will be capable of responding adaptively to their novel or disrupted environments depends on their immediate ability to actively disperse to suitable habitat as larvae and YOY and return to intact spawning habitats as adults; to forage efficiently in the presence of competitors and predators; to specialize if need be and/or be flexible enough in their personalities to withstand environmental variation; and to locate shoaling kin-conspecifics while avoiding cannibalistic cohorts. These perch behaviours will then drive social group and population dynamics, affect interspecific interactions, and influence how animals cope with environmental changes. To sum, the overall challenge now for research on perch is for continued empirical and modeling work on how different state variables (e.g., age, size, energy level, residual reproductive value) interact with each other and with the multiplicative effects of multiple stressors to affect individual behaviour, population dynamics and even whole-lake community structure in the face of human-induced, rapid environmental change (Montiglio and Royauté 2014). Such discoveries will advance both fundamental behavioural-ecological research, in addition to enhancing its applied value.

9.8 References

Aalto, S.K. and G.E. Newsome. 1990. Additional evidence supporting demic behaviour of a yellow perch (*Perca flavescens*) population. Can. J. Fish. Aquat. Sci. 47(10): 1959–1962.

Abrams, P.A. and L. Rowe. 1996. The effects of predation on the age and size of maturity of prey. Evolution 50(3): 1052–1061. doi: Doi 10.2307/2410646.

Araújo, M.S., P.R. Guimaraes, Jr., R. Svanbäck, A. Pinheiro, P. Guimaraes, S.F.d. Reis and D.I. Bolnick. 2008. Network analysis reveals contrasting effects of intraspecific competition on individual vs. population diets. Ecology 89(7): 1981–1993.

Attayde, J.L. and L.-A. Hansson. 2001. Fish-mediated nutrient recycling and the trophic cascade in lakes. Can. J. Fish. Aquat. Sci. 58(10): 1924–1931. doi: 10.1139/cjfas-58-10-1924.

Azizishirazi, A., W. Dew, H. Forsyth and G. Pyle. 2013. Olfactory recovery of wild yellow perch from metal contaminated lakes. Ecotox. Environ. Safe. 88: 42–47. doi: 10.1016/j.ecoenv.2012.10.015.

Azizishirazi, A., W.A. Dew, B. Bougas, M. Dashtband, L. Bernatchez and G.G. Pyle. 2014. Chemosensory mediated behaviors and gene transcription profiles in wild yellow perch (*Perca flavescens*) from metal contaminated lakes. Ecotox. Environ. Safe. 106: 239–245. doi: 10.1016/j.ecoenv.2014.04.045.

Bauer, W.F., N.B. Radabaugh and M.L. Brown. 2009. Diel movement patterns of yellow perch in a simple and a complex lake basin. N. Am. J. Fish. Manage. 29(1): 64–71. doi: 10.1577/m07-087.1.

Beeck, P. 2003. The early piscivory of European perch (*Perca fluviatilis*)—a neglected phenomenon with notable consequences for the population structure and fish community in lake ecosystems. (PhD), Köln: Universität zu Köln.

Begon, M., M. Mortimer and D.J. Thompson. 2009. Interspecific competition. *In*: Population Ecology: A Unified Study of Animals and Plants, Third Edition. Blackwell Science Ltd., Oxford, UK.

Behrmann-Godel, J., G. Gerlach and R. Eckmann. 2006. Kin and population recognition in sympatric Lake Constance perch (*Perca fluviatilis* L.): can assortative shoaling drive population divergence? Behav. Ecol. Sociobiol. 59(4): 461–468. doi: 10.1007/s00265-005-0070-3.

Behrmann-Godel, J. and G. Gerlach. 2008. First evidence for postzygotic reproductive isolation between two populations of Eurasian perch (*Perca fluviatilis* L.) within Lake Constance. Front Zool. 5(1): 3.

Beisner, B.E., A.R. Ives and S.R. Carpenter. 2003. The effects of an exotic fish invasion on the prey communities of two lakes. J. Anim. Ecol. 72(2): 331–342.

Bennetts, R.E., J.D. Nichols, J.-D. Lebreton, R. Pradel, J.E. Hines and W.M. Kitchens. 2001. Methods for estimating dispersal probabilities and related parameters using marked animals. pp. 3–17. J. Clobert, E. Danchin, A.A. Dhondt and J.D. Nichols (eds.). *In*: Dispersal. Oxford University Press, Oxford, UK.

Bergek, S. and M. Björklund. 2007. Cryptic barriers to dispersal within a lake allow genetic differentiation of Eurasian perch. Evolution 61(8): 2035–2041.

Bergek, S., G. Sundblad and M. Björklund. 2010. Population differentiation in perch *Perca fluviatilis*: environmental effects on gene flow? J. Fish Biol. 76(5): 1159–1172.

Bergman, E. 1990. Effects of roach *Rutilus rutilus* on two percids, *Perca fluviatilis* and *Gymnocephalus cernua*: importance of species interactions for diet shifts. Oikos 241–249.

Bertolo, A., F.G. Blanchet, P. Magnan, P. Brodeur, M. Mingelbier and P. Legendre. 2012. Inferring processes from spatial patterns: the role of directional and non-directional forces in shaping fish larvae distribution in a freshwater lake system. PLoS One 7(11): e50239. doi: 10.1371/journal.pone.0050239.

Biro, P.A., M.V. Abrahams, J.R. Post and E.A. Parkinson. 2006. Behavioural trade-offs between growth and mortality explain evolution of submaximal growth rates. J. Anim. Ecol. 75(5): 1165–1171.

Blair, B.D., J.P. Crago, C.J. Hedman and R.D. Klaper. 2013. Pharmaceuticals and personal care products found in the Great Lakes above concentrations of environmental concern. Chemosphere 93(9): 2116–2123.

Bolnick, D.I., R. Svanbäck, J.A. Fordyce, L.H. Yang, J.M. Davis, C.D. Hulsey and M.L. Forister. 2003. The ecology of individuals: incidence and implications of individual specialization. Am. Nat. 161(1): 1–28.

Borcherding, J. 2006. Prey or predator: 0+ perch (*Perca fluviatilis*) in the trade-off between food and shelter. Environ. Biol. Fishes 77(1): 87–96. doi: 10.1007/s10641-006-9057-9.

Borcherding, J., P. Beeck, D.L. Deangelis and W.R. Scharf. 2010. Match or mismatch: the influence of phenology on size-dependent life history and divergence in population structure. J. Anim. Ecol. 79(5): 1101–1112. doi: 10.1111/j.1365-2656.2010.01704.x.

Borcherding, J. and C. Magnhagen. 2008. Food abundance affects both morphology and behaviour of juvenile perch. Ecology of Freshwater Fish 17(2): 207–218. doi: 10.1111/j.1600-0633.2007.00272.x.

Brabrand, A. 1995. Intra-cohort cannibalism among larval stages of perch (*Perca fluviatilis*). Ecol. Freshw. Fish 4(2): 70–76.

Brosse, S. and S. Lek. 2002. Relationships between environmental characteristics and the density of age-0 Eurasian perch *Perca fluviatilis* in the littoral zone of a lake: a nonlinear approach. T. Am. Fish. Soc. 131(6): 1033–1043.

Brown, A., A. Riddle, I. Winfield, J. Fletcher and J. James. 2005. Predicting the effects of endocrine disrupting chemicals on healthy and disease impacted populations of perch (*Perca fluviatilis*). Ecol. Model. 189(3): 377–395.

Bruylants, B., A. Vandelannoote and R. Verheyen. 1986. The movement pattern and density distribution of perch, *Perca fluviatilis* L., in a channelized lowland river. Aquacult. Res. 17(1): 49–57.

Budaev, S. and C. Brown. 2011. Personality traits and behaviour. pp. 135–165. *In*: C. Brown, K. Laland and J. Krause (eds.). Fish Cognition and Behavior. Wiley-Blackwell, Oxford.

Byström, P., M. Huss and L. Persson. 2012. Ontogenetic constraints and diet shifts in perch (*Perca fluviatilis*): mechanisms and consequences for intra-cohort cannibalism. Freshwater Biol. 57(4): 847–857. doi: 10.1111/j.1365-2427.2012.02752.x.

Byström, P., L. Persson and E. Wahlström. 1998. Competing predators and prey: juvenile bottlenecks in whole-lake experiments. Ecology 79(6): 2153–2167.

Christensen, B. and L. Persson. 1993. Species-specific antipredatory behaviours: effects on prey choice in different habitats. Behav. Ecol. Sociobiol. 32(1): 1–9.

Claessen, D., A.M. de Roos and L. Persson. 2000. Dwarfs and giants: cannibalism and competition in size-structured populations. Am. Nat. 155(2): 219–237.

Conrad, J.L., K.L. Weinersmith, T. Brodin, J.B. Saltz and A. Sih. 2011. Behavioural syndromes in fishes: a review with implications for ecology and fisheries management. J. Fish Biol. 78(2): 395–435. doi: 10.1111/j.1095-8649.2010.02874.x.

Craig, J.F. 2008. Percid Fishes: Systematics, Ecology and Exploitation. John Wiley & Sons.

Dall, S.R., L.A. Giraldeau, O. Olsson, J.M. McNamara and D.W. Stephens. 2005. Information and its use by animals in evolutionary ecology. Trends Ecol. Evol. 20(4): 187–193.

Dettmers, J.M., J. Janssen, B. Pientka, R.S. Fulford and D.J. Jude 2005. Evidence across multiple scales for offshore transport of yellow perch (*Perca flavescens*) larvae in Lake Michigan. Can. J. Fish. Aquat. Sci. 62(12): 2683–2693.

Dew, W.A., A. Azizishirazi and G.G. Pyle. 2014. Contaminant-specific targeting of olfactory sensory neuron classes: Connecting neuron class impairment with behavioural deficits. Chemosphere. Available only online doi: 10.1016/j.chemosphere.2014.02.047.

Diehl, S. and P. Eklöv. 1995. Effects of piscivore-mediated habitat use on resources, diet, and growth of perch. Ecology 76(6): 1712–1726. doi: Doi 10.2307/1940705.

Dieterich, A., M. Mörtl and R. Eckmann. 2004. The effects of zebra mussels (*Dreissena polymorpha*) on the foraging success of Eurasian perch (*Perca fluviatilis*) and ruffe (*Gymnocephalus cernuus*). Int. Rev. Hydrobiol. 89(3): 229–237. doi: 10.1002/iroh.200310693.

Eadie, J.M. and A. Keast. 1982. Do goldeneye and perch compete for food? Oecologia 55(2): 225–230.

Eklöv, P. 1992. Group foraging versus solitary foraging efficiency in piscivorous predators—the perch, *Perca fluviatilis* and pike, *Esox lucius*. Patterns Anim. Behav. 44(2): 313–326. doi: 10.1016/0003-3472(92)90037-a.

Eklöv, P. 1997. Effects of habitat complexity and prey abundance on the spatial and temporal distributions of perch (*Perca fluviatilis*) and pike (*Esox lucius*). Can. J. Fish. Aquat. Sci. 54(7): 1520–1531.

Eklöv, P. and L. Persson. 1995. Species-specific antipredator capacities and prey refuges: interactions between piscivorous perch (*Perca fluviatilis*) and juvenile perch and roach (*Rutilus rutilus*). Behav. Ecol. Sociobiol. 37(3): 169–178.

Eklöv, P. and R. Svanbäck. 2006. Predation risk influences adaptive morphological variation in fish populations. Am. Nat. 167(3): 440–452. doi: 10.1086/499544.

Eklöv, P. and T. VanKooten. 2001. Facilitation among piscivorous predators: effects of prey habitat use. Ecology 82(9): 2486–2494.

Elmberg, J., L. Dessborn and G. Englund. 2010. Presence of fish affects lake use and breeding success in ducks. Hydrobiologia 641(1): 215–223. doi: 10.1007/s10750-009-0085-2.

Estlander, S., L. Nurminen, M. Olin, M. Vinni, S. Immonen, M. Rask, J. Ruuhijärvi, J. Horppila and H. Lehtonen. 2010. Diet shifts and food selection of perch *Perca fluviatilis* and roach *Rutilus rutilus* in humic lakes of varying water colour. J. Fish Biol. 77(1): 241–256. doi: 10.1111/j.1095-8649.2010.02682.x.

Fobert, E., M.G. Fox, M. Ridgway and G.H. Copp. 2011. Heated competition: how climate change will affect non-native pumpkinseed *Lepomis gibbosus* and native perch *Perca fluviatilis* interactions in the U.K. J. Fish Biol. 79(6): 1592–1607. doi: 10.1111/j.1095-8649.2011.03083.x.

Fullerton, A.H. and G.A. Lamberti. 2006. A comparison of habitat use and habitat-specific feeding efficiency by Eurasian ruffe (*Gymnocephalus cernuus*) and yellow perch (*Perca flavescens*). Ecol. Freshw. Fish 15(1): 1–9. doi: 10.1111/j.1600-0633.2005.00114.x.

Fullerton, A.H., G.A. Lamberti, D.M. Lodge and F.W. Goetz. 2000. Potential for resource competition between Eurasian ruffe and yellow perch: growth and RNA responses in laboratory experiments. T. Am. Fish. Soc. 129(6): 1331–1339. doi: 10.1577/1548-8659(2000)129<1331:pfrcbe>2.0.co;2.

Gerlach, G., U. Schardt, R. Eckmann and A. Meyer. 2001. Kin-structured subpopulations in Eurasian perch (*Perca fluviatilis* L.). Heredity 86: 213–221. doi: 10.1046/j.1365-2540.2001.00825.x.

Glover, D.C., J.M. Dettmers, D.H. Wahl and D.F. Clapp. 2008. Yellow perch (*Perca flavescens*) stock structure in Lake Michigan: an analysis using mark–recapture data. C. J. Fish. Aquat. Sci. 65(9): 1919–1930. doi: 10.1139/f08-100.

Graeb, B.D., T. Galarowicz, D.H. Wahl, J.M. Dettmers and M.J. Simpson. 2005. Foraging behavior, morphology, and life history variation determine the ontogeny of piscivory in two closely related predators. Can. J. Fish. Aquat. Sci. 62(9): 2010–2020.

Grant, J.W. 1993. Whether or not to defend? The influence of resource distribution. Marine & Freshwater Behaviour & Physiology 23(1-4): 137–153.

Heermann, L., W. Scharf, G. van der Velde and J. Borcherding. 2013. Does the use of alternative food resources induce cannibalism in a size-structured fish population? Ecol. Freshw. Fish 23(2): 129–140. doi: 10.1111/eff.12060.

Helfman, G.S. 1979. Twilight activities of yellow perch, *Perca flavescens*. J. Fish. Res. Board Can. 36(2): 173–179.

Helfman, G.S. 1984. School fidelity in fishes—the yellow perch pattern. Anim. Behav. 32(AUG): 663–672. doi: 10.1016/s0003-3472(84)80142-6.

Hellström, G. and C. Magnhagen. 2011. The influence of experience on risk taking: results from a common-garden experiment on populations of Eurasian perch. Behav. Ecol. Sociobiol. 65(10): 1917–1926. doi: 10.1007/s00265-011-1201-7.

Heynen, M., L. Heermann and J. Borcherding. 2011. Does the consumption of divergent resources influence risk taking behaviour in juvenile perch (*Perca fluviatilis* L.)? Ecol. Freshw. Fish 20(1): 1–4. doi: 10.1111/j.1600-0633.2010.00473.x.

Heynen, M., I. Rentrop and J. Borcherding. 2014. Age matters—Experienced predation risk affects behavior and morphology of juvenile 0+ and 1+ perch. Limnologica—Ecol. Manage. Inl. Wat. 44: 32–39. doi: 10.1016/j.limno.2013.06.003.

Holyoak, M., R. Casagrandi, R. Nathan, E. Revilla and O. Spiegel. 2008. Trends and missing parts in the study of movement ecology. Proc. Natl. Acad. Sci. B 105(49): 19060–19065.

Hrabik, T.R., J.J. Magnuson and A.S. McLain. 1998. Predicting the effects of rainbow smelt on native fishes in small lakes: evidence from long-term research on two lakes. C. J. Fish. Aquat. Sci.55(6): 1364–1371.

Humphries, P. 2005. Spawning time and early life history of Murray cod, *Maccullochella peelii* (Mitchell) in an Australian river. Environ. Biol. Fishes 72(4): 393–407.

Huss, M., P. Byström and L. Persson. 2008. Resource heterogeneity, diet shifts and intra-cohort competition: effects on size divergence in YOY fish. Oecologia 158(2): 249–257. doi: 10.1007/s00442-008-1140-9.

Jacobsen, L., S. Berg, H. Baktoft, P.A. Nilsson and C. Skov. 2014. The effect of turbidity and prey fish density on consumption rates of piscivorous Eurasian perch *Perca fluviatilis*. J. Limnol. 73(1): doi: 10.4081/jlimnol.2014.837.

Jacobsen, L. and M. Perrow. 1998. Predation risk from piscivorous fish influencing the diel use of macrophytes by planktivorous fish in experimental ponds. Ecol. Freshw. Fish 7(2): 78–86.

Jamet, J.L. and N. Lair. 1991. An example of diel feeding cycle of 2 percids, perch (*Perca fluviatilis*) and ruffe (*Gymnocephalus cernuus*) in eutrophic lake Aydat (France). Ann. Sci. Nat. Zool. Biol. Anim. 12(3): 99–105.

Kaemingk, M.A., J.C. Jolley, D.W. Willis and S.R. Chipps. 2012a. Priority effects among young-of-the-year fish: reduced growth of bluegill sunfish (*Lepomis macrochirus*) caused by yellow perch (*Perca flavescens*)? Freshwater Biol. 57(4): 654–665. doi: 10.1111/j.1365-2427.2011.02728.x.

Kaemingk, M.A. and D.W. Willis. 2012b. Mensurative approach to examine potential interactions between age-0 yellow perch (*Perca flavescens*) and bluegill (*Lepomis macrochirus*). Aquat. Ecol. 46(3): 353–362. doi: 10.1007/s10452-012-9406-z.

Kahl, U. and R.J. Radke. 2006. Habitat and food resource use of perch and roach in a deep mesotrophic reservoir: enough space to avoid competition? Ecol. Freshw. Fish 15(1): 48–56. doi: 10.1111/j.1600-0633.2005.00120.x.

Kaufman, S.D., J.M. Gunn, G.E. Morgan and P. Couture. 2006. Muscle enzymes reveal walleye (*Sander vitreus*) are less active when larger prey (cisco, Coregonus artedi) are present. Can. J. Fish. Aquat. Sci. 63(5): 970–979.

Kocovsky, P., T. Sullivan, C. Knight and C. Stepien. 2013. Genetic and morphometric differences demonstrate fine-scale population substructure of the yellow perch *Perca flavescens*: need for redefined management units. J. Fish Biol. 82(6): 2015–2030.

Koolhaas, J.M., S.M. Korte, S.F. De Boer, B.J. Van Der Vegt, C.G. Van Reenen, H. Hopster, I.C. De Jong, M.A.W. Ruis and H.J. Blokhuis. 1999. Coping styles in animals: current status in behavior and stress-physiology. Neurosci. Biobehav. R. 23(7): 925–935.

Krause, J. and G.D. Ruxton. 2002. Living in Groups. Oxford University Press, Oxford.

Lass, S. and P. Spaak. 2003. Chemically induced anti-predator defences in plankton: a review. Hydrobiologia 491(1-3): 221–239.

Lechner, A., H. Keckeis, E. Schludermann, P. Humphries, N. McCasker and M. Tritthart. 2013. Hydraulic forces impact larval fish drift in the free flowing section of a large European river. Ecohydrology 7(2): 648–658. doi: 10.1002/eco.1386.

Lechner, A., H. Keckeis, E. Schludermann, F. Loisl, P. Humphries, M. Glas, M. Tritthart and H. Habersack. 2013. Shoreline configurations affect dispersal patterns of fish larvae in a large river. ICES J. Mar. Sci. 71(4): 930–942. doi: 10.1093/icesjms/fst139.

Leclerc, E., Y. Mailhot, M. Mingelbier and L. Bernatchez. 2008. The landscape genetics of yellow perch (*Perca flavescens*) in a large fluvial ecosystem. Mol. Ecol. 17(7): 1702–1717.

LeCren, E.D. 1992. Exceptionally big individual perch (*Perca fluviatilis* L.) and their growth. J. Fish Biol. 40(4): 599–625.

Leung, C., P. Magnan and B. Angers. 2012. Genetic evidence for sympatric populations of yellow perch (*Perca flavescens*) in Lake Saint-Pierre (Canada): the crucial first step in developing a fishery management plan. J. Aquacult. Res. Devel. S6: 001. doi:10.4172/2155-9546.S6-001.

Linehan, J.E., R.S. Gregory and D.C. Schneider. 2001. Predation risk of age-0 cod (*Gadus*) relative to depth and substrate in coastal waters. J. Exp. Mar. Biol. Ecol. 263(1): 25–44. doi: 10.1016/s0022-0981(01)00287-8.

Lott, J.P., D.W. Willis and D.O. Lucchesi. 1996. Relationship of food habits to yellow perch growth and population structure in South Dakota lakes. J. Freshw. Ecol. 11(1): 27–37.

MacGregor, R.B. and L.D. Witzel. 1987. A Twelve Year Study of the Fish Community in the Nanticoke Region of Long Point Bay, Lake Erie. Lake Erie Fisheries Assessment Unit Report 1987-3. Port Dover, ON: Ontario Ministry of Natural Resources, Lake Erie Fisheries Assessment Unit.

Magnhagen, C. 2007. Social influence on the correlation between behaviours in young-of-the-year perch. Behav. Ecol. Sociobiol. 61(4): 525–531.

Magnhagen, C. 2012. Personalities in a crowd: What shapes the behaviour of Eurasian perch and other shoaling fishes? Curr. Zool. 58(1).

Magnhagen, C. and J. Borcherding. 2008. Risk-taking behaviour in foraging perch: does predation pressure influence age-specific boldness? Anim. Behav. 75(2): 509–517. doi: 10.1016/j.anbehav.2007.06.007.

Magnhagen, C. and N. Bunnefeld. 2009. Express your personality or go along with the group: what determines the behaviour of shoaling perch? Proc. Natl. Acad. Sci. B 276(1671): 3369–3375. doi: 10.1098/rspb.2009.0851.

Magnhagen, C. and F. Staffan. 2003. Social learning in young-of-the-year perch encountering a novel food type. J. Fish Biol. 63(3): 824–829.

Martin, C.W., F.J. Fodrie, K.L. Heck, Jr. and J. Mattila. 2010. Differential habitat use and antipredator response of juvenile roach (*Rutilus rutilus*) to olfactory and visual cues from multiple predators. Oecologia 162(4): 893–902. doi: 10.1007/s00442-010-1564-x.

McDowall, R.M. 1990. New Zealand freshwater fishes: a natural history and guide. Heinemann Reed, Auckland. 553 p.

Miehls, S.M. and J.M. Dettmers. 2011. Factors influencing habitat shifts of age-0 yellow perch in southwestern Lake Michigan. T. Am. Fish. Soc. 140(5): 1317–1329. doi: 10.1080/00028487.2011.620484.

Mikheev, V.N., A.F. Pasternak, G. Tischler and J. Wanzenbock. 2005. Contestable shelters provoke aggression among 0+ perch, *Perca fluviatilis*. Environ. Biol. Fishes 73(2): 227–231. doi: 10.1007/s10641-005-0558-8.

Mikheev, V.N., J. Wanzenbock and A.F. Pasternak. 2006. Effects of predator-induced visual and olfactory cues on 0+ perch (*Perca fluviatilis* L.) foraging behaviour. Ecol. Freshw. Fish 15(2): 111–117. doi: 10.1111/eff.2006.15.issue-2.

Mirza, R.S. and D.P. Chivers. 2001. Do juvenile yellow perch use diet cues to assess the level of threat posed by intraspecific predators? Behaviour 138: 1249–1258.

Mirza, R.S., W.W. Green, S. Connor, A.C.W. Weeks, C.M. Wood and G.G. Pyle. 2009. Do you smell what I smell? Olfactory impairment in wild yellow perch from metal-contaminated waters. Ecotox. Environ. Safe. 72(3): 677–683.

Mitchell, J.C. and S.C. Walls. 2008. Cannibalism. pp. 513–517. *In*: S.E. Jørgensen and B.D. Fath (eds.). Population Dynamics, Vol. 1. Elsevier, Oxford.

Montiglio, P.-O. and R. Royauté. 2014. Contaminants as a neglected source of behavioural variation. Anim. Behav. 88: 29–35.

Mustamäki, N., T. Cederberg and J. Mattila. 2014. Diet, stable isotopes and morphology of Eurasian perch (*Perca fluviatilis*) in littoral and pelagic habitats in the northern Baltic Proper. Environ. Biol. Fishes 97(6): 675–689.

Nislow, K.H., J.D. Armstrong and J.W.A. Grant. 2010. The role of competition in the ecology of juvenile Atlantic salmon. *In*:Ø. Aas, S. Einum, A. Klemetsen and J. Skurdal (eds.). Atlantic Salmon Ecology. Wiley-Blackwell, Oxford, UK. doi: 10.1002/9781444327755.ch7.

Nummi, P., V.-M. Väänänen, M. Rask, K. Nyberg and K. Taskinen. 2011. Competitive effects of fish in structurally simple habitats: perch, invertebrates, and goldeneye in small boreal lakes. Aquatic Sciences 74(2): 343–350. doi: 10.1007/s00027-011-0225-4.

Olson, M.H., D.M. Green and L.G. Rudstam. 2001. Changes in yellow perch (*Perca flavescens*) growth associated with the establishment of a walleye (*Stizostedion vitreum*) population in Canadarago Lake, New York (USA). Ecology of Freshwater Fish 10(1): 11–20.

Olsson, J., R. Svanback and P. Eklov. 2007. Effects of resource level and habitat type on behavioral and morphological plasticity in Eurasian perch. Oecologia 152(1): 48–56. doi: 10.1007/s00442-006-0588-8.

Parker, A.D., C.A. Stepien, O.J. Sepulveda-Villet, C.B. Ruehl and D.G. Uzarski. 2009. The interplay of morphology, habitat, resource use, and genetic relationships in young yellow perch. Transactions of the American Fisheries Society 138(4): 899–914. doi: 10.1577/t08-093.1.

Persson, L. 1983. Effects of intra- and interspecific competition on dynamics and size structure of a perch *Perca fluviatilis* and a roach *Rutilus rutilus* population. Oikos 41(1): 126–132.

Persson, L. 1987a. The effects of resource availability and distribution on size class interactions in perch, *Perca fluviatilis*. Oikos 48(2): 148–160.

Persson, L. 1987b. Effects of habitat and season on competitive interactions between roach (*Rutilus rutilus*) and perch (*Perca fluviatilis*). Oecologia 73(2): 170–177.

Persson, L. 1991. Behavioral response to predators reverses the outcome of competition between prey species. Behav. Ecol. Sociobiol. 28(2): 101–105.

Persson, L. 1993. Predator-mediated competition in prey refuges: the importance of habitat dependent prey resources. Oikos 68(1): 12–22.

Persson, L., D. Claessen, A.M. De Roos, P. Byström, S. Sjögren, R. Svanbäck, E. Wahlström and E. Westman. 2004. Cannibalism in a size-structured population: energy extraction and control. Ecol. Monograph. 74(1): 135–157.

Persson, L. and A.M. de Roos. 2013. Symmetry breaking in ecological systems through different energy efficiencies of juveniles and adults. Ecology 94(7): 1487–1498.

Persson, L., A.M. De Roos, D. Claessen, P. Byström, J. Lövgren, S. Sjögren, R. Svänback, E. Wahlström and E. Westman. 2003. Gigantic cannibals driving a whole-lake trophic cascade. Proc. Natl. Acad. Sci. USA 100(7): 4035–4039. doi: 10.1073/pnas.0636404100.

Persson, L. and P. Eklov. 1995. Prey refuges affecting interactions between piscivorous perch and juvenile perch and roach. Ecology 76(1): 70–81.

Persson, L., A. Van Leeuwen and A.M. De Roos. 2014. The ecological foundation for ecosystem-based management of fisheries: mechanistic linkages between the individual-, population-, and community-level dynamics. ICES J. Mar. Sci. doi: 10.1093/icesjms/fst231.

Pitcher, T.J., A.E. Magurran and I.J. Winfield. 1982. Fish in larger shoals find food faster. Behav. Ecol. Sociobiol. 10(2): 149–151. doi: 10.1007/bf00300175.

Pitcher, T.J. and J.K. Parrish. 1993. Functions of shoaling behaviour in teleosts. pp. 363–439. *In*: T.J. Pitcher (ed.). Behaviour of Teleost Fishes. Chapman & Hall, London.

Polis, G. 1988. Exploitation competition and the evolution of interference, cannibalism, and intraguild predation in age/size-structured populations. pp. 185–202. *In*: B. Ebenman and L. Persson (eds.). Size Structured Populations. Springer-Verlag, Berlin, Heidelberg.

Post, J.R., M.R.S. Johannes and D.J. McQueen. 1997. Evidence of density-dependent cohort splitting in age-0 yellow perch (*Perca flavescens*): Potential behavioural mechanisms and population-level consequences. Can. J. Fish. Aquat. Sci. 54(4): 867–875. doi: 10.1139/cjfas-54-4-867.

Post, J.R. and D.J. McQueen. 1988. Ontogenetic changes in the distribution of larval and juvenile yellow perch (Perca flavescens): a response to prey or predators? Can. J. Fish. Aquat. Sci. 45(10): 1820–1826.

Post, J.R., E.A. Parkinson and N.T. Johnston. 1998. Spatial and temporal variation in risk to piscivory of age-0 rainbow trout: Patterns and population level consequences. T. American Fish. Soc. 127(6): 932–942. doi: 10.1577/1548-8659(1998)127<0932:satvir>2.0.co;2.

Probst, W.N., G. Thomas and R. Eckmann. 2009. Hydroacoustic observations of surface shoaling behaviour of young-of-the-year perch *Perca fluviatilis* (Linnaeus 1758) with a towed upward-facing transducer. Fish. Res. 96(2-3): 133–138. doi: 10.1016/j.fishres.2008.10.009.

Radabaugh, N.B., W.F. Bauer and M.L. Brown. 2010. A Comparison of Seasonal Movement Patterns of Yellow Perch in Simple and Complex Lake Basins. N. Am. J. Fish. Manage. 30(1): 179–190. doi: 10.1577/m08-243.1.

Réale, D., S.M. Reader, D. Sol, P.T. McDougall and N.J. Dingemanse. 2007. Integrating animal temperament within ecology and evolution. Biol. Rev. 82(2): 291–318.

Rennie, M.D., N.C. Collins, B.J. Shuter, J.W. Rajotte and P. Couture. 2005. A comparison of methods for estimating activity costs of wild fish populations: more active fish observed to grow slower. Canadian J. Fish. Aquat. Sci. 62(4): 767–780.

Rennie, M.D., C.F. Purchase, B.J. Shuter, N.C. Collins, P.A. Abrams and G.E. Morgan. 2010. Prey life-history and bioenergetic responses across a predation gradient. J. Fish Biol. 77(6): 1230–1251. doi: 10.1111/j.1095-8649.2010.02735.x.

Revilla, E. and T. Wiegand. 2008. Individual movement behavior, matrix heterogeneity, and the dynamics of spatially structured populations. Proc. Natl. Acad. Sci. 105(49): 19120–19125.

Robillard, S.R. and J.E. Marsden. 2001. Spawning substrate preferences of yellow perch along a sand-cobble shoreline in southwestern Lake Michigan. T. Am. Fish. Soc. 21(1): 208–215.

Robb, T. and M.V. Abrahams. 2002. The influence of hypoxia on risk of predation and habitat choice by the fathead minnow, *Pimephales promelas*. Behav. Ecol. Sociobiol. 52(1): 25–30.

Rosier, R. and T. Langkilde. 2011. Behavior under risk: how animals avoid becoming dinner. Nature Education Knowledge 2(11): 8.

Roswell, C.R., S.A. Pothoven and T.O. Höök. 2013. Spatio-temporal, ontogenetic and interindividual variation of age-0 diets in a population of yellow perch. Ecol. Freshw. Fish 22(3): 479–493. doi: 10.1111/eff.12041.

Ryer, C.H. and B.L. Olla. 1998. Effect of light on juvenile walleye pollock shoaling and their interaction with predators. Mar. Ecol. Prog. Series 167: 215–226. doi: 10.3354/meps167215.

Sale, P.F. 2004. Connectivity, recruitment variation, and the structure of reef fish communities. Integr. Comp. Biol. 44(5): 390–399.

Sanderson, B.L., T.R. Hrabik, J.J. Magnuson and D.M. Post. 1999. Cyclic dynamics of a yellow perch (*Perca flavescens*) population in an oligotrophic lake: evidence for the role of intraspecific interactions. Can. J. Fish. Aquat. Sci. 56(9): 1534–1542.

Sanderson, B.L., K.A. Barnas and A.M.W. Rub. 2009. Nonindigenous species of the Pacific Northwest: an overlooked risk to endangered salmon? BioScience 59(3): 245–256.

Schleuter, D. and R. Eckmann. 2006. Competition between perch (*Perca fluviatilis*) and ruffe (*Gymnocephalus cernuus*): the advantage of turning night into day. Freshw. Biol. 51(2): 287–297. doi: 10.1111/j.1365-2427.2005.01495.x.

Schleuter, D., S. Haertel-Borer, P. Fischer and R. Eckmann. 2007. Respiration rates of Eurasian perch *Perca fluviatilis* and ruffe: Lower energy costs in groups. T. Am. Fish. Soc. 136(1): 43–55. doi: 10.1577/t06-123.1.

Schoenebeck, C.W. and M.L. Brown. 2010. Potential importance of competition, predation, and prey on yellow perch growth from two dissimilar population types. Prairie Natural. 42: 32–37.

Schulze, T., U. Baade, H. Dörner, R. Eckmann, S.S. Haertel-Borer, F. Hölker and T. Mehner. 2006. Response of the residential piscivorous fish community to introduction of a new predator type in a mesotrophic lake. Can. J. Fish. Aquat. Sci. 63(10): 2202–2212. doi: 10.1139/f06-099.

Sepulveda-Villet, O.J., C.A. Stepien and R. Vinebrooke. 2011. Fine-scale population genetic structure of the yellow perch *Perca flavescens* in Lake Erie. Can. J. Fish. Aquat. Sci. 68(8): 1435–1453.

Sepulveda-Villet, O.J. and C.A. Stepien. 2012. Waterscape genetics of the yellow perch (*Perca flavescens*): patterns across large connected ecosystems and isolated relict populations. Mol. Ecol. 21(23): 5795–5826.

Sih, A., A. Bell and J.C. Johnson. 2004. Behavioral syndromes: an ecological and evolutionary overview. Trends Ecol. Evol. 19(7): 372–378.

Sih, A., G. Englund and D. Wooster. 1998. Emergent impacts of multiple predators on prey. Trends Ecol. Evol. 13(9): 350–355.

Snickars, M., A. Sandström and J. Mattila. 2004. Antipredator behaviour of 0+ year *Perca fluviatilis*: effect of vegetation density and turbidity. J. Fish Biol. 65: 1604–1613.

Staffan, F., C. Magnhagen and A. Alanärä. 2002. Variation in food intake within groups of juvenile perch. J. Fish Biol. 60(3): 771–774.

Staffan, F., C. Magnhagen and A. Alanärä. 2005. Individual feeding success of juvenile perch is consistent over time in aquaria and under farming conditions. J. Fish Biol. 66(3): 798–809.

Stephens, D.W. and J.R. Krebs. 1986. Foraging theory. Princeton University Press, Princeton, New Jersey, USA.

Strand, A., A. Alanärä and C. Magnhagen. 2007. Effect of group size on feed intake, growth and feed efficiency of juvenile perch. J. Fish Biol. 71(2): 615–619. doi: 10.1111/j.1095-8649.2007.01497.x.

Sullivan, T. 2013. A Fine-scale Analysis of Spatial and Temporal Population Genetic Patterns in the Yellow Perch (*Perca flavescens*). Electronic Thesis or Dissertation. University of Toledo. Retrieved from https://etd.ohiolink.edu/.

Sumpter, D.J.T., J. Krause, R. James, I.D. Couzin and A.J.W. Ward. 2008. Consensus decision making by fish. Curr. Biol. 18(22): 1773–1777.

Svanbäck, R. and D.I. Bolnick. 2005. Intraspecific competition affects the strength of individual specialization: an optimal diet theory method. Evol. Ecol. Res. 7(7): 993–1012.

Svanbäck, R. and L. Persson. 2004. Individual diet specialization, niche width and population dynamics: implications for trophic polymorphisms. J. Anim. Ecol. 73(5): 973–982.

Swanson, B.O., A.C. Gibb, J.C. Marks and D.A. Hendrickson. 2003. Trophic polymorphism and behavioral differences decrease intraspecific competition in a cichlid, Herichthys minckleyi. Ecology 84(6): 1441–1446.

Urbatzka, R., P. Beeck, G. van der Velde and J. Borcherding. 2008. Alternative use of food resources causes intra-cohort variation in the size distribution of young-of-the-year perch (Perca fluviatilis). Ecol. Freshw. Fish 17(3): 475–480. doi: 10.1111/j.1600-0633.2008.00300.x.

Urho, L. 1996. Habitat shifts of perch larvae as survival strategy. Annales Zoologi Fennici 33(3): 329–340.

Utne, A., E. Brännäs and C. Magnhagen. 1997. Individual responses to predation risk and food density in perch (*Perca fluviatilis* L.). Can. J. Zool. 75(12): 2027–2035.

Väänänen, V.-M., P. Nummi, H. Pöysä, M. Rask and K. Nyberg. 2012. Fish–duck interactions in boreal lakes in Finland as reflected by abundance correlations. Hydrobiologia 697(1): 85–93. doi: 10.1007/s10750-012-1172-3.

Wahlström, E., L. Persson, S. Diehl and P. Byström. 2000. Size-dependent foraging efficiency, cannibalism and zooplankton community structure. Oecologia 123(1): 138–148.

Wanzenbock, J., V.N. Mikheev and A.F. Pasternak. 2006. Modification of 0+ perch foraging behaviour by indirect cues of predation risk. Ecol. Freshw. Fish 15(2): 118–124. doi: 10.1111/j.1600-0633.2006.00139.x.

Ward, A.J., M.M. Webster and P.J. Hart. 2006. Intraspecific food competition in fishes. Fish Fish. 7(4): 231–261.

Ward, A.J.W., J.E. Herbert-Read, D.J.T. Sumpter and J. Krause. 2011. Fast and accurate decisions through collective vigilance in fish shoals. Proc. Natl. Acad. Sci. USA 108(6): 2312–2315. doi: 10.1073/pnas.1007102108.

Westerberg, M., F. Staffan and C. Magnhagen. 2004. Influence of predation risk on individual competitive ability and growth in Eurasian perch, *Perca fluviatilis*. Anim. Behav. 67(2): 273–279.

Wisenden, B.D., J. Klitzke, R. Nelson, D. Friedl and P.C. Jacobson. 2004. Predator-recognition training of hatchery-reared walleye (*Stizostedion vitreum*) and a field test of a training method using yellow perch (*Perca flavescens*). Can. J. Fish. Aquat. Sci. 61(11): 2144–2150. doi: 10.1139/f04-164.

Zamora, L. and R. Moreno-Amich. 2002. Quantifying the activity and movement of perch in a temperate lake by integrating acoustic telemetry and a geographic information system. Hydrobiologia 483(1-3): 209–218. doi: 10.1023/a:1021396016424.

10

Using *Perca* as Biomonitors in Ecotoxicological Studies

Patrice Couture,[1,*] Greg Pyle,[2] Peter G.C. Campbell[1] and Alice Hontela[2]

ABSTRACT

Perch have been used for ecotoxicological research for nearly a half-century owing to their overlapping geographic distribution with large-scale industrial activities in the Northern Hemisphere. Because they can withstand relatively high concentrations of environmental contaminants, perch are often found in more contaminated habitats than those that can be tolerated by more sensitive species. In this chapter we survey the ecotoxicological literature for studies involving perch, and present two case studies of our own that illustrate the value of perch in aquatic toxicology. The case studies focus on wild yellow perch in northeastern Ontario and northern Québec, Canada, from lakes affected by metal mining and copper smelting, respectively. Using both field and laboratory-based approaches, these studies provide important information on metal uptake, accumulation, and depuration, sub-cellular metal partitioning and toxicity, and metal effects on growth and condition, metabolism and aerobic capacity, and longevity. Newly developed molecular tools are now being used to study ecotoxicogenomic mechanisms of long-term perch adaptation to contaminated environments. Together, these studies inform ecological risk assessments of metal-contaminated environments, which can be used to develop strong, scientifically based policies to protect at-risk perch populations.

[1] Institut national de la recherche scientifique, Centre Eau Terre Environnement, Québec, QC G1K 9A9 Canada.

[2] Dept. of Biological Sciences, Water Institute for Sustainable Environments (WISE), University of Lethbridge, Lethbridge, AB T1K 3M4 Canada.

* Corresponding author: patrice.couture@ete.inrs.ca

Keywords: Ecotoxicology, yellow perch, European perch, metals, organic contaminants, historical review, mechanisms of toxicity, indirect effects of contaminants

10.1 Introduction

The geographic distribution of yellow perch and European perch (*Perca flavescens* and *P. fluviatilis*, respectively) naturally overlaps with some of the most productive industrial regions in the world. Over the past decade or so, these two species have received an increasing amount of research attention by ecotoxicologists and freshwater ecologists. These species, given their sedentary and non-migratory lifestyle and their considerable tolerance to environmental contaminants—, especially metals, have demonstrated their value as environmental sentinels.

In this chapter, we survey the scientific literature for ecotoxicological studies involving *Perca* spp. We provide a brief historical overview of some of the earlier studies with *Perca* spp., and how their utility in ecotoxicological research has greatly expanded recently to include areas affected by most major industrial centres in the northern hemisphere. Two case studies of our own work are examples of the kind of information yellow perch can provide about the receiving waters of two major Canadian industrial operations in northern Ontario and Québec. We end the chapter with a section describing some cutting edge work in perch ecotoxicogenomics that has already shed light on local population adaptations to long-term metal exposure from industrial contamination.

Although perch are difficult to maintain under laboratory conditions, they show great promise for investigating population-level effects in native populations inhabiting contaminated receiving waters. With this chapter, we aim to combine historical context with our own experiences to demonstrate the value of *Perca* spp. in ecotoxicological studies, as well as to provide readers with sufficient background to consider these species in their own research.

10.2 Historical Review of Field and Laboratory Toxicological Studies

The peer-reviewed scientific literature was surveyed for published toxicological and ecotoxicological research involving the three *Perca* species. Of the 181 published papers[1] identified by our survey, only three reported on contaminant (i.e., radionuclide) accumulation in *Perca schrenkii* (Lind et al. 2013; Salbu et al. 2013; Strømman et al. 2013). The remaining papers reported on uptake, accumulation, and effects of contaminants on the other two perch species, *P. fluviatilis* and *P. flavescens*. Among these remaining papers, 86 focused on *P. flavescens*, and 90 papers on *P. fluviatilis*. Consequently, both the North American yellow perch and European perch are represented approximately equally in the ecotoxicological literature. The geographic

[1] An annotated database of these studies has been compiled by the authors and is available on request.

distribution of sites sampled in these studies (Fig. 1) covers only a small portion of the worldwide distribution range of the *Perca* genus, presented in Chapter 1, and clearly reflects the density of occidental research centres and universities publishing ecotoxicological studies in the mainstream scientific literature.

Early ecotoxicological research on perch began in the 1970s with a study examining the incidence of *Aeromonashydrophila* in *P. flavescens* in several lakes in southern Québec (Vezina and Desrochers 1971). This study was the first to investigate the links between environmental contamination and parasitism in perch (see Section 10.3 for a brief discussion of this topic). It took another five years before the first paper that characterized the uptake and accumulation of an anthropogenic contaminant (mercury released from a wood pulp factory in southern Norway) into the tissues of *P. fluviatilis* was published (Steinnes et al. 1976). Over the next decade, only a few ecotoxicological papers focusing on perch appeared in the scientific literature. Of those that did appear, the primary contaminants of concern were metals and the metalloid selenium in both *P. flavescens* and *P. fluviatilis* (Kearns and Atchison 1979; Edgren and Notter 1980; Badsha and Goldspink 1982; Klaverkamp et al. 1983) and acid rain. The study by Ryan and Harvey (1980) in the La Cloche area in Ontario is one of the earliest reports of reduced population densities, together with changes in growth rates, of yellow perch in acidified lakes. Concurrent studies in Scandinavia supported the findings from Ontario and provided extensive evidence for additional impacts of acidification on perch, including egg mortality, reproductive failure and changes in population structure (Lappalainen et al. 1988; Linløkken et al. 1991). Experimental acidification of a lake by David Schindler's group in the Experimental Lakes Area

Fig. 1. Worldwide distribution of ecotoxicology field studies involving the three species of *Perca*. Yellow points represent studies involving yellow perch (*Perca flavescens*), green points represent European perch (*P. fluviatilis*), and purple points represent Balkhash perch (*P. schrenkii*). Location data are approximate from the papers listed in Table 10.1 (where available). Map prepared using Google Maps; Imagery ©2015, TerraMetrics, map data ©2015, Google, INEGI.

(Schindler et al. 1985) greatly increased the understanding of the effects of acidification on freshwater systems, even though perch were not present in these lakes. More recent studies (e.g., Lippert et al. 2007) examined the combined effects of acidification and metal stress in perch, and focused on recovery and remediation treatments.

Publication rates for ecotoxicological studies involving perch have been generally low, but steadily increasing since the 1970s (Table 1). The publication rate for papers involving *P. fluviatilis* has always exceeded that observed for *P. flavescens* except for the 2000s, during which period the latter saw a publication rate of 5.2 papers per year, relative to 3.4 for the former. The increasing publication rates, easily observed from Table 1, probably represent an acknowledgement by the research community of the ecological value these species bring to ecotoxicological research and environmental risk assessments owing to their relative widespread geographic distributions, abundance, and physiological tolerance (see Sections 10.3 and 10.4).

The vast majority of investigations identified by the literature survey involve the study of metals (129 of 181 studies). There are more yellow perch studies involving metals (76) than European perch (50), and the three ecotoxicological studies on the Balkhash perch focus on metals, specifically radionuclide uptake and tissue accumulation from populations around uranium mining operations in central Asia (Lind et al. 2013; Salbu et al. 2013; Strømman et al. 2013). The only other perch studies to focus on radionuclides were conducted using European perch (Kryshev et al. 1993; Zalewska and Suplinska 2013). Field studies examining metals are exclusively dominated by *P. flavescens*, but laboratory studies are divided equally between *P. flavescens* and *P. fluviatilis*. The dominance of *P. flavescens* in these field studies clearly reflects the abundance of the species in Canada where the mining industry is thriving. Mining is a major industrial activity in Russia as well, where *P. fluviatilis* is widely distributed, but the paucity in the mainstream literature of Russian-based environmental studies using this species highlights important gaps to fill.

Only 23 of the 181 studies identified by the literature survey are strictly laboratory-based studies. Adding those studies that had both a laboratory and field component increases the count by only 11 for a total of 34 studies that have at least a field component, regardless of species. Consequently, the vast majority of ecotoxicology studies involving *Perca* spp. (155 of 181; or 82%) are field-based. The lack of laboratory-based studies may be a reflection of the difficulty to maintain these species under laboratory conditions.

Table 1. Average number of ecotoxicological publications per year for each decade since the 1970s for the three species of *Perca*.

Decade	*P. flavescens*	*P. fluviatilis*	*P. schrenkii*	Total
1970s	0.2	0.1	0	0.3
1980s	0.6	1.2	0	1.8
1990s	1.3	2.2	0	3.5
2000s	5.2	3.4	0	8.6
2010s*	2.8	4.4	0.6	7.8

* The apparent drop off in publication rate for *P. flavescens* may simply reflect that these data were compiled before the end of the decade. Moreover, the apparent increase in publications involving *P. schrenkii* reflects three papers from one special issue of J. Environ. Radioactivity.

The worldwide distribution of the field studies is presented in Fig. 1. In North America, most yellow perch studies have been conducted in the Laurentian Great Lakes and St. Lawrence River systems, including a disproportionately large number around the mining district of Sudbury, Ontario, and the copper-smelting region of Rouyn-Noranda, Québec. Only a few perch studies are available west of Ontario (notwithstanding those studies conducted in the Experimental Lakes Area in northwestern Ontario; Klaverkamp et al. (1983)), and those involving contaminants associated with bitumen extraction from the oil sands district of northern Alberta (van den Heuvel et al. 1999a; van den Heuvel et al. 1999b). The distribution of ecotoxicological field studies with *P. fluviatilis* seems to be more reflective of its entire natural range across most of Europe and western Asia than for *P. flavescens* in North America. However, despite the more even distribution of studies across their natural range, field studies from Scandinavian countries involving *P. fluviatilis* dominate the distribution. In comparison to *P. flavescens* or *P. fluviatilis*, few studies have been conducted on Balkhash perch; all three that were published in English were conducted in central Asia near the border region between Kazakhstan, Kyrgyzstan, and Tajikistan (Lind et al. 2013; Salbu et al. 2013; Strømman et al. 2013).

The ecotoxicology literature survey identified 23 variables or endpoints in studies involving *Perca* spp. (Table 2). The only endpoint measured with *Perca schrenkii* is tissue radionuclide accumulation (see above). Over 70% of all papers in the survey measure the uptake and (or) accumulation of contaminants into fish tissues, regardless of species. Studies examining yellow perch measured uptake and accumulation more frequently (79%) than those examining European perch (64%). Nevertheless, the high proportion of studies for both species that included measurements of contaminant uptake and accumulation suggests that the genus *Perca* is an appropriate group of fish for monitoring environmental contamination because of its lifestyle, tolerance to environmental contaminants, and distribution. Other criteria favouring perch in contaminant accumulation studies include ease of capture with common fishing gear like gill nets, seines, fishing rods and minnow traps. The mean size of fish from this genus is also favourable, being generally larger than the smaller cyprinids, which are sometimes too small for accurate measures of contaminant accumulation in individual fish, and much smaller than the less abundant and commonly sympatric species like northern pike (*Esox lucius*).

A wide range of variables is commonly examined in perch in relation to environmental contamination and tissue accumulation. Not surprisingly, simple measurements of weight, length and age and related calculation of indices of condition (globally termed "biometrics") are the most common, and are reported in 29% of all the studies in the literature survey. Other than accumulation of contaminants and biometrics, the variables most studied in each species differ somewhat. For example, in yellow perch, effects of contaminants on metabolic indicators (25%), ecological and indirect effects (22%), and endocrinology (16%) were studied most frequently, whereas detoxification (18%), reproduction (12%), and oxidative stress (ROS production) and antioxidant production (11%) were studied most frequently in European perch. This difference reflects the fact that impacts of organic contaminants have been more extensively studied in European perch than the other *Perca* species, justifying the relevance of the corresponding endpoints.

Table 2. Proportion of the total number of ecotoxicological studies for each species that focus on a particular topic. Proportions are calculated from data from Table 10.1.

Variable, Estimate, or Endpoint	Yellow perch	European perch	Balkhash perch	Total
Accumulation	0.79	0.64	1.00	0.72
Biometrics	0.39	0.21	0.00	0.29
Partitioning/detoxification	0.15	0.18	0.00	0.16
Metabolism	0.25	0.08	0.00	0.16
Ecology/indirect effects	0.22	0.09	0.00	0.15
Endocrinology	0.16	0.07	0.00	0.11
Reproduction	0.09	0.12	0.00	0.10
ROS/antioxidant	0.09	0.11	0.00	0.10
Modeling	0.13	0.03	0.00	0.08
Histopathology	0.05	0.10	0.00	0.07
DNA damage/Apoptosis	0.01	0.10	0.00	0.06
Parasites	0.02	0.09	0.00	0.06
Bioenergetics	0.11	0.00	0.00	0.06
Immune system	0.06	0.03	0.00	0.04
BLM	0.07	0.00	0.00	0.03
Ion regulation	0.02	0.03	0.00	0.03
Evolutionary ecotoxicology	0.03	0.02	0.00	0.03
Behaviour	0.01	0.02	0.00	0.02
Olfaction	0.02	0.01	0.00	0.02
Toxicogenomics	0.03	0.00	0.00	0.02
Acute toxicity	0.02	0.00	0.00	0.01
Digestion	0.00	0.01	0.00	0.01
Epigenetics	0.01	0.00	0.00	0.01

The other types of endpoints examined in perch ecotoxicological studies represent a very small proportion (1% or less) of the literature on the subject. These include studies of acute toxicity and effects of contaminants on digestion, bioenergetics or behaviour. Only a handful of studies (nine in total) have taken advanced approaches involving molecular biology, including toxicogenomics, epigenetics and evolutionary ecotoxicology. These are reviewed in Section 10.5.

Considerable effort has gone into establishing Biotic Ligand Model (BLM) parameters for yellow perch to improve the BLM's relevance for the industrial receiving waters of Ontario and Québec, and to improve its ecological relevance to the area (Niyogi et al. 2004; Taylor et al. 2004; Klinck et al. 2007; Niyogi et al. 2007; Mirza et al. 2009). The BLM is a mechanistic model used to establish site-specific water quality criteria for metals using freshwater fish (Niyogi and Wood 2004). The model makes use of established relationships among site-specific water quality variables, such as hardness, pH, and dissolved organic carbon, and the amount of metal bound to and/or taken up by a physiologically-sensitive 'biotic ligand' (usually a gill) that can induce toxicity. Metal binding and subsequent uptake dynamics are often species dependent, which necessitates derivation of model parameters for each species that cannot be represented by existing BLMs. At present, BLMs are available for model species, such as rainbow trout and fathead minnows but no *Perca* species; however, yellow perch BLM development is ongoing. In the productive mining districts of northern Ontario and Québec, yellow perch are dominant species in lakes around industrial operations. However, yellow perch are considerably less sensitive to metals than either rainbow trout or fathead minnows (Taylor et al. 2004). Whereas yellow perch are nearly ubiquitous in these mining receiving waters, rainbow trout and fathead minnows are all but absent. Development of BLMs specifically for yellow perch will allow for the development of ecologically-relevant site specific water quality criteria in some of the most important industrial regions of Canada and north eastern United States.

10.3 Accumulation and Effects of Organic Contaminants, Municipal Effluents and Radionuclides in Wild Perch

The first study attempting to link exposure to contaminants with effects on perch that we could identify (Vézina and Desrochers 1971) is a brief paper reporting on the rate of infection by *Aeromonashydrophila*, a dangerous bacterium responsible for hemorrhagic septicaemia in populations of yellow perch from four lakes in Eastern Canada. There was no evidence that the infection rate was influenced by pollution, even though some of their study lakes were strongly contaminated by municipal effluents leading to algal blooms and hypoxia. In contrast, Marcogliese and colleagues (Marcogliese et al. 2005; Dautremepuits et al. 2009) measured indicators of oxidative stress and parasitism in yellow perch from the St. Lawrence River in Québec (Canada) collected at clean sites and sites contaminated by municipal effluents and also metals. They concluded that a combination of contamination and parasitism exacerbate oxidative stress, making fish from contaminated sites more vulnerable to the detrimental health effects of parasitism. In *Perca fluviatilis*, two studies (Valtonen et al. 1997; Valtonen et al. 2003) revealed the complexities involved in attempting to link parasitism and pollution in perch. Environmental contamination can weaken the fish's immune system, partly as a consequence of the induced oxidative stress described above. The same contamination can affect parasitism if intermediate hosts, such as bivalves, are lost from a contaminant-induced reduction in biodiversity. Furthermore, and as also supported by Marcogliese et al. (2012) experimentally, several parasites are also quite sensitive to contaminants. Overall, these perch studies reveal that the relationships between parasitism and water quality are complex, and that synergistic effects between environmental contamination and parasitism can be either detrimental or beneficial to perch health, despite the relatively high tolerance of perch to contamination exposure, as we discuss below.

Studies of accumulation and effects of organic contaminants in yellow perch are largely limited to the Great Lakes-St. Lawrence River continuum, where industrial and urban pollution from major Canadian and American cities and industries is concentrated. In a series of studies, Hontela and collaborators demonstrated that chronic exposure to complex mixtures of organic and inorganic contaminants lead to an impairment of the normal endocrine stress response involving the release of cortisol (Hontela et al. 1992, 1995; Girard et al. 1998; Hontela 1998). The contaminant-induced impairment of the cortisol stress response is further discussed in Section 10.4.1.

Yellow perch are also present in western Canada, where the oil sands industry is an important source of environmental contaminants in freshwater systems (van den Heuvel et al. 1999a, 1999b). A series of studies using yellow perch transplanted from clean lakes nearby has examined accumulation and effects of exposure to a complex mixture of polycyclic aromatic hydrocarbons (PAHs) and other organic and inorganic contaminants from process water in experimental ponds simulating reclamation ponds. After 5 to 11 months in these contaminated ponds, gross biometric and reproductive indicators such as condition factor, gonad size or fecundity were higher in fish from the contaminated pond, compared to reference fish (van den Heuvel et al. 1999b). Furthermore, although an increase in liver mixed-function oxidase, an indicator of exposure to petroleum hydrocarbons, was correlated with increased PAH metabolites in bile, a normal response of PAH accumulation and detoxification in oil-contaminated fish, levels of female sex steroids remained within the normal range (van den Heuvel et al. 1999a). Severe gill lesions and fin erosion were observed initially, however, after a population decline, survivors recovered and the rates of histopathologies decreased. Nevertheless, over time, the presence of endocrine disruptors in oil sands was confirmed by lower levels of sex hormones in males leading to a reduction in testicular size (van den Heuvel et al. 2012). Overall, the authors found little evidence of a chronic chemical stress after long-term exposure in contaminated experimental ponds, and concluded that favourable nutritional conditions in these ponds had a stronger influence on their condition than did contamination. To us, this demonstrates that yellow perch are remarkably opportunistic and tolerant of contamination.

The 45 studies of accumulation and effects of organic contaminants in European perch that we identified in our literature survey cover a wider range of contamination types and habitats than those involving yellow perch, investigated essentially in the Great Lakes-St. Lawrence region, with the exception of a few studies in the oil sands district, described above. The list of contaminants investigated in European perch is much more extensive than for yellow perch and reflects the wide geographic distribution of the species (Chapter 1) and the overlapping anthropogenic impacts of an intensely urbanized continent. These include agricultural contaminants (pesticides, herbicides), chemicals from industrial sources (e.g., pulp and paper mill effluents, plasticizers, petroleum and derivatives), urban effluents, radioisotopes and metals. European perch in these studies were captured in lakes, rivers and estuaries, in England, France, Germany, Czech Republic, Poland, Estonia, Latvia, Moldova, Finland, Sweden and Russia (Fig. 1).

Literature suggests that the European perch is a good bioindicator of environmental contamination, probably due to its tolerance to contaminants, abundance and relative sedentary nature. A remarkable example is the analysis of cesium radionuclides in non-predatory fish and in three species of predatory fish, northern pike, pike perch (*Sander lucioperca*, a fish in the Percidae family that is closely related to members of the genus *Perca*) and European perch samples collected in the Chernobyl area following the 1986 nuclear accident (Kryshev et al. 1993). Radionuclide concentrations peaked in non-predatory fish in the year following the accident, but maximum concentrations were observed up to three years later in perch and other predatory fish. A Polish study that measured cesium concentrations in five fish species collected in and around the Baltic sea between 2000 and 2010 confirmed the findings of Kryshev et al. (1993): ^{137}Cs concentrations in fish muscle were the signature of the Chernobyl accident and reliably tracked the decrease of radioisotope contamination in the salt waters of the Baltic Sea over time (Zalewska and Suplinska 2013). In contrast to Kryshev et al. (1993) however, concentrations in European perch muscle were lower than in most other species examined, an observation that the authors attributed to the important dilution of the ^{137}Cs–contaminated sea water by freshwater in the Szczecin Lagoon where perch were sampled, in contrast to the other species which were sampled along the shores of the Baltic Sea.

European perch are also raised for human consumption in ponds throughout Europe, and can be used as a biomonitor of watershed contamination. For example, Lazartigues et al. (2013) measured the concentrations of thirteen pesticides in the muscle of carp (*Cyprinus carpio*), roach (*Rutilus rutilus*) and European perch raised in five dam ponds from watersheds varying in agricultural pressure in France, and compared tissue pesticide residues with concentrations measured in water and sediment. Their study confirmed that fish contamination reflected watershed and pond contamination.

Studies such as these, in addition to their contribution to our understanding of the fate of contaminants in aquatic foodwebs, have the added ecological value of providing indirect evidence of the trophic level occupied by perch in the wide range of environments in which they occur. For instance, in the Lazartigues et al. (2013) study, examination of relationships between pesticide concentrations in sediment, water and muscle suggested that in the ponds studied, perch were mostly pelagic and planktivorous, whereas carp occupied a more benthic niche. In contrast, two studies investigating biomagnification of organic contaminants in fishes, one in a Latvian lake (Olsson et al. 2000) and the other in the Baltic Sea (Burreau et al. 2004), concluded that European perch are piscivorous and at the top of the food chain as they reach adult size. Clearly, if the biological characteristics of perch make them useful for ecotoxicological studies, conversely, these studies are also helpful in improving our understanding of perch biology.

10.4 Accumulation and Effects of Metals in Wild Perch

As reviewed in Section 10.2, metals are the most studied contaminants in perch, and a large proportion of these studies were carried out on yellow perch from Canadian lakes varying in metal contamination caused by metal mining and smelting activities. Studies by individual teams examining metal accumulation and effects in yellow perch collected along metal contamination gradients from two major mining regions of eastern Canada (e.g., Brodeur et al. 1997; Laflamme et al. 2000; Eastwood and Couture 2002; Rajotte and Couture 2002; Audet and Couture 2003; Couture and Kumar

2003) revealed that yellow perch were suitable for investigations of the mechanisms of metal toxicity in wild fish. Because elucidating mechanisms of metal toxicity was a main objective of the *Metals in the Environment Research Network*[2] (MITE-RN), the network invested in two large-scale collaborative projects focused on yellow perch, which are presented as case studies below. These case studies adopted the yellow perch as its sentinel or biomonitor organism and were designed to explore the following questions:

1. Can bioaccumulated metal concentrations in chronically exposed yellow perch be predicted from the ambient water quality (e.g., dissolved metal concentrations, pH, hardness and dissolved organic carbon concentrations)?
2. What is the relative importance of water *versus* food as vectors of metal accumulation in chronically exposed juvenile yellow perch?
3. How are the bioaccumulated metals handled by the native yellow perch? Do the mechanisms of metal detoxification change along the metal contamination gradient, i.e., as the ambient metal levels increase? (This part of the study was designed to field test the threshold toxicity model that is commonly assumed to govern metal toxicity in aquatic organisms).
4. Do the bioaccumulated metals have demonstrable deleterious effects on fish condition, growth, longevity and metabolic capacities in chronically exposed yellow perch?
5. Can metal exposure lead to indirect (i.e., food-web mediated) effects on yellow perch?
6. What is the influence of season on tissue metal accumulation, fish condition, and metabolic capacities in wild yellow perch?

10.4.1 Case study 1

A collaborative field research project, focused on yellow perch and developed under the auspices of MITE-RN, was carried out between 1999–2004 (Campbell et al. 2003; Campbell et al. 2008; Rasmussen et al. 2008). In the spatial component of this field study, yellow perch were collected from a set of lakes located upwind and downwind from either a copper/nickel smelter (Sudbury, Ontario) or a copper/zinc smelter (Rouyn-Noranda, Quebec). The lakes exhibited contrasting dissolved metal concentrations and represented a marked gradient in exposure to metals (especially Cu and Ni in the case of the Sudbury lakes, Cd and Zn for the Rouyn-Noranda lakes). A description of the physical-chemical characteristics of the lakes, including pH, hardness, dissolved organic carbon concentrations and dissolved metal concentrations, can be found in Giguère et al. (2004). Metal accumulation was determined in the fish from each lake, with an emphasis on the liver as the target organ. In a more detailed study of metal bioaccumulation in fish from a single lake (Lake Osisko), we explored how metal accumulation in different target organs varied as a function of age and size.

[2] The *Metals in the Environment Research Network (MITE-RN)* received financial contributions from the Natural Sciences and Engineering Research Council of Canada (NSERC), the Mining Association of Canada, Ontario Power Generation Inc., the International Copper Association, the International Lead Zinc Research Organization, and the Nickel Producers Environmental Research Association.

The spatial component of the study focused on juvenile fish that were still in the planktivorous stage of development (nominally aged 1+); this choice was designed to minimize the effects of different food availabilities on metal accumulation (the zooplankton communities in the different lakes were expected to be less variable than the benthic communities—see Rasmussen et al. (2008) for a discussion of this point). In addition to identifying differences in metal bioaccumulation along the metal exposure gradient, we also looked for metal-induced effects in the yellow perch, at the biochemical and physiological levels (Levesque et al. 2003; Giguère et al. 2005; Gravel et al. 2005). Finally, as a complement to the sampling of resident fish populations, we carried out reciprocal transplant experiments, where juvenile perch were moved between a clean lake and a metal-contaminated lake, allowing us to follow the dynamics of metal uptake and metal depuration over a 70 d period (Kraemer et al. 2005a,b).

Detailed descriptions of the experimental approach can be found in the original papers, cited below, but the following brief summary should suffice for the present purposes. Ambient dissolved metal concentrations (Cd, Cu, Ni, Pb, Zn) were determined in each lake with *in situ* diffusion passive samplers; these samplers also provided samples for determination of the general water quality (pH, [Ca^{2+}], [Mg^{2+}], [Cl^{-}], [SO_4^{2-}] and dissolved organic carbon or DOC). Metal exposure was evaluated on the basis of the free metal ion concentrations ([M^{2+}]), as calculated from chemical equilibrium simulations using the ambient water chemistry data (Giguère et al. 2004). Total hepatic metal concentrations were determined as a measure of metal bioaccumulation and various markers of oxidative stress were measured to assess metal effects in liver cells (Giguère et al. 2005). Blood, liver, gonad, gill and kidney samples were collected in fish captured in lakes located along the metal gradient in summer and fall, to evaluate the physiological and hormonal status in relation to tissue metal concentrations. Thyroid hormones (triiodothyronine T3, thyroxine T4), cortisol, a key corticosteroid released by the interrenal tissue in teleost fish, and reproductive hormones (estradiol, testosterone) were measured in plasma. Liver glycogen and triglyceride reserves, plasma free fatty acids and glucose, and the activity of key gluconeogenic and glycolytic enzymes were also measured in the fish to assess their metabolic status. Thyroid tissue, gills, gonads and interrenal tissue were collected for histopathology. To further characterize the potential effects of metals on the functional integrity of the endocrine axis, a stress test challenge was used to evaluate the capacity of the fish from lakes situated along the metal gradient to mount the normal hormonal stress response (Laflamme et al. 2000). Fish were captured and kept overnight in *in situ* enclosures to recover from the capture. The next day, at a time standardized across lakes, fish were confined for 1 h in 20-L containers (10 fish/container) filled with water from the sampled lake. Plasma cortisol was measured following the confinement to compare the stress response among fish from different lakes. Furthermore, head kidney, the tissue that secretes cortisol, was isolated and tested for its ability to secrete cortisol *in vitro* (Laflamme et al. 2000). To investigate metal detoxification strategies at the subcellular level, metal partitioning among potentially metal-sensitive fractions (cytosolic enzymes, organelles) and detoxified metal fractions (metallothionein, granules) was determined after differential centrifugation of fish liver homogenates (Giguère et al. 2006).

In the reciprocal transplant experiments, juvenile yellow perch were caught in a reference lake and transplanted to cages held within a lake impacted by mining activities, with elevated levels of aqueous bioavailable Cd, Cu and Zn. The mesh size of the cages was chosen to be large enough to permit the entry of zooplankton from the lake into the cages, thus obviating need to feed the captive fish. Fish were sampled from the cages over 70 d and changes in metal concentrations were followed over time in the gills, gut, liver and kidney (Kraemer et al. 2005a). Similar experiments were carried out in the opposite direction, from the contaminated lake to the reference lake, to follow the depuration dynamics (Kraemer et al. 2005a,b). The data from the transplant experiments were then used to develop a biodynamic model for Cd uptake and loss from juvenile yellow perch and the results were compared to the measured metal levels in native fish in variably contaminated lakes (Kraemer et al. 2008).

Principal Results

Metal Accumulation in Yellow Perch (Questions 1, 2 and 6)

In our initial studies in which metal concentrations were determined in a range of fish organs (kidney, liver, gills, gastrointestinal tract, carcass), it was shown that the accumulated metal concentrations were consistently higher in the kidney and liver than in the other organs (Giguère et al. 2004). Similar patterns of tissue accumulation were reported in other studies from this area (Laflamme et al. 2000; Levesque et al. 2002). For subsequent work, we focused on the liver as the internal organ offering the optimal combination of a wide range of concentrations along the exposure gradient and a large mass of tissue for analysis.

In the spatial (inter-lake) studies, important variations in hepatic metal concentrations were observed. To illustrate this variability, we calculated the ratio of the highest observed concentration ($[M]^{max}$) divided by the lowest ($[M]^{min}$. For example, for juvenile yellow perch (1+) collected from eight different lakes in the Rouyn-Noranda and Sudbury areas, ratios of $[M]_{liver}^{max} / [M]_{liver}^{min}$ decreased in the order Ni (36) > Cd (14) > Cu (13) > Zn (2). As expected, hepatic concentrations of the nonessential metals (Cd, Ni) tended to increase in response to increases in the estimated waterborne free metal concentrations ($[Cd^{2+}]$, $[Ni^{2+}]$), but only in the case of Ni was the relationship between $[M]_{liver}$ and $[M^{2+}]$ statistically significant. The proton (H^+) and the hardness cations (Ca^{2+}, Mg^{2+}) are known to compete with cationic trace metals and reduce metal uptake. Accordingly, equilibrium modelling (analogous to the BLM (Campbell and Fortin 2013)) was used to account for differences in pH and water hardness in the sampled lakes, but this approach failed to improve the relationships; based on regression analyses, neither lake water pH nor hardness significantly influenced the steady-state metal concentrations in various body parts of yellow perch. Possible reasons for the apparent failure of BLM approach are discussed by Giguère et al. (2005). For example, Cu and Zn are essential elements and yellow perch presumably exert a certain degree of homeostatic control over the internal concentrations of these metals, especially in the case of Zn, weakening possible relationships with the aqueous metal concentrations. For all four metals, if diet-borne metals contributed to hepatic metal burdens, uptake from food might well obscure relationships between aqueous exposure and metal bioaccumulation.

The possible role of diet-borne metals was investigated in the case of Cd, using a diet manipulation approach with caged yellow perch. To determine the relative importance of water and food as Cd sources for juvenile yellow perch, fish were caged for up to 30 days in either a reference lake or a Cd-contaminated lake, and offered prey (zooplankton) from one of these lakes. Four Cd-exposure regimes were tested: (i) a control, (ii) Cd-contaminated water only, (iii) Cd-contaminated food only, and (iv) Cd-contaminated food and water. Analysis of the results suggested that Cd in the gills and kidney of the caged perch was taken up largely from lake water, whereas liver and gut Cd appeared to come from both dietary and aqueous sources (Kraemer et al. 2006b). Note that in field-collected yellow perch, Cd concentrations are generally higher in the gastrointestinal tract than in the gills, also suggesting that uptake of this metal from food is important (Giguère et al. 2004). This hypothesis is further discussed later in the chapter (see "Relative contribution of water and diet to tissue metal accumulation (Question 2)" in Case Study 2).

To explore possible effects of fish age on metal bioaccumulation, we collected fish ranging in age from 1 to 10 years from a single lake (Lake Osisko) and determined Cd, Cu and Zn concentrations in the kidney, liver, gills, gastrointestinal tract and carcass. Cadmium concentrations in the liver and kidney increased with fish age, which we interpret to be linked to changes in fish growth rate with age (i.e., the effect of growth dilution on tissue metal concentrations decreases in older fish). Relationships between age and Cu or Zn accumulations were weak or nonexistent, a result consistent with the essential nature of these metals (Giguère et al. 2004). Finally, in a small study of seasonal variations in metal concentrations, fluctuations in metal (Cd, Cu) concentrations were monitored over four months (May to August) in the livers of juvenile yellow perch collected from four Rouyn-Noranda lakes situated along a metal concentration. Cadmium and Cu concentrations varied most, in both absolute and relative values, in fish with the highest hepatic metal concentrations, whereas fish sampled from the reference lake did not show any significant variation (Kraemer et al. 2006a).

Metal Dynamics in Yellow Perch (Question 2)

The results discussed in the preceding section were all derived from native yellow perch collected in the field from their home lake. In such studies, the steady-state metal concentrations in the liver (or any other organ) reflect a balance between metal uptake (from water or food or both) and metal loss by excretion and elimination; steady-state concentrations are also affected by the growth of the fish, i.e., by growth dilution. These various processes are represented in equation (1), for Cd:

$$\frac{d[Cd]_{organ}}{dt} = k_u\left[Cd^{2+}\right] + (AE \times IR \times [Cd]_{food}) - k_e\left[Cd\right]_{organ} - k_g\left[Cd\right]_{organ} \qquad (1)$$

where k_u, k_e and k_g are rate constants for Cd-influx from water ($L^{-1}\cdot g^{-1}\cdot d^{-1}$), Cd-efflux ($d^{-1}$) and growth ($d^{-1}$), respectively; $[Cd^{2+}]$ is the free Cd ion concentration in lake water (nmol·L^{-1}); $[Cd]_{organ}$ is the Cd concentration in the organ of interest (nmol·g^{-1} dry weight); $[Cd]_{food}$ is the Cd concentration in the food of perch (nmol·g^{-1} dry weight); AE is the efficiency (%) with which perch assimilate Cd from their food; and IR is the rate at which food is ingested by perch (g prey·g^{-1} perch·d^{-1}) (Kraemer et al. 2008). To parameterize this equation, organ-specific values of k_e were obtained from the 70-day Cd-efflux experiment in which Cd-contaminated juvenile yellow perch were transplanted to a lake with low ambient $[Cd^{2+}]$ (Kraemer et al. 2005b). Values of k_u were derived from the 75-d Cd-uptake experiment involving the transplantation of juvenile yellow perch from a clean lake to a metal-contaminated lake (Kraemer et al. 2005a) and from the 30-d diet manipulation study described earlier (Kraemer et al. 2006b).

When juvenile yellow perch were transplanted up-gradient, Cd concentrations increased in all the organs sampled, whereas Cu mainly increased in the gills, gut and liver but not the kidney; only some slight accumulation of Zn occurred in the kidneys and gills of the transplanted fish. When the experiment was performed in the opposite direction, from the contaminated lake to the clean lake, Cd concentrations decreased most rapidly in the gills and gut, i.e., from organs in contact with the ambient water and food; biological half-lives ($t_{½}$) were 18 and 37 d, respectively, for each organ. Longer half-lives were observed in the liver (75 d) and kidney (52 d) for this metal. Elimination of excess Cu by the liver and gut occurred much more rapidly, with estimated half-lives of labile Cu being 8 and 4 d, respectively, for these two organs. In contrast to Cd and Cu, there was little Zn elimination.

Biodynamic modelling with equation (1) was then used to simulate Cd accumulation in yellow perch (Kraemer et al. 2008). Model simulations for the liver agreed well with the observed time course of hepatic Cd accumulation, but organs in contact with the external environment (i.e., the gills and gut) responded very differently than did the liver; predicted accumulations in these organs after 70 d were about three times higher than the observed values. These "interface" organs presumably have the capacity to adjust their ability to take up and/or eliminate Cd, depending on the ambient conditions. Recall that the fish used for the Cd uptake experiments were collected in the reference lake and thus were metal-naïve; the proposed adjustment of their ability to take up and/or eliminate Cd must have taken place early within the 70-day time frame. The model was also used to predict steady-state Cd concentrations in the gills, gut, and liver of juvenile yellow perch living in lakes along a Cd gradient. Agreement between predicted and observed steady-state Cd concentrations was reasonable in lakes with low to moderate Cd concentrations, but in lakes with high dissolved Cd (>1.5 nM), the model overestimated Cd accumulation, particularly in the gills and gut (Kraemer et al. 2008). These results again suggest that yellow perch are able to acclimate to Cd-contaminated environments by decreasing their uptake of waterborne or diet-borne Cd, or by increasing their rates of Cd elimination. We conclude that biodynamic modelling is not yet sufficiently advanced to allow accurate predictions of tissue metal concentrations in free-ranging perch.

Metal Detoxification—Subcellular Partitioning (Question 3)

The metal bioaccumulation data discussed to this point are all total metal concentrations, and thus include both detoxified metal (e.g., metal sequestered in insoluble granules or bound to intracellular peptides like metallothioneins) and labile metal (e.g., metal bound to metal-sensitive target molecules such as enzymes or nucleic acids in the cytosol or in organelles within the cell). To be able to predict whether the accumulated metals are actually causing deleterious effects, it would obviously be helpful to know to what extent the metals have been detoxified, or conversely to what extent they remain labile and available within the organ of interest. To this end, with juvenile yellow perch collected from the Rouyn-Noranda and Sudbury lakes, using the liver as the target organ and employing a differential centrifugation procedure adapted from that described by Wallace et al. (2003), we determined the partitioning of the metals of interest among the following sub-cellular compartments: (1) a fraction comprising nuclei, cell membranes, intact cells and connective tissue, termed "nuclei/debris" hereafter, (2) a granule-like or resistant fraction, (3) mitochondria, (4) a fraction combining microsomes and lysosomes, (5) cytosolic heat-stable proteins, including metallothioneins, termed "HSP" hereafter, and (6) cytosolic heat-denatured proteins, termed "HDP" hereafter.

In such a fractionation scheme, fractions (2), (3), (5) and (6) are of particular importance from a diagnostic point of view: fractions (2) and (5) can be considered to be "biologically detoxified metals" or "BDM", whereas fractions (3) and (6) correspond to potentially "metal-sensitive fractions" or "MSF", as originally defined by Wallace et al. (2003). Based on the threshold model of metal detoxification (Mason and Jenkins 1995), we hypothesized that the non-essential metals (Cd, Ni) would be effectively detoxified in fish collected from lakes at the low end of the metal exposure gradient, but that above a certain threshold metal accumulation—to be determined—these metals would "spill over" into the potentially metal-sensitive sub-cellular pools (3) and (6). Since we are interested here in linking the partitioning of accumulated metal concentrations to the onset of deleterious effects, we shall limit the discussion to the nonessential metals (Cd and Ni).

In juvenile yellow perch collected from variably contaminated lakes in the Rouyn-Noranda and Sudbury areas, major proportions of hepatic Cd were found in the heat-stable cytosolic peptides and proteins fraction (HSP), a fraction including metallothioneins, whereas the potentially metal-sensitive heat-denaturable proteins fraction (HDP) was the largest contributor to the total Ni burdens (Giguère et al. 2006). The concentrations of Cd and Ni in each sub-cellular fraction increased along the metal contamination gradient, but the relative contributions of each fraction to the total burden of each of these metals remained generally constant. In other words, for these chronically exposed fish there was no threshold hepatic metal concentration below which binding of Cd or Ni to the heat-denaturable protein fraction did not occur (Fig. 2). The presence of Cd and Ni in this HDP fraction, even for low chronic exposure concentrations, suggests that metal detoxification was imperfect, i.e., that yellow perch were subject to some metal-related stress even under these conditions. One might thus anticipate finding evidence of metal-induced effects in the native fish (see below and Section 10.4.2).

The subcellular partitioning of Cd was also studied in a temporal context. For example, in the diet manipulation study where juvenile yellow perch were exposed to waterborne or diet-borne Cd over a 30-d period, the majority of the newly accumulated hepatic Cd was found in the HSP fraction, regardless of the exposure source (Kraemer

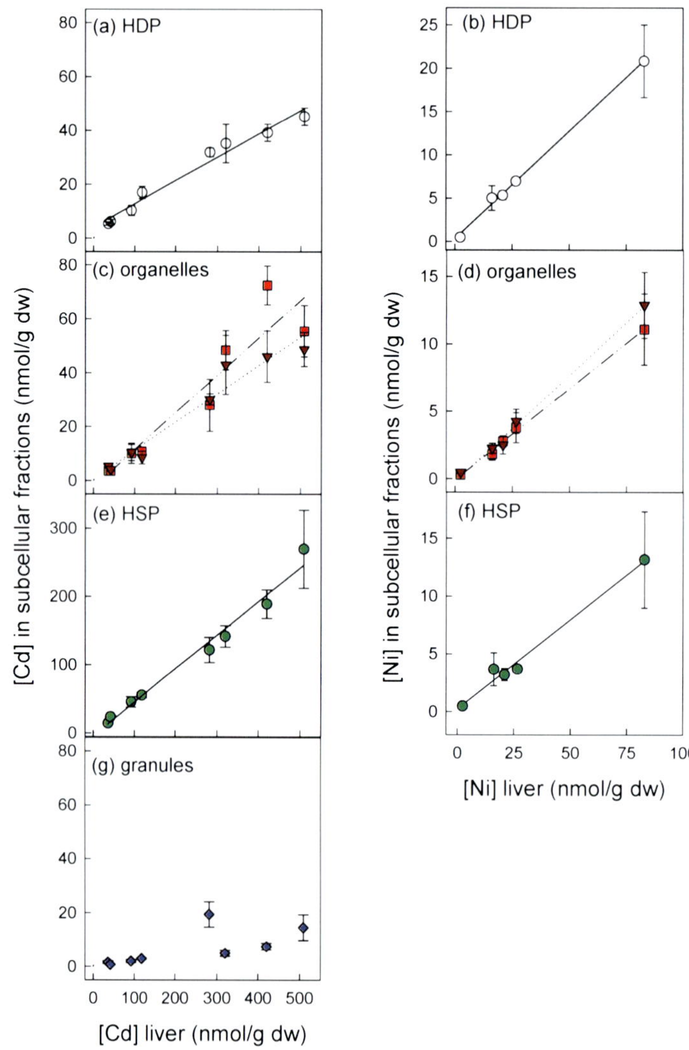

Fig. 2. Cadmium and nickel partitioning in the liver of yellow perch, *Perca flavescens*, collected from eight (Cd) or five (Ni) lakes along a metal contamination gradient (see Table 3). Cadmium and Ni concentrations in the various fractions are compared to Cd and Ni concentrations in the whole organ: (a,b) = heat denaturable proteins (HDP); (c,d) = organelles (open squares = mitochondria; inverted triangles = lysosomes + microsomes); (e,f) = heat-stable proteins (HSP); (g) = granules. Each point represents data for juvenile perch collected in a single lake (mean ± SD; N = 3 composite samples). Results for Ni in the granules fraction are unavailable. Metal concentrations for the metal-sensitive fractions are plotted in panels a-d; metals in the metal-detoxified fractions are plotted in panels e-h. Curves are lines of best fit. Reproduced from Campbell and Hare (2009) with permission from The Royal Society of Chemistry.

et al. 2006b). Similarly, in the metal uptake experiment where juvenile yellow perch were transplanted from a clean lake to a metal-contaminated lake and held there in cages for 70 days (Kraemer et al. 2005a), sequestration of Cd in the HSP fraction again proved to be the dominant strategy used by these fish to detoxify the incoming metal. Note however that this binding did not completely protect the mitochondria (fraction 3) or the heat-denatured proteins (fraction 6); even early in the metal accumulation phase of the transplantation there was no hint of a threshold response (Campbell and Hare 2009).

The metal depuration study, in which juvenile yellow perch were moved between the same two lakes but in the down-gradient direction, afforded an opportunity to determine how the perch liver handled Cd during the 75-d metal elimination phase (Kraemer et al. 2005b). Consistent with the long half-life observed for Cd at the whole organ level, there was no significant movement of Cd among the subcellular fractions and no significant decrease in any individual fraction. Copper, on the other hand, was lost from both the organelle and cellular debris fractions. As these fractions likely contain structures such as lysosomes, we suggest that Cu is depurated from the liver by direct elimination of these sub-cellular vesicles. The results from both time-course studies, uptake and depuration, demonstrate that the liver in juvenile perch handles essential (Cu, Zn) and non-essential metals (Cd) differently.

Evidence of Direct Metal Toxicity (Question 4)

In the preceding sections, we have clearly demonstrated that yellow perch can accumulate nonessential elements metals such as Cd and Ni when they are exposed to these metals in their environment, and we have just shown that the detoxification of these elements by binding to metallothionein or sequestration in granules appears to be incomplete, as judged by the presence of Cd and Ni in potentially metal-sensitive subcellular fractions. In the present section we summarize the evidence for metal-induced deleterious effects in the chronically exposed yellow perch, at the cellular, organ and individual levels.

Biochemical endpoints associated with oxidative stress have long been considered as markers for environmental impact assessment and have been used in the identification of subtle and early effects of contaminants (Lushchak 2011). To evaluate the importance of oxidative stress in juvenile yellow perch in the Rouyn-Noranda and Sudbury lakes, we investigated the status of their hepatic antioxidant defence system. Biochemical measurements included an indicator of lipid peroxidation (malondialdehyde), the activities of two enzymatic antioxidants (glutathione reductase and glutathione peroxidase), along with the concentration of an antioxidant peptide, glutathione (Giguère et al. 2005).

Overall, oxidative stress seems to be well controlled in indigenous yellow perch exposed to metals. Although the levels of two antioxidants, glutathione reductase and glutathione, tended to decrease along the metal contamination gradient, there was no attendant increase in the concentrations of malondialdehyde. Indeed, concentrations of this biomarker actually decreased as a function of hepatic metal concentrations, suggesting that any oxidative stress must have been controlled by other antioxidants at the cellular level (e.g., metallothionein, which is known to have the capacity to inactivate reactive oxygen species—see Viarengo et al. (2000)). We conclude that reactive oxygen species were probably not directly responsible for major deleterious effects at the whole-fish level. The added metabolic cost for controlling oxidative stress levels could, however, be indirectly responsible for deleterious effects in the exposed fish (Sherwood et al. 2000).

The assessment of the relationship between the metal tissue concentrations of fish sampled along the metal gradient and their physiological and hormonal status revealed significant concentration-dependent effects of metals on several functional endpoints (Fig. 3). Metabolic anomalies were diagnosed in fish sampled from lakes where the

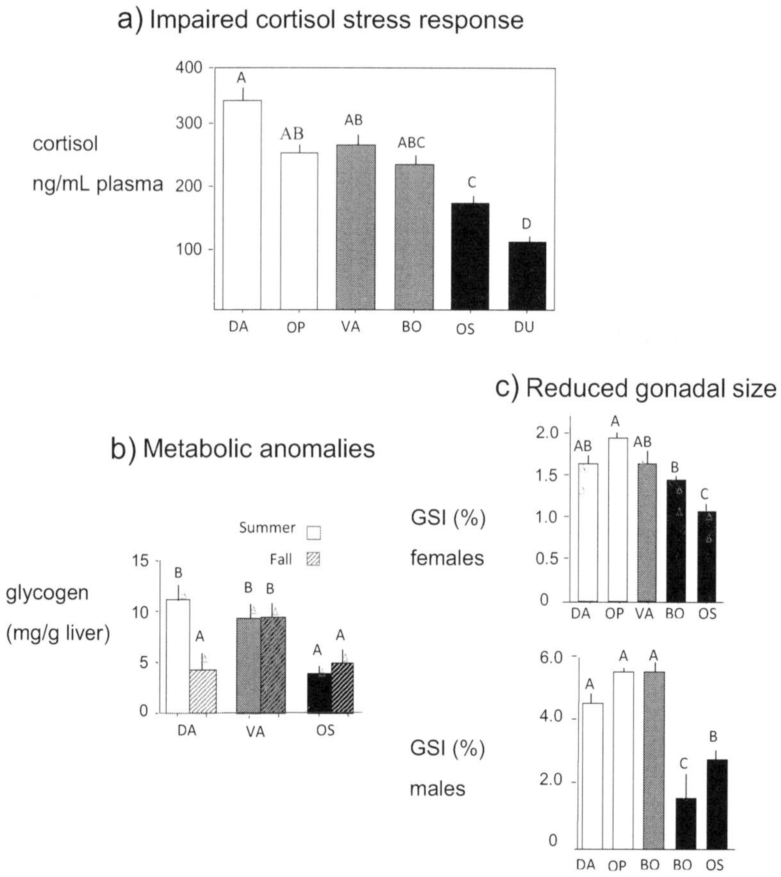

Fig. 3. Yellow perch from metal-contaminated lakes exhibit (a) endocrine anomalies such as a reduced capacity to elevate plasma cortisol levels in response to a standardized confinement stress, (b) metabolic anomalies such as differences in seasonal cycling of liver glycogen reserves, and (c) reduced gonadal size (GSI, gonadosomatic index). Fish were sampled from reference lakes, Opasatica and Dasserat (white bars), intermediate Bousquet and Vaudray (gray) and highly contaminated lakes Osisko and Dufault (black). Letters that are different indicate significantly different means ($P < 0.01$, Tukey-Kramer test). Modified from Campbell et al. (2003) and reproduced with permission of Taylor and Francis Group.

exposures led to tissue burdens of Cd, Zn and Cu above those measured in reference lakes (Levesque et al. 2002). These chronic exposures led to abnormal seasonal cycling of energy substrates such as glycogen and triglycerides, and altered the activity of key enzymes, including phosphoenolpyruvate carboxykinase, malate dehydrogenase and glucose 6-phosphate dehydrogenase, that regulate the metabolism of glucose and free fatty acids. The results provided evidence that chronic environmental exposures to metals, specifically Cd, Zn and Cu, result in a decreased ability to store energy reserves in the fall and mobilize these reserves in early summer. Also linked to high metal tissue concentrations were alterations of the thyroid function, assessed both by plasma thyroid hormone levels and thyroid histopathology, and impaired function of the steroidogenic interrenal tissue, diagnosed both *in vivo* and *in vitro* with functional stress tests (Levesque et al. 2002; Levesque et al. 2003; Gravel et al. 2005). Delayed gonadal development, detected using gonadal histology and plasma concentrations of the reproductive hormones estradiol and testosterone, was also linked to higher tissue metal concentrations. Gill histopathology, characterized by thickened secondary gill lamellae and filling of the interlamellar spaces, was observed in fish sampled at the high end of the metal gradient (Levesque et al. 2003). Taken together, these results provided clear evidence that chronic environmental exposures to Cd, Zn and Cu in lakes impacted by metal contamination have the potential to disrupt, at multiple target sites, the endocrine and physiological function of yellow perch.

Clearly, any direct toxic effect on individuals will have consequences at the population level. Further discussion of effects on populations can be found in Section 10.5.2, where we explore the consequences of metal contamination on population genetics and how these impacts may affect the long-term survival of local populations. In addition to direct toxic effects, contaminant-induced fluctuations in perch populations may also be due to indirect effects on more sensitive components of food webs. This aspect is discussed in the next section.

Indirect Effects of Metal Contamination (Question 5)

The collaborative research field project executed within the framework of MITE-RN also provided important new evidence for indirect food-web mediated effects of metals on fish growth and bioenergetics (Rasmussen et al. 2008). For these ecological studies in lakes situated along a metal contamination gradient, specific daily consumption rates, growth rates and conversion efficiencies were determined in yellow perch using ^{137}Cs, a globally dispersed trophically transferred radiotracer (Sherwood et al. 2000; Sherwood et al. 2002a), along with detailed characterization of foodwebs (Kövecses et al. 2005).

Yellow perch were captured by hook and line, gill nets or seine, and body mass and length were measured for calculations of the condition index [(wt (g)/length^3(cm)) x 100]. Scales, otoliths and opercular bones were collected for age determination. Stomach contents were sampled for detailed analysis of prey items consumed by the fish, and to further characterize the food web accessible to the fish in each lake, the invertebrate and plankton communities were described and quantified, using Shannon's index and identifications (Kövecses et al. 2005). Specific daily consumption rates were estimated using ^{137}Cs in fish muscle and prey items, in a mass balance model validated *in situ* and in the laboratory (Rowan and Rasmussen 1996). Muscle lactate dehydrogenase (LDH) activity was also measured, as a marker of fish activity (Sherwood et al. 2002b).

These field studies provided evidence for greatly simplified prey bases in metal-contaminated lakes, characterized by less diverse populations of benthic macroinvertebrates, occurring at lower densities (Kövecses et al. 2005). Yellow perch in these lakes fed on smaller prey items compared to fish from less contaminated lakes, and they continued to consume small prey as they aged, while fish in reference lakes consumed increasingly larger items. These results suggested that fish in metal-contaminated lakes, feeding on small items and depending on zooplankton throughout their life, incur increased costs of foraging. Estimates of activity levels using muscle LDH as a marker confirmed this observation (Sherwood et al. 2002b). Field estimates of specific daily fish growth and *in situ* consumption rates using ^{137}Cs to calculate conversion efficiency and total energy budget, revealed that the conversion efficiency was lower in perch from the most metal-contaminated lakes, indicating that these fish experience greater energetic costs (Sherwood et al. 2000). The term "energetic bottlenecks" was used to describe the phenomenon of reduced fish growth resulting from greater energetic expense and lower conversion efficiency when perch feed on small prey items in the simplified food web of metal-contaminated lakes (Sherwood et al. 2002a; Campbell et al. 2003; Rasmussen et al. 2008). The food web-mediated indirect effects of metals, caused by effects on the metal-sensitive invertebrates, had significant effects on fish growth and population dynamics, as these effects reduced the efficiency of energy transfer to higher trophic levels in metal-contaminated lakes. The relative partitioning of energetic expense associated with direct effects of metal toxicity, including detoxification costs (Giguère et al. 2006) and indirect food-web mediated effects, has not been determined thus far.

10.4.2 Case Study 2

Another investigation carried out in the context of MITE-RN, focused exclusively on yellow perch, examined metal accumulation and effects in fish from the same two metal contamination gradients (Sudbury and Rouyn-Noranda) described in 10.4.1. The objectives of Case Study 2 overlapped somewhat with those of the Case Study 1. However, the approaches taken by each study were very different. Whereas the first study focused on mechanisms of metal accumulation, modes of toxic action and effects, this second study aimed at providing a large database that would allow for an evaluation of the influences of environmental metal contamination and the confounding influences of natural factors on tissue metal accumulation and metabolic impairment. Data collection was designed to yield information that would allow the formulation of recommendations and guidelines for field sampling using yellow perch as a biomonitor in environmental effects monitoring studies such as those developed for Canadian mines (Environment Canada 2013). In this study, we sampled 2400 yellow perch from five lakes in each of two metal-contamination gradients (Sudbury and Rouyn-Noranda; i.e., 10 lakes total) over two seasons (spring and summer) using gear that caught the full size range of perch from every population.

Relative Contribution of Water and Diet to Tissue Metal Accumulation (Question 2)

Our first objective was to estimate the relative contributions of water and diet as metal sources and their influence on liver and kidney metal accumulation. Whereas the first study examined this question using caging and cross-transplantation between one clean and one contaminated lake from the Rouyn-Noranda region, the present study used a correlative approach to investigate relationships between environmental contamination (metal concentrations in lake water and sediments and fish diet) and fish contamination (liver and kidney metal concentrations) for 2400 yellow perch collected as follows: 120 fish from the full size range available were captured in spring and summer in each of five lakes forming a gradient of metal contamination in the Sudbury region in 2002; this design was repeated in Rouyn-Noranda in 2003 (Couture et al. 2008a). Since the diet of yellow perch varies as they grow, from zooplankton to benthic invertebrates to fish, we chose to measure the metal contamination of fish gut contents. Although this approach has its limitations compared to sampling a range of potential food items for determination of metal concentrations (Couture et al. 2008a), it has the advantage of not having to make assumptions about perch dietary preferences, which will vary greatly depending on the lake, season and fish size. Diet metal contamination reflected environmental contamination gradients, supporting the use of gut contents in dietary contamination studies. The relative influences of water and diet on tissue metal accumulation depended on the metal, particularly for Cd, which was strongly correlated to a dietary source. At the other end of the spectrum, diet appeared to have little influence on zinc tissue accumulation. However, given that $[M]_{diet}$ covaried with $[M]_{water}$, the correlative approach that we took must be interpreted cautiously. Nevertheless, evidence from this and other studies strongly suggest that water is a constant source of metal contamination for yellow perch, and that, in periods when fish eat, diet can also contribute to uptake.

Beyond investigation of Question 2, the extensive dataset collected in our study and its sampling design allowed us to verify earlier suggestions that, in a given lake, Cd concentrations in yellow perch increase with fish age (Giguère et al. 2004). We concluded that there is no evidence that the tissue concentrations of metals other than Cd increase with age. On the contrary, tissue metal concentrations simply reflected environmental contamination and varied seasonally. For Cd, patterns of increasing concentrations with age were common in yellow perch from Rouyn-Noranda but non-existent in fish from Sudbury, despite the similarities in environmental Cd concentrations in both gradients. Patterns of Cu and Ni accumulation and the influence of environmental contamination also differed between regions and suggested that Sudbury perch may have evolved more effective strategies for limiting uptake and/or stimulating elimination. Together, these data supported the idea that the two populations may be genetically distinct, a hypothesis that was later tested and confirmed (see Section 10.5).

Seasonal and Regional Effects (Question 6)

A major objective of Case Study 2 was to examine seasonal and regional variations in tissue metal accumulation, fish condition and metabolic capacities (Couture et al. 2008a; Couture and Pyle 2008; Couture et al. 2008b; Pyle et al. 2008). Earlier studies in the Sudbury area had already revealed that seasonal variations influenced tissue metal concentrations and metabolic capacities within populations, especially in fish from contaminated lakes (Eastwood and Couture 2002; Audet and Couture 2003). Similarly, in the Rouyn-Noranda region, two studies had reported seasonal variations in both perch tissue metal contamination (Kraemer et al. 2006a) and metabolic endpoints (Levesque et al. 2002). Combining these studies, we demonstrated that seasonal variations of fish contamination and condition were substantial, but largely unpredictable and variable inter-annually. Although the knowledge gained with these and other studies (reviewed in Case study 1 above) may eventually allow us to better predict variations in tissue metal concentrations in chronically exposed wild perch exposed to complex metal mixtures, to date, even predicting accumulation of a single non-essential metal like Cd from a single source (water) in one lake within a season remains a challenge, as detailed in Section 10.4.1 Question 2 above. Presently, the only way to assess metal accumulation in perch from metal-contaminated environments requires a rigorous sampling of local populations at different periods of the year.

In addition to differences in metal handling capacities, our study also highlighted other regional variations providing evidence of potential genetic differences between yellow perch from the two regions. Growth patterns deduced from age-to-size relationships established by pooling the fish by region revealed that growth trajectories varied substantially between populations: fish from Rouyn-Noranda showed rapid early growth compared to Sudbury fish (Fig. 4). Tissue metabolic capacities also appeared strongly affected by region: fish sampled at the same time of the year in similar lakes separated by a few hundred kilometres differed in metabolic capacities (Couture et al. 2008b).

Effects of Metal Accumulation (Question 4)

A major objective of Case Study 2 was to investigate the extent to which metal contamination affects perch condition, growth, longevity and metabolic capacities, and to provide evidence of direct metal toxicity in contaminated fish. In one of the earliest studies investigating the effects of metals on wild yellow perch, Kearns and Atchison (1979) demonstrated a negative relationship between young-of-the-year yellow perch growth and Cd concentration. Sjobeck et al. (1984) demonstrated that wild European perch from Cd-contaminated sites were anaemic and showed signs of disturbed carbohydrate metabolism. Our study built upon the results of these earlier studies to determine if we could observe direct effects on wild yellow perch populations inhabiting metal-contaminated sites.

In our study, fish from contaminated lakes grew faster but died younger than those from clean lakes (Pyle et al. 2008). Aerobic capacities in tissues of fish chronically exposed to Cu and Cd were impaired, possibly through direct enzyme inhibition, which corroborated earlier results (Couture and Rajotte 2003). The Tissue Residue

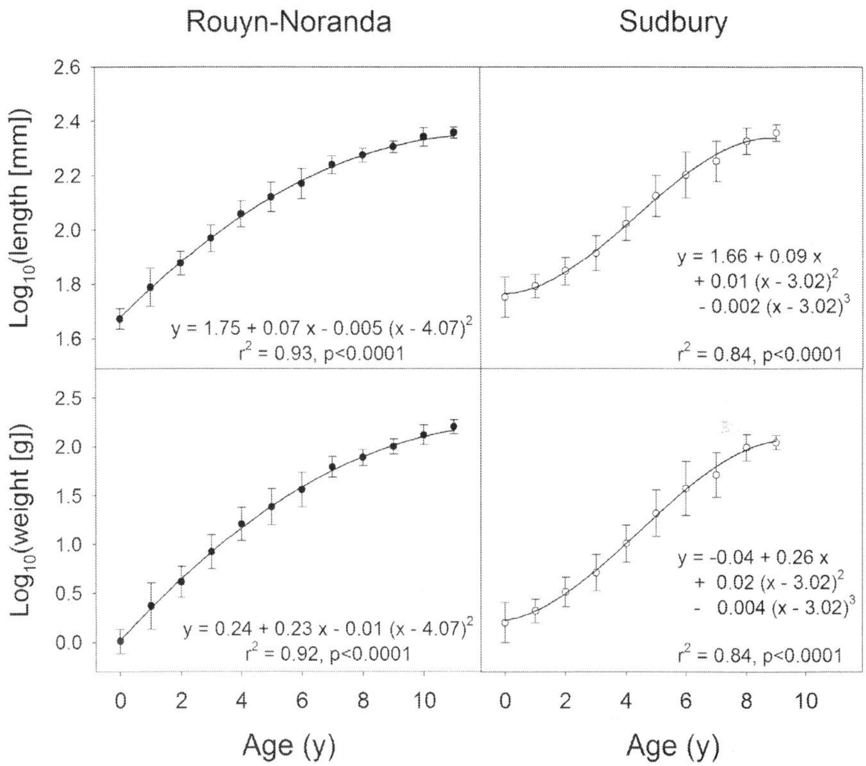

Fig. 4. Length and weight at age relationships for wild yellow perch sampled from five lakes along two metal contamination gradients in Sudbury, ON (n = 1324, for the entire relationship) and Rouyn-Noranda, QC (n = 1125), Canada. Points represent means ± SD for each age group. Black dots represent fish from Rouyn-Noranda lakes, white dots represent fish from Sudbury lakes. From Pyle et al. 2008. Reprinted with permission.

Approach was developed to investigate the relationships between tissue concentrations of contaminants in tolerant species and toxicity in the same species or in more sensitive species from the same environment that may not readily accumulate contaminants (Adams et al. 2011). We developed a simplified version of this approach and proposed thresholds of tissue metal concentrations in yellow perch that allowed us to discriminate between fish from clean and contaminated environments and to predict direct metabolic consequences (Couture and Pyle 2008). Thresholds of liver and kidney Cu and Cd concentrations were defined as concentrations exceeded by only 10% of fish from clean lakes from the two metal contamination gradients investigated (Sudbury and Rouyn-Noranda). On average, 72% of fish from contaminated lakes exceeded the Cd and Cu thresholds in their liver, and 30 to 70% in the kidney. In order to examine whether exceeding the tissue metal concentration thresholds led to deleterious metabolic consequences, tissue protein concentrations and metabolic capacities of fish above or below the thresholds were compared. The comparison revealed clear differences between fish from the two regions. Exceeding the threshold did not affect tissue protein concentrations in Sudbury fish, but was associated with a significant decrease in the muscle of Rouyn-Noranda fish and with an increase in their liver, suggesting that in the latter fish metal contamination reduced energy accumulation (decreased muscle protein) and diverted it towards detoxification (increased liver protein). The tissue threshold approach further corroborated previous studies reviewed above that metal contamination affected tissue aerobic capacities. However the response differed between liver and muscle and, as in the case of tissue protein, between regions, as discussed in more detail in Couture and Pyle (2008). Regional differences in metal handling and tolerance, as highlighted by the tissue threshold approach, were a major incentive to examine genetic differences between the two populations, described in Section 10.5.2.

Regardless of the evidence reviewed above suggesting direct toxicity of metals in wild perch, indirect effects of metal contamination on metabolic capacities cannot be discounted. Indeed, as discussed in Section 10.4.1 above, metal contamination was demonstrated to lead to impoverished food webs (Rasmussen et al. 2008). Furthermore, we later demonstrated experimentally that feeding rate affects aerobic, anaerobic and biosynthetic capacities as well as growth and condition factor (Gauthier et al. 2008). The mechanisms through which metal contamination can lead to metabolic impairment are clearly multiple and complex, and likely involve a combination of direct toxicity and environmental impacts leading to indirect consequences, both positive and negative. New approaches involving evolutionary ecotoxicology and toxicogenomics offer the tantalizing possibility to decipher signals and mechanisms of direct metal toxicity in these adaptable fish having to cope with a complex, varied and variable environments.

Recommendations for Perch Sampling in Ecotoxicological Studies

A last objective of Case Study 2, facilitated by the extensive database it had generated, was to propose approaches for yellow perch sampling in environmental effects monitoring studies that take into account seasonal variations in fish condition, metal contamination and indirect (food-web mediated) effects of metal contamination. Our results suggest that several factors need to be taken into consideration when sampling yellow perch if one wishes to tease apart the complex effects of metal contamination, region, and season on metal uptake, accumulation, physiological effects and fish condition (Couture and Pyle 2008). For example, fish from different populations should be size-matched or allometric corrections should be made to properly interpret metabolic effects or population-level biometric data. Because fish condition and metabolic capacities vary substantially both inter- and intra-seasonally, fish must be sampled during a narrow temporal window within a single season, or over seasons, to allow for comparisons within and between populations. Furthermore, that "temporal window" is site-specific and depends on the depth of water body, latitude, altitude and any other factor influencing temperature. Most importantly, selecting appropriate reference populations is required for any sampling regime designed to investigate the effects of environmental contaminants on perch populations. Ideally, more than one reference population should be sampled and they should be in relatively close proximity to the contaminated populations of interest, to reduce the probability of sampling different regional (genetic) subpopulations (Couture et al. 2008a; Couture and Pyle 2008).

10.5 Omics and Population Genetics in the Context of Perch Ecotoxicology

Molecular approaches to investigate the effects of contaminants in wild fish are recent and to date, only a handful of studies (we identified a mere eight of them) have used these novel approaches for the study of *Perca*. All of them were carried out with yellow perch, among which seven focused on metal contamination. Investigations described in the two case studies above (Sections 10.4.1 and 10.4.2) prompted us to develop molecular approaches to examine two questions: (1) Can examination of gene transcription levels facilitate our understanding of the mechanisms of metal toxicity in wild yellow perch? (2) Can long-term metal contamination affect genetic diversity and direct selection? These questions are discussed in the sections below.

10.5.1 Ecotoxicogenomics

Two decades of studies examining direct effects of chronic metal exposure have demonstrated that these contaminants induce toxicity and metabolic costs in wild yellow perch. However, our traditional effects biomarkers tend to be generic and lack contaminant specificity. Ecotoxicogenomics offers the possibility to determine modes of toxic action in the field, in chronically exposed fish, and from this information to design new and contaminant-specific biomarkers. In other words, examining the response of gene transcription to contaminant exposure broadens the perspective of ecotoxicological studies.

Pierron et al. (2009) used a quantitative reverse transcriptase polymerase chain reaction (RT-qPCR) approach to measure the transcription levels of genes encoding for proteins involved in metal detoxification (metallothionein), protection of protein structure (heat shock protein-70), growth (insulin-like growth factor), aerobic capacities (cytochrome C oxidase) and protection against oxidative stress (superoxide dismutase) in yellow perch collected along the same metal contamination gradients described in Section 10.4. The approach also included examination of three corresponding physiological biomarkers (activity of superoxide dismutase and cytochrome C oxidase enzymes, and tissue total protein concentrations), in order to examine to what extent genomic and physiological responses to chronic metal contamination were consistent. The mechanisms of toxicity of cadmium and copper differed. Gene transcription levels were more affected by cadmium in liver than in muscle, whereas the opposite trend was observed for copper. For instance, accumulation of cadmium in liver was correlated with a decrease of the transcription level of the genes encoding for cytochrome C oxidase but did not affect the activity of the corresponding enzyme in the liver. In contrast, liver copper accumulation was positively correlated with the transcription level of the cytochrome C oxidase gene in muscle, but the activity of the enzyme was negatively correlated with liver copper concentrations. We hypothesized that chronic

Cu accumulation can lead to inhibition or even breakdown of muscle enzymes, including cytochrome C oxidase, a phenomenon that has been reported in rats fed a Cu-contaminated diet (Sokol et al. 1993). Parallel increases in total muscle protein suggest that the increase in the transcription level of the cytochrome C oxidase gene was part of a global compensatory mechanism aimed at counteracting the inhibitory effects of Cu on muscle enzymes in contaminated perch. In addition to providing new insights into mechanisms of metal toxicity in wild perch, this study demonstrated the potential of combining physiological and genomic endpoints to identify the effects of individual contaminants present in environmental mixtures. Nevertheless, PCR and physiological biomarkers have in common that they must be selected *a priori*.

In investigations of mechanisms of toxicity in wild fish, an *a posteriori* approach would be advantageous. Progress in technology now allows the sequencing of a large number of genes and examining at once hundreds, or even thousands, of genes involved in a wide range of metabolic pathways and their response to environmental stressors. These responses can then be used *a posteriori* to orient mechanistic studies involving both genomic and physiological endpoints. Using this approach, we sampled yellow perch from four lakes in the Rouyn-Noranda region representing a strong metal contamination gradient, extracted their liver RNA and identified over 1500 contigs representing different genes (Pierron et al. 2011). The transcription levels of almost 200 genes were affected by metal contamination and the majority of these correlations were negative, a result that is consistent with the inhibition of metabolism reported in metal-contaminated perch from earlier studies (Section 10.4). Furthermore, we identified several metabolic pathways that were particularly affected by metal contamination, including those involved in protein biosynthesis, immune response, and lipid and energy metabolism.

The preceding study oriented further investigations of mechanisms of metal toxicity in wild yellow perch in directions that only such a large-scale genomic approach could have dictated. In particular, we identified a suite of genes related to retinol (vitamin A) metabolism that appeared to be particularly sensitive to chronic metal exposure (Pierron et al. 2011). This result triggered a study that confirmed that retinol metabolism is indeed affected by cadmium contamination in wild yellow perch (Defo et al. 2012). Retinoids play several physiological functions, including performing as antioxidants and participating in cell differentiation, growth and reproduction.

DNA microarrays allow the measurement of transcription levels for a large number of genes simultaneously. Convinced by the potential of genomics in fish ecotoxicological studies and using the sequencing results reported by Pierron et al. (2009), we designed a one thousand gene microarray specifically for the yellow perch (Bougas et al. 2013). Genes that responded to metal contamination (196 genes, see Pierron et al. 2011) were selected for inclusion on the microarray. We completed the microarray with genes representing each major metabolic pathway. This tool was then applied to investigate mechanisms of cadmium and nickel toxicity in laboratory-exposed perch and the combined effects of natural stressors (temperature, hypoxia and food restriction) on gene transcription levels. The microarray allowed us to distinguish stress caused by metals from that arising from natural stressors. This approach shows great promise in environmental effects monitoring studies, since wild fish are continuously exposed to multiple stressors which affect to variable extents most effects biomarkers.

Although the yellow perch microarray has not been tested on the other *Perca* species, given the close phylogenetic relationship among the three species, it is likely to be applicable to all species of the genus *Perca*. Given the distribution of this genus around the globe, we suggest that perch may well prove useful in ecotoxicogenomic approaches to investigate global environmental issues.

10.5.2 Evolutionary Ecotoxicology

The field of evolutionary ecotoxicology examines how contaminants influence natural selection, by selectively eliminating the more vulnerable phenotypes within a population and/or by randomly reducing genetic diversity following population crashes. Among the first such studies on perch is that of Kopp et al. (1994), who compared the effects of acidity and associated elevated aluminum concentrations on genetic diversity in yellow perch and four other species sampled from locations varying in pH within a watershed. Based on its distribution and abundance in both clean and contaminated sites, the authors classified yellow perch as relatively tolerant to acid-stressed environments. Hence, they expected a high degree of genetic variability, characteristic of tolerant, generalist species. However, although yellow perch were abundant and ubiquitous in the study area and demonstrated a high degree of tolerance to acid deposition, their genetic diversity was low among sampling sites. Although the authors concluded that population genetics may not be suitable to detect tolerance to environmental stressors, we can argue that the design of their study may not have been suitable to properly investigate the question. As Kopp et al. (1994) mention, yellow perch were introduced to their study area, inducing a founder effect that may be responsible for a low genetic diversity among study sites. Furthermore, yellow perch are recognized as quite tolerant to acidity. Consequently, it is unlikely that the low variation in pH among their study sites (ranging from 6.4 to 8.0, which is typical of the habitat for the species) was sufficient to induce a reduction in genetic diversity. Moreover, one cannot rule out the possibility that effective population sizes (N_e) vary naturally among populations, independent of any environmental stressors, which would blur a signal of association with the latter. Perhaps more importantly, since yellow perch as a whole harbour very low levels of genetic diversity most likely due to historical bottlenecks associated with the last glaciations (Stepien et al., Chapter 2 of this volume), it is plausible that the range of genetic variation in this species limits the possibility to detect any association with an environmental stressor that likely imposes a relatively weak selective pressure, as would be the case with acidity, given perch's tolerance to it. However, this interpretative constraint imposed by limited genetic variation in perch would be less important if selective pressures are very strong, as is likely for metals (see below).

Indeed, in a more recent study, we sampled yellow perch from ten lakes in each of the two metal contamination gradients described in Section 10.4 and examined genetic diversity using neutral microsatellites. Overall gene diversity decreased with increasing cadmium, but not copper, contamination (Bourret et al. 2008). This study suggested that more than a century of environmental degradation and metal contamination may have compromised gene diversity in the most contaminated yellow perch populations, leading to contamination-driven microevolutionary processes favouring metal tolerance. Tremblay et al. (2008) also examined genetic variability in yellow perch collected in eight lakes along a gradient of metal contamination. Their study revealed that contamination may have affected genetic diversity indirectly, by modifying fish communities. Indeed, when yellow perch are the only species present, they occupy both pelagic and benthic niches, and their body shape is more elongated than when predators are present. In the latter situation, yellow perch populations tend to become benthic and adopt a more robust morphology (Lippert et al. 2007). The abundance and diversity of predators and competitors being higher in cleaner lakes, contamination may indirectly affect yellow perch genetic diversity. Nevertheless, the Tremblay et al. (2008) study also concluded that metal contamination had a direct influence on population genetic variability, although they did not report on the direction of these changes.

This hypothesis was tested in an investigation examining single nucleotide polymorphism (SNP) in non-neutral alleles (Bélanger-Deschênes et al. 2013). Our study was designed to identify polymorphism in functional regions of genes under directional selection, for which the frequency is linked to contamination intensity. We identified two such genes. Although the adaptive implications of SNPs in NADH dehydrogenase subunit 2, a gene with several roles, cannot be convincingly explained to date, polymorphism in the functional region of the cyclin G1 gene, a gene involved in cell growth and stress response, suggests that contaminant-induced selection has favoured a rapid early growth and reduced longevity. This hypothesis is supported by an earlier study in which we reported a lower longevity for contaminated yellow perch (Pyle et al. 2008).

To our knowledge, there is no published study in evolutionary ecotoxicology examining *Perca fluviatilis*, other than a modeling exercise investigating the theoretical consequences of endocrine disruptors on the population dynamics of Windermere perch, a population to which Chapter 6 is dedicated (Brown et al. 2005). Given its broad distribution in habitats varying widely in levels of environmental stressors and the sedentary nature of perch, the genus presents outstanding research opportunities to investigate the effects of natural and anthropogenic stressors on perch evolution.

10.6 Conclusions

The aquatic ecotoxicology literature is dominated by studies involving model species, such as rainbow trout (*Oncorhynchus mykiss*), fathead minnows (*Pimephales promelas*), zebrafish (*Danio rerio*) and others. There is no doubt that these species have been instrumental in developing our understanding of contaminant effects on fish. However, these species often do not inhabit environments receiving contaminants of industrial origin, which could lead to questions about the ecological relevance of their use to predict effects in natural environments where they do not occur.

Over the past four decades researchers have progressively realized the value of studying indigenous populations of *Perca* spp. in receiving waters affected by industrial activities because of their non-migratory nature and relative tolerance to environmental contaminants. Most research has been on wild populations of *Perca* spp., especially *P. flavescens* and *P. fluviatilis*, and has been dominated by studies investigating the effects of metals. The work that has been conducted to date has demonstrated that *Perca* spp. can provide environmental managers with ecologically relevant data demonstrating physiological effects of contamination as well as a mechanistic understanding of these effects, such as subcellular contaminant partitioning and mechanisms of toxic action. Recently, research on *Perca flavescens* has demonstrated that molecular and ecotoxicogenomic approaches can provide meaningful insights into microevolutionary processes that can lead to local, population-level adaptations after long-term, multigeneration contaminant exposures. Given the widespread geographic distribution of *Perca* spp. over the northern hemisphere, the potential for similar insights over their entire natural range is promising.

10.7 References

Adams, W.J., R. Blust, U. Borgmann, K.V. Brix, D.K. DeForest, A.S. Green, J.S. Meyer, J.C. McGeer, P.R. Paquin, P.S. Rainbow and C.M. Wood. 2011. Utility of tissue residues for predicting effects of metals on aquatic organisms. Integr. Environ. Assess. Mgt. 7(1): 75–98.

Audet, D. and P. Couture. 2003. Seasonal variations in tissue metabolic capacities of yellow perch (*Perca flavescens*) from clean and metal-contaminated environments. Can. J. Fish. Aquat. Sci. 60: 269–278.

Badsha, K.S. and C.R. Goldspink. 1982. Preliminary observations on the heavy metal content of four species of freshwater fish in NW England. J. Fish Biol. 21: 251–267.

Bélanger-Deschênes, S., P. Couture, P.G.C. Campbell and L. Bernatchez. 2013. Evolutionary change driven by metal exposure as revealed by coding SNP genome scan in wild yellow perch (*Perca flavescens*). Ecotoxicol. 22: 938–957.

Bougas, B., E. Normandeau, F. Pierron, P.G.C. Campbell, L. Bernatchez and P. Couture. 2013. How does exposure to nickel and cadmium affect the transcriptome of yellow perch (*Perca flavescens*)—Results from a 1000 candidate-gene microarray. Aquat. Toxicol. 142-143: 355–364.

Bourret, V., P. Couture, P.G.C. Campbell and L. Bernatchez. 2008. Evolutionary ecotoxicology of wild yellow perch (*Perca flavescens*) populations chronically exposed to a polymetallic gradient. Aquat. Toxicol. 86: 76–90.

Brodeur, J.C., G. Sherwood, J.B. Rasmussen and A. Hontela. 1997. Impaired cortisol secretion in yellow perch (*Perca flavescens*) from lakes contaminated by heavy metals: *in vivo* and *in vitro* assessment. Can. J. Fish. Aquat. Sci. 54: 2752–2758.

Brown, A.R., A.M. Riddle, I.J. Winfield, J.M. Fletcher and J.B. James. 2005. Predicting the effects of endocrine disrupting chemicals on healthy and disease impacted populations of perch (*Perca fluviatilis*). Ecol. Mod. 189(3-4): 377–395.

Burreau, S., Y. Zebühr, D. Broman and R. Ishaq. 2004. Biomagnification of polychlorinated biphenyls (PCBs) and polybrominated diphenyl ethers (PBDEs) studied in pike (*Esox lucius*), perch (*Perca fluviatilis*) and roach (*Rutilus rutilus*) from the Baltic Sea. Chemosphere 55(7): 1043–1052.

Campbell, P.G.C., L. Kraemer, A. Giguere, L. Hare and A. Hontela. 2008. Subcellular distribution of cadmium and nickel in chronically exposed wild fish: inferences regarding metal detoxification strategies and implications for setting water quality guidelines for dissolved metals. Hum. Ecol. Risk Assess. 14(2): 290–316.

Campbell, P.G.C. and C. Fortin. 2013. Biotic ligand model. pp. 237–245. *In*: J.F. Férard and C. Blais (eds.). Encyclopedia of Aquatic Ecotoxicology. Springer-Verlag, Heidelberg.

Campbell, P.G.C. and L. Hare. 2009. Metal detoxification in freshwater animals. pp. 239–277. *In*: A. Sigel, H. Sigel and R.K.O. Sigel (eds.). Metallothioneins and Related Chelators. Royal Society of Chemistry, Cambridge, UK.

Campbell, P.G.C., A. Hontela, J.B. Rasmussen, A. Giguère, A. Gravel, L. Kraemer, J. Kovesces, A. Lacroix, H. Levesque and G. Sherwood. 2003. Differentiating between direct (physiological) and food-chain mediated (bioenergetic) effects on fish in metal-impacted lakes. Hum. Ecol. Risk Assess. 9(4): 847–866.

Couture, P., P. Busby, J.W. Rajotte, C. Gauthier and G.C. Pyle. 2008a. Seasonal and regional variations of metal contamination and condition indicators in yellow perch (*Perca flavescens*) along two polymetallic gradients. I. Factors influencing tissue metal concentrations. Hum. Ecol. Risk Assess. 14(1): 97–125.

Couture, P. and P.R. Kumar. 2003. Impairment of metabolic capacities in copper and cadmium contaminated wild yellow perch (*Perca flavescens*). Aquat. Toxicol. 64: 107–120.

Couture, P. and G.G. Pyle. 2008. Live fast and die young: metal effects on the condition and physiology of wild yellow perch from along two metal contamination gradients. Hum. Ecol. Risk Assess. 14(1): 73–96.

Couture, P. and J.W. Rajotte. 2003. Morphometric and metabolic indicators of metal stress in wild yellow perch (*Perca flavescens*) from Sudbury Ontario: A review. J. Environ. Monit. 5: 216–221.

Couture, P., J.W. Rajotte and G.C. Pyle. 2008b. Seasonal and regional variations of metal contamination and condition indicators in yellow perch (*Perca flavescens*) along two polymetallic gradients. III. Energetic and physiological indicators. Hum. Ecol. Risk Assess. 14(1): 146–165.

Dautremepuits, C., D.J. Marcogliese, A.D. Gendron and M. Fournier. 2009. Gill and head kidney antioxidant processes and innate immune system responses of yellow perch (*Perca flavescens*) exposed to different contaminants in the St. Lawrence River, Canada. Sci. Tot. Environ. 407(3): 1055–1064.

Defo, M.A., F. Pierron, P.A. Spear, L. Bernatchez, P.G.C. Campbell and P. Couture. 2012. Evidence for metabolic imbalance of vitamin A2 in wild fish chronically exposed to metals. Ecotoxicol. Environ. Saf. 85: 88–95.

Eastwood, S. and P. Couture. 2002. Seasonal variations in condition and liver metal concentrations of yellow perch (*Perca flavescens*) from a metal-contaminated environment. Aquat. Toxicol. 58: 43–46.

Edgren, M. and M. Notter. 1980. Cadmium uptake by fingerlings of perch (*Perca fluviatilis*) studied by Cd-115m at two different temperatures. Bull. Environ. Contam. Toxicol. 24(5): 647–651.

Environment Canada. 2013. Pollution and Waste - Metal Mining - DEV. Available from http://www.ec.gc.ca/esee-eem/default.asp?lang=En&n=996CDD5D-1 [accessed 02/15/14 2014].

Gauthier, C., P.G.C. Campbell and P. Couture. 2008. Physiological correlates of growth and condition in the yellow perch (*Perca flavescens*). Comp. Biochem. Physiol. A. 151(4): 526–532.

Giguère, A., P.G.C. Campbell, L. Hare and C. Cossu-Leguille. 2005. Metal bioaccumulation and oxidative stress in yellow perch (*Perca flavescens*) collected from eight lakes along a metal contamination gradient (Cd, Cu, Zn, Ni). Can. J. Fish. Aquat. Sci. 62: 563–577.

Giguère, A., P.G.C. Campbell, L. Hare and P. Couture. 2006. Sub-cellular partitioning of cadmium, copper, nickel and zinc in indigenous yellow perch (*Perca flavescens*) sampled along a polymetallic gradient. Aquat. Toxicol. 77: 178–189.

Giguère, A., P.G.C. Campbell, L. Hare, D.G. McDonald and J.B. Rasmussen. 2004. Influence of lake chemistry and fish age on cadmium, copper and zinc concentrations in various organs of indigenous yellow perch (*Perca flavescens*). Can. J. Fish. Aquat. Sci. 61: 1702–1716.

Girard, C., J.C. Brodeur and A. Hontela. 1998. Responsiveness of the interrenal tissue of yellow perch (*Perca flavescens*) from contaminated sites to an ACTH challenge test *in vivo*. Can. J. Fish. Aquat. Sci. 55(2): 438–450.

Gravel, A., P.G.C. Campbell and A. Hontela. 2005. Disruption of the hypothalamo-pituitary-interrenal axis in 1+ yellow perch (*Perca flavescens*) chronically exposed to metals in the environment. Can. J. Fish. Aquat. Sci. 62: 982–990.

Harvey, H.H. 1980. Widespread and diverse changes in the biota of North American lakes and rivers coincident with acidification. pp. 93–98. *In*: D. Drabløs and A. Tollan (eds.). Proceedings of the International Conference on Ecological Impact of Acid Precipitation. SNSF Project, Norway.

Hontela, A. 1998. Interrenal dysfunction in fish from contaminated sites: *in vivo* and *in vitro* assessment. Environ. Toxicol. Chem. 17(1): 44–48.

Hontela, A., P. Dumont, D. Duclos and R. Fortin. 1995. Endocrine and metabolic dysfunction in yellow perch, *Perca flavescens*, exposed to organic contaminants and heavy metals in the St. Lawrence River. Environ. Toxicol. Chem. 14(4): 725–731.

Hontela, A., J.B. Rasmussen, C. Audet and G. Chevalier. 1992. Impaired cortisol stress response in fish from environments polluted by PAHs, PCBs and mercury. Arch. Environ. Contam. Toxicol. 22: 278–283.

Kearns, P.K. and G.J. Atchison. 1979. Effects of trace metals on growth of yellow perch (*Perca flavescens*) as measured by RNA-DNA ratios. Environ. Biol. Fish. 4(4): 383–387.

Klaverkamp, J.F., D.A. Hodgins and A. Lutz. 1983. Selenite toxicity and mercury-selenium interactions in juvenile fish. Arch. Environ. Contam. Toxicol. 12: 405–413.

Klinck, J.S., W.W. Green, R.S. Mirza, S.R. Nadella, M.J. Chowdhury, C.M. Wood and G.G. Pyle. 2007. Branchial cadmium and copper binding and intestinal cadmium uptake in wild yellow perch (*Perca flavescens*) from clean and metal-contaminated lakes. Aquat. Toxicol. 84(2): 198–207.

Kopp, R.L., T.E. Wissing and S.I. Guttman. 1994. Genetic indicators of environmental tolerance among fish populations exposed to acid deposition. Biochem. Syst. Ecol. 22(5): 459–475.

Kövecses, J., G.D. Sherwood and J.B. Rasmussen. 2005. Impacts of altered benthic invertebrate communities on the feeding ecology of yellow perch (*Perca flavescens*) in metal-contaminated lakes. Can. J. Fish. Aquat. Sci. 62: 153–162.

Kraemer, L., P.G.C. Campbell and L. Hare. 2006a. Seasonal variations in hepatic Cd and Cu concentrations and in the sub-cellular distribution of these metals in juvenile yellow perch (*Perca flavescens*). Environ. Pollut. 142: 313–325.

Kraemer, L., P.G.C. Campbell and L. Hare. 2008. Modeling cadmium accumulation in indigenous yellow perch (*Perca flavescens*). Can. J. Fish. Aquat. Sci. 65(8): 1623–1634.

Kraemer, L., P.G.C. Campbell, L. Hare and J.-C. Auclair. 2006b. A field study examining the relative importance of food and water as sources of cadmium for juvenile yellow perch (*Perca flavescens*). Can. J. Fish. Aquat. Sci. 63: 549–557.

Kraemer, L.D., P.G.C. Campbell and L. Hare. 2005a. Dynamics of Cd, Cu and Zn accumulation in organs and sub-cellular fractions in field transplanted juvenile yellow perch (*Perca flavescens*). Environ. Pollut. 138: 324–337.

Kraemer, L.D., P.G.C. Campbell and L. Hare. 2005b. A field study examining metal elimination kinetics in juvenile yellow perch (*Perca flavescens*). Aquat. Toxicol. 75: 108–126.

Kryshev, I.I., I.N. Ryabov and T.G. Sazykina. 1993. Using a bank of predatory fish samples for bioindication of radioactive contamination of aquatic food chains in the area affected by the Chernobyl accident. Sci. Tot. Environ. 139-140: 279–285.

Laflamme, J.-S., Y. Couillard, P.G.C. Campbell and A. Hontela. 2000. Interrenal metallothionein and cortisol secretion in relation to Cd, Cu, and Zn exposure in yellow perch, *Perca flavescens*, from Abitibi lakes. Can. J. Fish. Aquat. Sci. 57: 1692–1700.

Lappalainen, A., M. Rask and P.J. Vuorinen. 1988. Acidification affects the perch, *Perca fluviatilis*, populations in small lakes of southern Finland. Environ. Biol. Fish. 21: 231–239.

Lazartigues, A., M. Thomas, C. Cren-Olive, J. Brun-Bellut, Y. Le Roux, D. Banas and C. Feidt. 2013. Pesticide pressure and fish farming in barrage pond in Northeastern France. Part II: residues of 13 pesticides in water, sediments, edible fish and their relationships. Environ. Sci. Pollut. Res. 20(1): 117–125.

Levesque, H., J. Dorval, A. Hontela, G. Van Der Kraak and P.G.C. Campbell. 2003. Hormonal, morphological, and physiological responses of yellow perch (*Perca flavescens*) to chronic environmental metal exposures. J. Toxicol. Environ. Health 66(Part A): 657–676.

Levesque, H.W., T.W. Moon, P.G.C. Campbell and A. Hontela. 2002. Seasonal variation in carbohydrate and lipid metabolism of yellow perch (*Perca flavescens*) chronically exposed to metals in the field. Aquat. Toxicol. 60: 257–267.

Lind, O.C., P. Stegnar, B. Tolongutov, B.O. Rosseland, G. Strømman, B. Uralbekov, A. Usubalieva, A. Solomaina, J.P. Gwynn, E. Lespukh and B. Salbu. 2013. Environmental impact assessment of radionuclide and metal contamination at the former U site at Kadji Sai, Kyrgyzstan. J. Environ. Radioact. 123: 37–49.

Linløkken, A., E. Kleiven and D. Matzow. 1991. Population structure, growth and fecundity of perch (*Perca fluviatilis* L.) in an acidified river system in Southern Norway. Hydrologia 229: 179–188.

Lippert, K.A., J.M. Gunn and G.E. Morgan. 2007. Effects of colonizing predators on yellow perch (*Perca flavescens*) populations in lakes recovering from acidification and metal stress. Can. J. Fish. Aquat. Sci. 64: 1413–1428.

Lushchak, V.I. 2011. Environmentally induced oxidative stress in aquatic animals. Aquat. Toxicol. 10(1): 13–30.

Marcogliese, D.J., L.G. Brambilla, F. Gagne and A.D. Gendron. 2005. Joint effects of parasitism and pollution on oxidative stress biomarkers in yellow perch *Perca flavescens*. Dis. Aquat. Org. 63(1): 77–84.

Marcogliese, D.J., K. Pulkkinen and E.T. Valtonen. 2012. Trichodinid (Ciliophora: Trichodinidae) infections in perch (*Perca fluviatilis*) experimentally exposed to pulp and paper mill effluents. Arch. Environ. Contam. Toxicol. 62(4): 650–656.

Mason, A.Z. and K.D. Jenkins. 1995. Metal detoxification in aquatic organisms. pp. 479–608. *In*: A. Tessier and J. Turner (eds.). Metal Speciation and Bioavailability in Aquatic Systems. Wiley & Sons, Chichester, UK.

Mirza, R.S., W.W. Green, S. Connor, A.C.W. Weeks, C.M. Wood and G.C. Pyle. 2009. Do you smell what I smell? Olfactory impairment in wild yellow perch from metal-contaminated waters. Ecotoxicol. Environ. Saf. 72(3): 677–683.

Niyogi, S., P. Couture, G. Pyle, D.G. McDonald and C.M. Wood. 2004. Acute cadmium biotic ligand model characteristics of laboratory-reared and wild yellow perch (*Perca flavescens*) relative to rainbow trout (*Oncorhynchus mykiss*). Can. J. Fish. Aquat. Sci. 61: 942–953.

Niyogi, S., G.C. Pyle and C.M. Wood. 2007. Branchial versus intestinal zinc uptake in wild yellow perch (*Perca flavescens*) from reference and metal-contaminated aquatic ecosystems. Can. J. Fish. Aquat. Sci. 64: 1605–1613.

Niyogi, S. and C.M. Wood. 2004. Biotic ligand model, a flexible tool for developing site-specific water quality guidelines for metals. Environ. Sci. Technol. 38: 6177–6192.

Olsson, A., K. Valters and S. Burreau. 2000. Concentrations of organochlorine substances in relation to fish size and trophic position: a study on perch (*Perca fluviatilis* L.). Environ. Sci. Technol. 34(23): 4878–4886.

Pierron, F., V. Bourret, J. St-Cyr, P.G.C. Campbell, L. Bernatchez and P. Couture. 2009. Transcriptional responses to environmental metal exposure in wild yellow perch (*Perca flavescens*) collected in lakes with differing environmental metal concentrations (Cd, Cu, Ni). Ecotoxicol. 18(5): 620–631.

Pierron, F., E. Normandeau, M. Defo, P. Campbell, L. Bernatchez and P. Couture. 2011. Effects of chronic metal exposure on wild fish populations revealed by high-throughput cDNA sequencing. Ecotoxicol. 20(6): 1388–1399.

Pyle, G.G., P. Busby, C. Gauthier, J.W. Rajotte and P. Couture. 2008. Seasonal and regional variations of metal contamination and condition indicators in yellow perch (*Perca flavescens*) along two polymetallic gradients. II. Growth patterns, longevity, and condition. Hum. Ecol. Risk Assess. 14(1): 126–145.

Rajotte, J.W. and P. Couture. 2002. Effects of environmental metal contamination on the condition, swimming performance, and tissue metabolic capacities of wild yellow perch (*Perca flavescens*). Can. J. Fish. Aquat. Sci. 59: 1296–1304.

Rasmussen, J.B., J.M. Gunn, G.D. Sherwood, A. Iles, A. Gagnon, P.G.C. Campbell and A. Hontela. 2008. Direct and indirect (foodweb mediated) effects of metal exposure on the growth of yellow perch (*Perca flavescens*): implications for ecological risk assessment. Hum. Ecol. Risk Assess. 14(2): 317–350.

Rowan, D.J. and J.B. Rasmussen. 1996. Measuring the bioenergetic cost of fish activity *in situ* using a globally dispersed radiotracer (137Cs). Can. J. Fish. Aquat. Sci. 53(4): 734–745.

Ryan, M.R. and H.H. Harvey. 1980. Growth responses of yellow perch, *Perca flavescens* (Mitchill), to lake acidification in the La Cloche Mountain lakes of Ontario. Environmental Biology of Fishes 5: 97–108.

Salbu, B., L.M. Burkitbaev, G. Strømman, I. Shishkov, P. Kayukov, B. Uralbekov and B.O. Rosseland. 2013. Environmental impact assessment of radionuclides and trace elements at the Kurday U mining site, Kazakhstan. J. Environ. Radioact. 123: 14–27.

Schindler, D.W., K.H. Mills, D.F. Malley, D.L. Findlay, J.A. Shearer, I.J. Davies, M.A. Turner, G.A. Linsey and D.R. Cruikshank. 1985. Long-term ecosystem stress: the effects of years of experimental acidification on a small lake. Science 228: 1395–1401.

Sherwood, G.D., J. Kövecses, A. Hontela and J.B. Rasmussen. 2002a. Simplified food webs lead to energetic bottlenecks in polluted lakes. Can. J. Fish. Aquat. Sci. 59: 1–5.

Sherwood, G.D., I. Pazzia, A. Moeser, A. Hontela and J.B. Rasmussen. 2002b. Shifting gears: enzymatic evidence for the energetic advantage of switching diet in wild-living fish. Can. J. Fish. Aquat. Sci. 59: 229–241.

Sherwood, G.D., J.B. Rasmussen, D.J. Rowan, J. Brodeur and A. Hontela. 2000. Bioenergetic costs of heavy metal exposure in yellow perch (*Perca flavescens*): *in situ* estimates with radiotracer (137Cs) technique. Can. J. Fish. Aquat. Sci. 57: 441–450.

Sjobeck, M.L., C. Haux, A. Larsson and G. Lithner. 1984. Biochemical and hematological studies on perch, *Perca fluviatilis*, from the cadmium-contaminated river Eman. Ecotoxicol. Environ. Saf. 8(3): 303–312.

Sokol, R., M. Devereaux, K. O'Brien, R. Khandwala and J. Loehr. 1993. Abnormal hepatic mitochondrial respiration and cytochrome c oxidase activity in rats with long term copper overload. Gastroenterology 105: 178–187.

Steinnes, E., T. Hastein, G. Norheim and A. Froslie. 1976. Mercury in various tissues of fish caught downstream of a wood pulp factory in the kammerfoss river, South Norway. Nordisk Veterinaer Medicin. 28(11): 557–563.

Strømman, G., B.O. Rosseland, L. Skipperud, L.M. Burkitbaev, B. Uralbekov, L.S. Heier and B. Salbu. 2013. Uranium activity ratio in water and fish from pit lakes in Kurday, Kazakhstan and Taboshar, Tajikistan. J. Environ. Radioact. 123: 71–81.

Taylor, L.N., W.J. McFarlane, G.G. Pyle, P. Couture and D.G. McDonald. 2004. Use of performance indicators in evaluating chronic metal exposure in wild yellow perch (*Perca flavescens*). Aquat. Toxicol. 67: 371–385.

Tremblay, A., D. Lesbarreres, T. Merritt, C. Wilson and J. Gunn. 2008. Genetic structure and phenotypic plasticity of yellow perch (*Perca flavescens*) populations influenced by habitat, predation, and contamination gradients. Integr. Environ. Assess. Mgt. 4(2): 264–266.

Valtonen, E.T., J.C. Holmes, J. Aronen and I. Rautalahti. 2003. Parasite communities as indicators of recovery from pollution: parasites of roach (*Rutilus rutilus*) and perch (*Perca fuviatilis*) in central Finland. Parasitol. 126 Suppl: S43–52.

Valtonen, E.T., J.C. Holmes and M. Koskivaara. 1997. Eutrophication, pollution and fragmentation: effects on the parasite communities in roach and perch in four lakes in central Finland. Paras sitologia 39(3): 233–236.

van den Heuvel, M.R., M. Power, M.D. MacKinnon and D.G. Dixon. 1999a. Effects of oil sands related aquatic reclamation on yellow perch (*Perca flavescens*). II. Chemical and biochemical indicators of exposure to oil sands related waters. Can. J. Fish. Aquat. Sci. 56(7): 1226–1233.

van den Heuvel, M.R., M. Power, M.D. MacKinnon, T.V. Meer, E.P. Dobson and D.G. Dixon. 1999b. Effects of oil sands related aquatic reclamation on yellow perch (*Perca flavescens*). I. Water quality characteristics and yellow perch physiological and population responses. Can. J. Fish. Aquat. Sci. 56(7): 1213–1225.

Vezina, R. and R. Desrochers. 1971. Occurrence of *Aeromonas hydrophila* in the perch, *Perca flavescens* Mitchill. Can. J. Micobiol. 17(8): 1101–1103.

Viarengo, A., B. Burlando, N. Ceratto and I. Panfoli. 2000. Antioxidant role of metallothioneins: a comparative overview. Cell. Mol. Biol. (Noisy-le-grand) 46(2): 407–417.

Wallace, W.G., L. Byeong-Gweon and S.N. Luoma. 2003. Subcellular compartmentalization of Cd and Zn in two bivalves. I. Significance of metal-sensitive fractions (MSF) and biologically detoxified metal (BDM). Marine Ecology Progress series 249: 183–197.

Zalewska, T. and M. Suplinska. 2013. Fish pollution with anthropogenic 137Cs in the southern Baltic Sea. Chemosphere 90(6): 1760–1766.

Index